MONOCYCLIC AZEPINES

The Syntheses and Chemical Properties
of the Monocyclic Azepines

This is the Fifty-sixth Volume in the Series
THE CHEMISTRY OF HETEROCYCLIC COMPOUNDS

THE CHEMISTRY OF HETEROCYCLIC COMPOUNDS

A SERIES OF MONOGRAPHS

EDWARD C. TAYLOR

Editor

MONOCYCLIC AZEPINES

The Syntheses and Chemical Properties
of the Monocyclic Azepines

GEORGE R. PROCTOR
University of Strathclyde

JAMES REDPATH
Organon Laboratories Ltd

A Wiley–Interscience Publication

JOHN WILEY & SONS
Chichester · New York · Brisbane · Singapore

Copyright © 1996 by John Wiley & Sons Ltd,
 Baffins Lane, Chichester,
 West Sussex PO19 1UD, England

National   01243 779777
International (+44) 1243 779777
e-mail (for orders and customer service enquiries): cs-books@wiley.co.uk
Visit our Home Page on http://www.wiley.co.uk
           or http://www.wiley.com

All Rights Reserved. No part of this publication may be reproduced, stored in a retrieval system, or transmitted, in any form or by any means, electronic, mechanical, photocopying, recording, scanning or otherwise, except under the terms of the Copyright, Designs and Patents Act 1988 or under the terms of a licence issued by the Copyright Licensing Agency, 90 Tottenham Court Road, London, UK W1P 9HE, without the permission in writing of the publisher.

Other Wiley Editorial Offices

John Wiley & Sons, Inc., 605 Third Avenue,
New York, NY 10158-0012, USA

Jacaranda Wiley Ltd, 33 Park Road, Milton,
Queensland 4064, Australia

John Wiley & Sons (Canada) Ltd, 22 Worcester Road,
Rexdale, Ontario M9W 1L1, Canada

John Wiley & Sons (Asia) Pte Ltd, 2 Clementi Loop #02-01,
Jin Xing Distripark, Singapore 0512

Library of Congress Cataloging-in-Publication Data

Proctor, George R.
  Monocyclic azepines : the syntheses and chemical properties of the monocyclic azepines / George R. Proctor, James Redpath.
    p.   cm. — (The chemistry of heterocyclic compounds ; v. 56)
  "A Wiley Interscience publication."
  Includes bibliographical references and index.
  ISBN 0-471-96372-0
  1. Azepines. I. Redpath, James. II. Title. III. Series.
QD401.P957   1996
547'.593—dc20   95-48039
                                                              CIP

British Library Cataloguing in Publication Data

A catalogue record for this book is available from the British Library

ISBN 0-471-96372-0

Typeset in 10/12pt Times by Alden Bookset, Didcot, Oxon
Printed and bound in Great Britain by Biddles Ltd, Guildford and King's Lynn

This book is printed on acid-free paper responsibly manufactured from sustainable forestation, for which at least two trees are planted for each one used for paper production.

Introduction to the Series

The chemistry of heterocyclic compounds is one of the most complex and intriguing branches of organic chemistry, of equal interest for its theoretical implications, for the diversity of its synthetic procedures, and for the physiological and industrial significance of heterocycles.

The Chemistry of Heterocyclic Compounds, published since 1950 under the initial editorship of Arnold Weissberger, and later, until Dr Weissberger's death in 1984, under our joint editorship, has attempted to make the extraordinarily complex and diverse field of heterocyclic chemistry as organized and readily accessible as possible. Each volume has traditionally dealt with syntheses, reactions, properties, structure, physical chemistry, and utility of compounds belonging to a specific ring system or class (e.g. pyridines, thiophenes, pyrimidines, three-membered ring systems). This series has become the basic reference collection for information on heterocyclic compounds.

Many broader aspects of heterocyclic chemistry are recognized as disciplines of general significance that impinge on almost all aspects of modern organic, medicinal and biochemistry, and for this reason we initiated several years ago a parallel series entitled *General Heterocyclic Chemistry* which treated such topics as nuclear magnetic resonance, mass spectra, and photochemistry of heterocyclic compounds, the utility of heterocycles in organic synthesis, and the synthesis of heterocycles by means of 1,3-dipolar cycloaddition reactions. These volumes were intended to be of interest to all organic, medicinal and biochemically oriented chemists, as well as to those whose particular concern is heterocyclic chemistry. It has, however, become increasingly clear that the above distinction between the two series was unnecessary and somewhat confusing, and we have therefore elected to discontinue *General Heterocyclic Chemistry* and to publish all forthcoming volumes in this general area in *The Chemistry of Heterocyclic Compounds* series.

The present volume extends our previous coverage of seven-membered nitrogen-containing rings to the parent monocyclic azepines. It is an authoritative, exhaustive and most welcome contribution to the literature of heterocyclic chemistry.

Edward C. Taylor

Department of Chemistry
Princeton University
Princeton, New Jersey 08544

Contents

Acknowledgements xi

List of Abbreviations xii

1. Introduction 1
2. Azepanes (Hexahydroazepines) 9
3. Azepanones 117
    - I. Azepan-2-ones (Caprolactams) 117
    - II. Azepan-3-ones 211
    - III. Azepan-4-ones 216
4. Azepanediones 251
    - I. Azepane-2,3-diones 251
    - II. Azepane-2,4-diones 260
    - III. Azepane-2,5-diones 266
    - IV. Azepane-2,6-diones 274
    - V. Azepane-2,7-diones 275
    - VI. Azepane-3,4-diones 281
    - VII. Azepane-3,5-diones 281
    - VIII. Azepane-3,6-diones* 282
    - IX. Azepane-4,5-diones 282
5. Tetrahydroazepines 288
    - I. 2,3,4,5-Tetrahydro[1H]azepines 288
    - II. 2,3,4,7-Tetrahydro[1H]azepines 304
    - III. 2,3,6,7-Tetrahydro[1H]azepines 314
    - IV. 3,4,5,6-Tetrahydro[2H]azepines 321

6.	Tetrahydroazepinones	359
	I. 1,3,4,5-Tetrahydro[2*H*]azepin-2-ones	359
	II. 1,3,4,7-Tetrahydro[2*H*]azepin-2-ones	367
	III. 1,3,6,7-Tetrahydro[2*H*]azepin-2-ones	372
	IV. 1,5,6,7-Tetrahydro[2*H*]azepin-2-ones	379
	V. 3,4,5,6-Tetrahydro[2*H*]azepin-2-ones*	–
	VI. 1,2,4,5-Tetrahydro[3*H*]azepin-3-ones*	–
	VII. 1,2,4,7-Tetrahydro[3*H*]azepin-3-ones*	–
	VIII. 1,2,6,7-Tetrahydro[3*H*]azepin-3-ones	392
	IX. 2,4,5,6-Tetrahydro[3*H*]azepin-3-ones*	–
	X. 4,5,6,7-Tetrahydro[3*H*]azepin-3-ones*	–
	XI. 1,2,3,5-Tetrahydro[4*H*]azepin-4-ones*	–
	XII. 1,2,3,7-Tetrahydro[4*H*]azepin-4-ones	393
	XIII. 1,5,6,7-Tetrahydro[4*H*]azepin-4-ones	395
	XIV. 2,3,5,6-Tetrahydro[4*H*]azepin-4-ones	399
	XV. 3,5,6,7-Tetrahydro[4*H*]azepin-4-ones	400
7.	Tetrahydroazepinediones	403
	I. 2,3,4,5-Tetrahydro[1*H*]azepine-2,3-diones	403
	II. 2,3,4,7-Tetrahydro[1*H*]azepine-2,3-diones*	–
	III. 2,3,6,7-Tetrahydro[1*H*]azepine-2,3-diones*	–
	IV. 3,4,5,6-Tetrahydro[2*H*]azepine-2,3-diones*	–
	V. 2,3,4,5-Tetrahydro[1*H*]azepine-2,4-diones	405
	VI. 2,3,4,7-Tetrahydro[1*H*]azepine-2,4-diones*	–
	VII. 3,4,5,6-Tetrahydro[2*H*]azepine-2,4-diones*	–
	VIII. 2,3,4,5-Tetrahydro[1*H*]azepine-2,5-diones	406
	IX. 2,5,6,7-Tetrahydro[1*H*]azepine-2,5-diones	407
	X. 3,4,5,6-Tetrahydro[2*H*]azepine-2,5-diones*	–
	XI. 2,3,6,7-Tetrahydro[1*H*]azepine-2,6-diones*	–
	XII. 2,5,6,7-Tetrahydro[1*H*]azepine-2,6-diones*	–
	XIII. 3,4,5,6-Tetrahydro[2*H*]azepine-2,6-diones*	–
	XIV. 2,3,4,7-Tetrahydro[1*H*]azepine-2,7-diones	408
	XV. 2,3,6,7-Tetrahydro[1*H*]azepine-2,7-diones	409
	XVI. 2,3,4,5-Tetrahydro[1*H*]azepine-3,4-diones*	–
	XVII. 2,3,4,7-Tetrahydro[1*H*]azepine-3,4-diones*	–
	XVIII. 3,4,5,6-Tetrahydro[2*H*]azepine-3,4-diones*	–
	XIX. 2,3,4,5-Tetrahydro[1*H*]azepine-3,5-diones*	–
	XX. 3,4,5,6-Tetrahydro[2*H*]azepine-3,5-diones*	–

Contents ix

XXI.	2,3,6,7-Tetrahydro[1H]azepine-3,6-diones*	–
XXII.	3,4,5,6-Tetrahydro[2H]azepine-3,6-diones*	–
XXIII.	2,3,4,5-Tetrahydro[1H]azepine-4,5-diones	410
XXIV.	3,4,5,6-Tetrahydro[2H]azepine-4,5-diones	411
XXV.	3,4,5,6-Tetrahydro[2H]azepine-4,6-diones*	–
XXVI.	3,4,5,6-Tetrahydro[2H]azepine-5,6-diones*	–

8. Dihydroazepines 412
   I. 2,3-Dihydro[1H]azepines 412
   II. 2,5-Dihydro[1H]azepines 423
   III. 2,7-Dihydro[1H]azepines 426
   IV. 4,5-Dihydro[1H]azepines 427
   V. 3,4-Dihydro[2H]azepines 442
   VI. 3,6-Dihydro[2H]azepines 444
   VII. 5,6-Dihydro[2H]azepines 444
   VIII. 4,5-Dihydro[3H]azepines 445

9. Dihydroazepinones 449
   I. 1,3-Dihydro[2H]azepin-2-ones 449
   II. 1,5-Dihydro[2H]azepin-2-ones 467
   III. 1,7-Dihydro[2H]azepin-2-ones 472
   IV. 3,4-Dihydro[2H]azepin-2-ones 473
   V. 3,6-Dihydro[2H]azepin-2-ones 474
   VI. 5,6-Dihydro[2H]azepin-2-ones* –
   VII. 1,2-Dihydro[3H]azepin-3-ones 474
   VIII. 2,4-Dihydro[3H]azepin-3-ones* –
   IX. 2,6-Dihydro[3H]azepin-3-ones* –
   X. 4,5-Dihydro[3H]azepin-3-ones* –
   XI. 4,7-Dihydro[3H]azepin-3-ones* –
   XII. 6,7-Dihydro[3H]azepin-3-ones* –
   XIII. 1,5-Dihydro[4H]azepin-4-ones* –
   XIV. 1,7-Dihydro[4H]azepin-4-ones* –
   XV. 2,3-Dihydro[4H]azepin-4-ones 478
   XVI. 3,5-Dihydro[4H]azepin-4-ones* –
   XVII. 3,7-Dihydro[4H]azepin-4-ones* –
   XVIII. 5,6-Dihydro[4H]azepin-4-ones* –

10. Dihydroazepinediones 481
    I. 2,3-Dihydro[1H]azepine-2,3-diones 482
    II. 3,4-Dihydro[2H]azepine-2,3-diones* –
    III. 3,6-Dihydro[2H]azepine-2,3-diones* –

IV.	3,4-Dihydro[2H]azepine-2,4-diones*	–
V.	2,5-Dihydro[1H]azepine-2,5-diones	482
VI.	5,6-Dihydro[2H]azepine-2,5-diones*	–
VII.	3,6-Dihydro[2H]azepine-2,6-diones*	–
VIII.	5,6-Dihydro[2H]azepine-2,6-diones*	–
IX.	2,7-Dihydro[1H]azepine-2,7-diones	485
X.	3,4-Dihydro[2H]azepine-3,4-diones*	–
XI.	4,5-Dihydro[3H]azepine-3,4-diones*	–
XII.	4,7-Dihydro[3H]azepine-3,4-diones*	–
XIII.	4,5-Dihydro[3H]azepine-3,5-diones*	–
XIV.	3,6-Dihydro[2H]azepine-3,6-diones*	–
XV.	4,5-Dihydro[1H]azepine-4,5-diones*	–
XVI.	4,5-Dihydro[3H]azepine-4,5-diones*	–

11. Azepines 487
    I. [1H]azepines 488
    II. [2H]azepines 521
    III. [3H]azepines 523
    IV. [4H]azepines 577

12. Azepinones 592
    I. [2H]azepin-2-ones 592
    II. [3H]azepin-3-ones 601
    III. [4H]azepin-4-ones 604

Index **607**

---

* Compounds marked with asterisks are not mentioned in the literature.

Acknowledgements

We cordially thank the following for helpful correspondence and discussions:

Professor L. A. Paquette, Ohio State University, Columbus, Ohio, USA
Professor T. Sano, Showa College, Tokyo, Japan
Dr H. McNab, University of Edinburgh, Scotland
Dr R. Hurnaus, Dr. Karl Thomae GmbH, Biberach, Germany
Dr I. R. Dunkin, University of Strathclyde, Glasgow, Scotland
Professor E. C. Taylor, Princeton University, Princeton, New Jersey, USA
Dr D. McFadzen, Organon Laboratories Ltd, Newhouse, Scotland
Ms L. Drummond, Organon Laboratories Ltd, Newhouse, Scotland
Ms L. Scott, Andersonian Library, University of Strathclyde, Glasgow, Scotland

List of Abbreviations

DMAD	Dimethyl acetylenedicarboxylate
DMSO	Dimethyl sulphoxide
DDQ	Dichlorodicyanobenzoquinone
DMF	N,N-Dimethylformamide
HMPA	Hexamethylphosphoric triamide
THF	Tetrahydrofuran
TCE	Tetracyanoethylene
GLC	Gas–liquid chromatography
Tos	Toluene-p-sulphonyl
PPA	Polyphosphoric acid
DBU	1,8-Diazabicyclo[5.4.0]undec-7-ene
NBS	N-Bromosuccinimide
Meerwein's reagent	Triethyloxonium borofluoride
LDA	Lithium diisopropylamine
Dibal	Diisobutyl aluminium hydride
DPPA	Diphenylphosphoryl azide
TMEDA	N,N,N',N'-Tetramethylethylenediamine

CHAPTER 1

Introduction

Although tricyclic and bicyclic azepines have been presented in this series,[1] monocyclic azepines have not been reviewed. Previous azepine compilations have included mono-, di- and tricyclic azepines[2,3] in some detail. In this volume we have collated all relevant data on monocyclic azepines in the style customary for this series. Where certain aspects of the subject have been satisfactorily reviewed previously,[3] we have summarised in an effort to avoid unhelpful duplication.

The subject is conveniently divided according to the state of oxidation of the heterocycle. Thus hexahydroazepines (azepanes) and their keto derivatives can be distinguished from tetrahydro- and dihydroazepines and their keto derivatives. Finally, azepines and azepinones provide examples at the highest state of oxidation (or dehydrogenation). Tables are provided where a significant number of examples justify this.

In order to systematise the presentation of the many possible structures containing carbonyl groups, we have referred to these compounds as di- or tetrahydroazepinones or -diones as appropriate, thereby maintaining the convention of naming the parent compounds according to the substituent with the highest oxidation state. For example, there are 26 distinct structures for tetrahydroazepinediones which are subdivided into 11 categories, depending on the position of the carbonyl groups: tetrahydroazepine-2,3-dione, tetrahydroazepine-2,4-dione, and so on. Of these possible structures, 18 are unknown at the present time. Similarly many of the possible dihydroazepinones and diones are not represented in the literature. For the sake of completeness, and to highlight the areas which have still to be explored, all the possible structural types have been included in the Contents list, but within the appropriate chapters we shall deal only with those structures for which compelling evidence has been adduced in the literature. Compounds for which no physical data were cited are only included in the tables when there is definite evidence of their existence.

Literature was searched via *Chemical Abstracts* up to late 1994 fairly exhaustively, and selectively for 1995. The nomenclature used is that currently seen in *Chemical Abstracts*.[4]

Table 1 Overview of monocyclic azepine structures

Chapter No.	Compound name	Structure
2	**Azepanes** (hexahydroazepines)	
3	**Azepanones**	
	I. Azepan-2-ones	
	II. Azepan-3-ones	
	III. Azepan-4-ones	
4	**Azepanediones**	
	I. Azepane-2,3-diones	
	II. Azepane-2,4-diones	
	III. Azepane-2,5-diones	
	IV. Azepane-2,6-diones	
	V. Azepane-2,7-diones	
	VI. Azepane-3,4-diones	
	VII. Azepane-3,5-diones	
	VIII. *azepane-3,6-diones*	
	IX. Azepane-4,5-diones	

Introduction

Table 1 (continued)

Chapter No.	Compound name	Structure
5	**Tetrahydroazepines**	
	I. 2,3,4,5-Tetrahydro[1*H*]-azepines	
	II. 2,3,4,7-Tetrahydro[1*H*]-azepines	
	III. 2,3,6,7-Tetrahydro[1*H*]-azepines	
	IV. 3,4,5,6-Tetrahydro[2*H*]-azepines	
6	**Tetrahydroazepinones**	
	I. 1,3,4,5-Tetrahydro[2*H*]-azepin-2-ones	
	II. 1,3,4,7-Tetrahydro[2*H*]-azepin-2-ones	
	III. 1,3,6,7-Tetrahydro[2*H*]-azepin-2-ones	
	IV. 1,5,6,7-Tetrahydro[2*H*]-azepin-2-ones	
	V. *3,4,5,6-Tetrahydro[2H]-azepin-2-ones*	
	VI. *1,2,4,5-Tetrahydro[3H]-azepin-3-ones*	
	VII. *1,2,4,7-Tetrahydro[3H]-azepin-3-ones*	
	VIII. 1,2,6,7-Tetrahydro[3*H*]-azepin-3-ones	
	IX. *2,4,5,6-Tetrahydro[3H]-azepin-3-ones*	
	X. *4,5,6,7-Tetrahydro[3H]-azepin-3-ones*	
	XI. *1,2,3,5-Tetrahydro[4H]-azepin-4-ones*	
	XII. 1,2,3,7-Tetrahydro[4*H*]-azepin-4-ones	
	XIII. 1,5,6,7-Tetrahydro[4*H*]-azepin-4-ones	
	XIV. 2,3,5,6-Tetrahydro[4*H*]-azepin-4-ones	
	XV. 3,5,6,7-Tetrahydro[4*H*]-azepin-4-ones	

Table 1 (continued)

Chapter No.	Compound name	Structure
7	**Tetrahydroazepinediones**	
	I. 2,3,4,5-Tetrahydro[1H]-azepine-2,3-diones
II. 2,3,4,7-Tetrahydro[1H]-azepine-2,3-diones
III. 2,3,6,7-Tetrahydro[1H]-azepine-2,3-diones
IV. 3,4,5,6-Tetrahydro[2H]-azepine-2,3-diones	
	V. 2,3,4,5-Tetrahydro[1H]-azepine-2,4-diones
VI. 2,3,4,7-Tetrahydro[1H]-azepine-2,4-diones
VII. 3,4,5,6-Tetrahydro[2H]-azepine-2,4-diones	
	VIII. 2,3,4,5-Tetrahydro[1H]-azepine-2,5-diones
IX. 2,5,6,7-Tetrahydro[1H]-azepine-2,5-diones
X. 3,4,5,6-Tetrahydro[2H]-azepine-2,5-diones	
	XI. 2,3,6,7-Tetrahydro[1H]-azepine-2,6-diones
XII. 2,5,6,7-Tetrahydro[1H]-azepine-2,6-diones
XIII. 3,4,5,6-Tetrahydro[2H]-azepine-2,6-diones	
	XIV. 2,3,4,7-Tetrahydro[1H]-azepine-2,7-diones
XV. 2,3,6,7-Tetrahydro[1H]-azepine-2,7-diones	
	XVI. 2,3,4,5-Tetrahydro[1H]-azepine-3,4-diones
XVII. 2,3,4,7-Tetrahydro[1H]-azepine-3,4-diones
XVIII. 3,4,5,6-Tetrahydro[2H]-azepine-3,4-diones	

Introduction

Table 1 (continued)

Chapter No.	Compound name	Structure
	XIX. 2,3,4,5-Tetrahydro[1H]-azepine-3,5-diones XX. 3,4,5,6-Tetrahydro[2H]-azepine-3,5-diones	
	XXI. 2,3,6,7-Tetrahydro[1H]-azepine-3,6-diones XXII. 3,4,5,6-Tetrahydro[2H]-azepine-3,6-diones	
	XXIII. 2,3,4,5-Tetrahydro[1H]-azepine-4,5-diones XXIV. 3,4,5,6-Tetrahydro[2H]-azepine-4,5-diones	
	XXV. 3,4,5,6-Tetrahydro[2H]-azepine-4,6-diones XXVI. 3,4,5,6-Tetrahydro[2H]-azepine-5,6-diones	
8	**Dihydroazepines**	
	I. 2,3-Dihydro[1H]-azepines II. 2,5-Dihydro[1H]-azepines III. 2,7-Dihydro[1H]-azepines IV. 4,5-Dihydro[1H]-azepines	
	V. 3,4-Dihydro[2H]-azepines VI. 3,6-Dihydro[2H]-azepines VII. 5,6-Dihydro[2H]-azepines	

Table 1 (continued)

Chapter No.		Compound name	Structure
	VIII.	4,5-Dihydro[3H]-azepines	
9		**Dihydroazepinones**	
	I.	1,3-Dihydro[2H]-azepin-2-ones	
	II.	1,5-Dihydro[2H]-azepin-2-ones	
	III.	1,7-Dihydro[2H]-azepin-2-ones	
	IV.	3,4-Dihydro[2H]-azepin-2-ones	
	V.	3,6-Dihydro[2H]-azepin-2-ones	
	VI.	*5,6-Dihydro[2H]-azepin-2-ones*	
	VII.	1,2-Dihydro[3H]-azepin-3-ones	
	VIII.	*2,4-Dihydro[3H]-azepin-3-ones*	
	IX.	*2,6-Dihydro[3H]-azepin-3-ones*	
	X.	4,5-Dihydro[3H]-azepin-3-ones	
	XI.	4,7-Dihydro[3H]-azepin-3-ones	
	XII.	6,7-Dihydro[3H]-azepin-3-ones	
	XIII.	*1,5-Dihydro[4H]-azepin-4-ones*	
	XIV.	*1,7-Dihydro[4H]-azepin-4-ones*	
	XV.	2,3-Dihydro[4H]-azepin-4-ones	
	XVI.	*3,5-Dihydro[4H]-azepin-4-ones*	
	XVII.	*3,7-Dihydro[4H]-azepin-4-ones*	
	XVIII.	*5,6-Dihydro[4H]-azepin-4-ones*	

Introduction

Table 1 (continued)

Chapter No.	Compound name	Structure
10	**Dihydroazepinediones**	
	I. 2,3-Dihydro[1H]-azepine-2,3-diones II. 3,4-Dihydro[2H]-azepine-2,3-diones III. 3,6-Dihydro[2H]-azepine-2,3-diones	
	IV. 3,4-Dihydro[2H]-azepine-2,4-diones V. 2,5-Dihydro[1H]-azepine-2,5-diones VI. 5,6-Dihydro[2H]-azepine-2,5-diones	
	VII. 3,6-Dihydro[2H]-azepine-2,6-diones VIII. 5,6-Dihydro[2H]-azepine-2,6-diones IX. 2,7-Dihydro[1H]-azepine-2,7-diones	
	X. 3,4-Dihydro[2H]-azepine-3,4-diones XI. 4,5-Dihydro[3H]-azepine-3,4-diones XII. 4,7-Dihydro[3H]-azepine-3,4-diones	
	XIII. 4,5-Dihydro[3H]-azepine-3,5-diones XIV. 3,6-Dihydro[2H]-azepine-3,6-diones	
	XV. 4,5-Dihydro[1H]-azepine-4,5-diones XVI. 4,5-Dihydro[3H]-azepine-4,5-diones	

Table 1 (continued)

Chapter No.	Compound name	Structure
11	**Azepines** I. [1H]azepines II. [2H]azepines III. [3H]azepines IV. [4H]azepines	
12	**Azepinones** I. [2H]Azepin-2-ones II. [3H]Azepin-3-ones III. [4H]Azepin-4-ones	

In Table 1, structures which are theoretically possible but which have not been found in the literature are shown in *italics*. For the sake of completeness, these structures are shown in the above diagrams, but they are not referred to further in the text. The authors hope that by highlighting the missing structures in this manner, other researchers who are involved in synthetic heterocyclic chemistry may be inspired to fill the gaps.

REFERENCES

1. B. Renfroe, C. Harrington and G.R. Proctor, 'Azepines, Part 1', in A. Rosowski (Ed.), *The Chemistry of Heterocyclic Compounds*, Wiley, Chichester, 1984.
2. J.A. Moore and E. Mitchell, in R.C. Elderfield (Ed.), *Heterocyclic Compounds*, Vol. 9, Wiley, Chichester, 1967.
3. R.K. Smalley, *Comprehensive Heterocyclic Chemistry*, Vol. 7, Pergamon Press, Oxford, 1984, pp. 491–546.
4. *Naming and Indexing of Chemical Substances for Chemical Abstracts, Appendix IV from the Chemical Abstracts 1977 Index guide*, American Chemical Society, Columbus, OH, 1977.

CHAPTER 2

Azepanes (Hexahydroazepines)

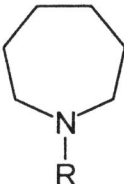

Originally the parent (R = H) of this series was referred to as 'hexamethyleneimine.' Later, and more correctly, these substances were seen as hexahydroazepines, but recently the name azepane has been preferred. The literature contains several thousand references to azepanes since it has been popular, particularly in the patent literature, to attach the azepane nucleus to very many structures in the same way as piperidine and pyrrolidine have been used. We are, therefore, obliged to restrict coverage to those cases in which the azepane portion comprises the major part of a molecule.

PREPARATION

Early work has been very adequately covered in a previous review.[1]

1. Reaction of 1,6-Difunctional Hexanes with Ammonia or Amines

This approach, introduced in 1928,[2] has continued to be employed from time to time with some variations.[3] In the latter case, 1,6-diaminohexane reacted with butane-1,4-diol in the presence of Raney nickel giving 1, although Adkins and co-workers had developed high-pressure treatment of various diols with amines in the presence of hydrogen and 'CuCrO' 20 years earlier.[4,5] From the appropriate diols azepanes such as 2 were obtained in moderate yields.[5] The use of 1,6-dihalohexanes[2] has limitations, side reactions tending to intervene.

However, the presence of electron-withdrawing groups α- to halogen atoms has a beneficial effect. Thus, following a precedent,[6] Italian chemists developed a mild reaction of benzylamine with the dibromo diester 3 which provided the azepane diester 4 in slightly impure form.[7] On the other hand, in Russia, α,ω-dihalo-monocarboxylic acids 5 reacted with a variety of amines yielding the α-amino acid derivatives 6.[8]

Organometallic chemistry has more recently had an impact on this type of azepane synthesis. Tetracarbonyl hydridoferrate [HFe(CO)$_4^-$], used stoichiometrically, promoted the reductive cyclisation of adipodialdehyde with primary amines,[9] but an improvement involves catalytic ruthenium halide plus triphenyl- or tributylphosphine promotion of reactions between primary amines and hexane-1,6-diol[10,11] at 180 °C under an inert atmosphere. This gives reasonable, somewhat variable, yields of 1-substituted azepanes: in general, RuCl$_3$.3H$_2$O + Ph$_3$P is advised for aromatic amines, whereas RuCl$_3$.3H$_2$O + Bu$_3$P is favoured for reactions involving aliphatic amines. A further useful development demonstrated the use of hexane-1,2,6-triol which gave passable yields of 3-hydroxy-*N*-arylazepanes **7**, but unfortunately failed for benzylamine and furfurylamine.[12]

2. Cyclisation of 1,6-Haloamines and Related Molecules

This has been a popular and successful approach which could be regarded as a stepwise variation of the previously discussed direct reaction of 1,6-dihalohexanes with amines. The first and relatively inefficient preparation of azepane,[13] later improved,[1] involved treating 1,6-haloamines with aqueous alkali, but the modification involving cyclisation of 6-bromo-*N*-tosylamine (**8**) to *N*-tosylazepane (**9**, R = tosyl) was an improvement.[14]

Access to useful intermediate bromoamino esters (**10**, R = CH_3, Ph, CH_2Ph) was gained by reaction of 1,5-dibromopentane with the isocyano esters (**11**, R = CH_3, Ph, CH_2Ph) and sodium hydride followed by treatment with hydrochloric acid. Cyclisation was effected with triethylamine, the overall yields of azepane esters (**12**) were good and the corresponding amino acids were obtained in good to excellent yield by alkaline hydrolysis.[15]

The amino acids **6** (R = H, CH_3, C_2H_5) have been obtained by cyclising the acyclic haloamino acids **5** (X = Cl, Y = NH_2, $NHCH_3$, NHC_2H_5 etc.) by boiling aqueous solutions of their alkali metal salts for 10 h along with some potassium iodide (~10%).[16] However, this approach did not appear to offer advantages over the direct method alluded to previously.[8] A chiral version has been reported[17] in which the chiral haloamino esters required for cyclisation (NaH/THF/RT) (RT = room temperature) were obtained by chirally induced substitution on imines **13** using (−)-2-hydroxypinan-3-one. The enantiomeric excess (ee) obtained was better than 95% in products **12**; in the case of **12** (R = CH_3) the absolute configuration was considered to be R.

Another interesting chiral synthesis proceeded by cyclisation on to a chiral imidazolidinone (14): aqueous acid treatment of the bicyclic product removed the auxiliary and provided the (S)-2-methylazepine-2-carboxylic acid (acid from 12, R = CH$_3$).[18]

The electrophilic alkyl halide component of these cyclisations can be replaced by an alcohol. Thus amino alcohols 15 (R = CH$_3$, C$_2$H$_5$, C$_3$H$_7$, C$_4$H$_9$) were transformed into the respective azepanes (9) by heating over alumina at 275 °C.[19] On occasion the alcohol portion required for cyclisation has been

RNH(CH$_2$)$_6$OH

(15)

(16) (17) (18)

generated *in situ* from a carbonyl component. For example, catalytic hydrogenation of 6-amino-6-desoxy-D-glucose (16) yielded the tetrahydroxyazepane (17).[20] The equivalence of furans to 1,4-diones has been exploited by catalytic (H$_2$/Pt–C/220 °C) reduction of some furan amines (18, R, R^1 = H or CH$_3$) which led to 2-propylazepanes (19, R, R^1 = H or CH$_3$).[21]

(19) (20) (21)

A further variation on this theme allowed reductive cyclisation of keto hydrazones (20, R = H, OCH$_3$) over nickel to produce 2-arylazepanes (21, R = H, OCH$_3$).[22]

The electrophilic iminium ion cyclisation technique of Hamersma and Speckamp[23,24] has been adapted most ingeniously for synthesis of azepinothiazolidine compounds (e.g. 23 from 22) from which sulphur removal leads to azepane 24.[25]

Azepanes (Hexahydroazepines)

(22) **(23)** **(24)**

Cyclisation of ω-halo-tertiary amines leads to the production of cyclised quaternary salts: this approach has been extensively exploited for azepane synthesis, particularly in connection with the search for analgesic substances.

(25) **(26)** **(27)**

The latter requirement provided the stimulus for syntheses of 3,3- and 4,4-disubstituted azepane derivatives. Accordingly the haloamine **25** (available by sequential alkylations of phenylacetonitrile) on heating gave **26**, which was demethylated on pyrolysis[26] to the corresponding 4-cyano-4-phenyl-1-methyl-azepane from which various esters (**27**) became accessible. In similar fashion, many analogous materials have been obtained.[27–30]

3. Reduction of Caprolactams and Other Ketoazepanes

This is arguably the most popular method for obtaining azepanes. Caprolactams (Chapter 3, Section I) being widely available, their chemistry, including various reductive processes, has been well studied. The reagent of choice for direct reduction of the lactams is LiAlH$_4$ and it was used more than 40 years ago for conversions (**28** → **29**,[31] **30** → **31**,[32] **32** → **33**[33]).

(28) **(29)** **(30)**

(31) **(32)** **(33)**

Simultaneous or consecutive reduction of two carbonyl functionalities is known: thus the lactam-ester **34** (R = H) was converted into the hydroxymethyl azepane **35** (R = H).[34] In another example the conversion **34** → **35** (R = Ph) allowed access eventually to a homobenzomorphan (**36**).[35]

(34) **(35)** **(36)** **(37)**

Since caprolactams can be alkylated at C-3, subsequent LiAlH$_4$ reduction of the products makes 3-substituted azepanes available. Thus alkylation of caprolactam gave **37** (R^1 = CH$_2$Ph, R^2 = CH$_3$)[36] and **37** (R^1 = CH$_3$, R^2 = CO$_2$C$_2$H$_5$),[37] which on reduction yielded **38** (R^1 = CH$_2$Ph, R^2 = CH$_3$ and R^1 = CH$_3$, R^2 = CO$_2$C$_2$H$_5$, respectively). A variation on this theme involves exploitation of the 2,3-diketoazepane **39** (from caprolactam), which underwent Grignard addition at C-3 followed by reduction to give **40**.[38]

(38) **(39)** **(40)**

LiAlH$_4$ reduction of chiral lactams produces chiral azepanes. The former have been obtained either by direct cyclisation as in the case of methyl L-lysinate, which led to **41**,[39] or by Beckmann rearrangement of a chiral oxime (**42**),[40] which gave **43**. Reductions proceeded to **44** and **45**, respectively, in fair optical purity.

Azepanes (Hexahydroazepines)

(41) (42) (43) (44) (45)

Sodium borohydride in acidified ethanol is known to reduce caprolactams to 2-ethoxyazepanes[41] provided the nitrogen atom was protected as a urethane, that is, lactams **46** (R = OC$_2$H$_5$, OCH$_2$Ph, O^{t}Bu) gave the products **47** (R = OC$_2$H$_5$, OCH$_2$Ph, O^{t}Bu) at −6 to 0 °C. This method complements those generally employed for these synthetically useful substances via imides[42] or by anodic oxidation.[43] Thus displacement of the ethoxy group at C-2 could be effected by various nucleophiles leading to structures **48** (R = CN, N$_3$ and 3-indolyl).[44]

(46) (47) (48) (49) (50)

Reduction of caprolactams using a borane–dimethyl sulphoxide combination has been reported.[45] Capitalising on the successful Schmidt reaction on chiral β-keto esters (**49**), the authors reduced the chiral products (lactams) to chiral azepane esters (**50**, R = PhCH$_2$, etc.; see Table 1) in generally good optical purity.

Very useful preparations of both 2-alkyl- and 2,2-dialkylazepanes have arisen from the work of Yamamoto and co-workers, who have studied combined Beckmann rearrangement–reductive alkylation protocols applied to the O-sulphonyl derivatives of cyclohexanone oximes.[46–48] In the first place it was

(51) (52) (53)

(54) (55)

shown that trialkylaluminium reagents converted oxime sulphonates to tetrahydroazepines alkylated at C-2 (51), which were reduced *in situ* by diisobutylaluminium hydride. A full recipe has been published for the case of 2-propylazepane (52).[49] It was furthermore demonstrated that if simple Grignard reagents in non-polar solvents replaced ^{i}Bu$_2$AlH in this sequence, then the products were 2,2-dialkylamines.[50] This has the clear advantage of giving access to substitution by two different alkyl groups (e.g. 53). It has long been known that caprolactams react with Grignard reagents to give 2,2-homodisubstituted azepanes.[32b]

Other azepanones have been reduced to azepanes. For example, the β-keto esters 54 and 55 have been catalytically hydrogenated over Raney Ni to give the corresponding hydroxy esters (56, 57)[51] and several 4-ketoazepanes (58) have been reacted with aryllithium or -magnesium reagents to yield products (59) of pharmacological interest.[52]

(56) (57) (58) (59)

4. Reduction of Azepines and Dihydro- and Tetrahydroazepines

The catalytic hydrogenation of azepines has been used on several occasions, principally to demonstrate that these molecules contained three double bonds in a seven-membered cyclic assembly. This was particularly useful in the early days of azepine synthesis by nitrene insertion to benzene by photolytic and thermal means.[53-55] The ring expansion work of Johnson and co-workers (Chapter 8) was supported by evidence obtained by catalytic hydrogenation.[56,57] All of these involved reduction of [1H]azepine derivatives, but the [3H]azepine 60 has also been likewise reduced to the azepanyl phosphonate 61.[58] Catalysts used have usually been platinum or palladium supported on carbon or occasionally Adam's platinum oxide.

Not surprisingly, dihydroazepines (e.g. **62**) can be reduced catalytically[59] and several examples of tetrahydroazepine hydrogenations are known. Both 2-alkyl-[60] and 2-aryl-[61] 3,4,5,6-tetrahydro[2H]azepines (**63**, R = alkyl, aryl) have given substituted azepanes, as have 2,3,6,7-tetrahydro[2H]azepines (**64**[62] and **65**[51,63]). 2,3,4,7-Tetrahydroazepine (**66**) was both reduced and hydrolysed using palladium on charcoal.[72]

Nitrones (e.g. N-oxide of **63**) prepared recently by hydroxylamine–alkyne cyclisations (Chapter 5, Section IV) react with Grignard reagents (e.g. vinylmagnesium bromide) to give 2,2-disubstituted N-hydroxyazepanes[234,235] which are convertible to azepanes with $TiCl_3$.

5. C–C Bond Cyclisations

Syntheses of azepanes by cyclisations involving carbon–carbon bond formation are of relatively recent origin for the most part. Apparently Cope and Burrows[64] were the first to recognise that iminium ions generated during

'Clarke–Eschweiler' methylation could lead to azepanes. In this way the alkenamine **67** (R = H) treated with formaldehyde and formic acid yielded the azepane **68** in 39% yield, along with the expected N,N-dimethyl starting

(67)   **(68)**   **(69)**   **(70)**

material (**67**, R = CH_3). This cyclisation would appear to proceed via the iminium ion **69** in the *endo* mode; it is surprising that none of the alternative *exo* mode product (**70**) was found. Twenty years later[65] a similar type of reaction was applied (CH_2O/CF_3CO_2H) to the silylamine **71** to give a very good yield of the azepane **72**. An alkynylsilane (**73**) was also shown[66] to participate via an iminium ion, leading to the interesting allenic structure **74**.

(71)   **(72)**

$(CH_3)_3SiCH_2C \equiv C(CH_2)_4NH^iPr$

(73)   **(74)**

Palladium chemistry has helped in promoting radical-like cyclisations of α-halo esters and ketones. Thus the iodoketone **75** gave the N-tosyl azepanes **76** along with isomeric piperidines (**77**) using $Pd(PPh_3)_4$ (10%) in dioxane at room temperature.[67] The radical-like features of this type of reaction were revealed by Curran and Chang,[68] who demonstrated that identical product mixtures were obtained from unsaturated α-iodocarbonyls using either Pd^0 or $(Bu_3Sn)_2$; compound **78** along with several others was isolated from the iodo

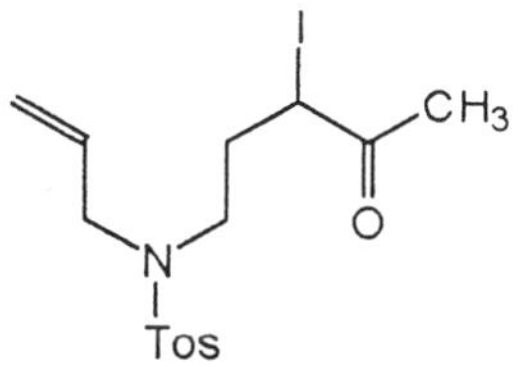

Azepanes (Hexahydroazepines)

(75) 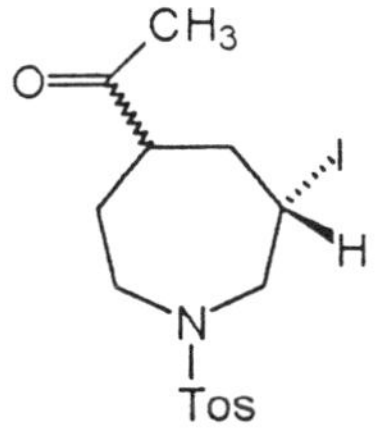 (76) 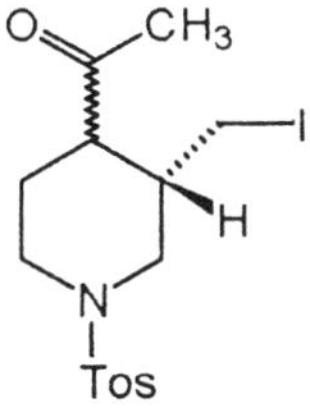 (77) 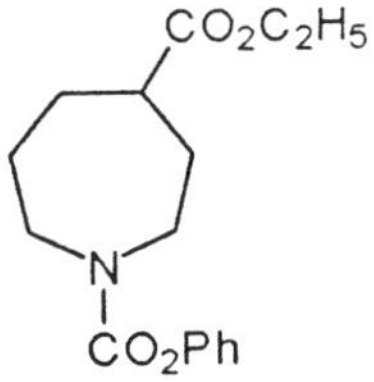 (78) 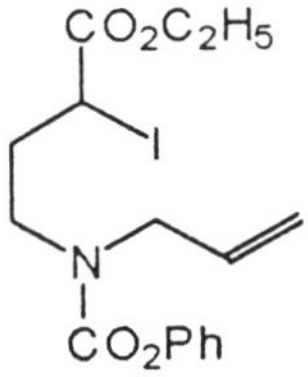 (79)

ester **79**. As yet these interesting studies could not be said to constitute viable syntheses for medium-scale work.

A [4 + 3] reaction is known[69] whereby the bis(methylthiomethyl)amine **80** reacted with diketene silyl acetal (**81**) in the presence of trimethylsilyl triflate to give an acceptable (63%) yield of the azepane **82**.

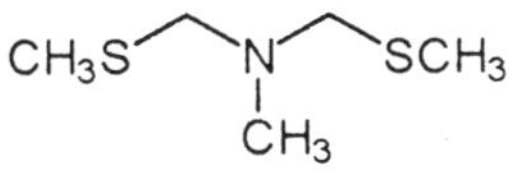

(80) 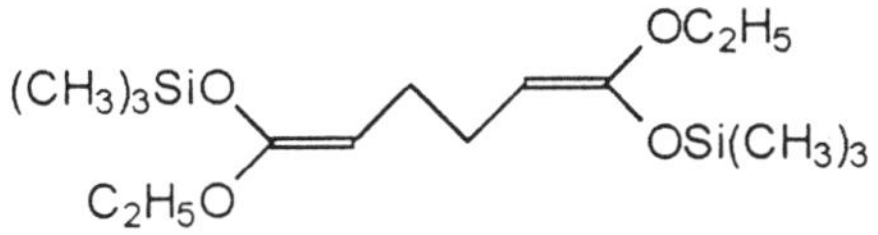 (81)

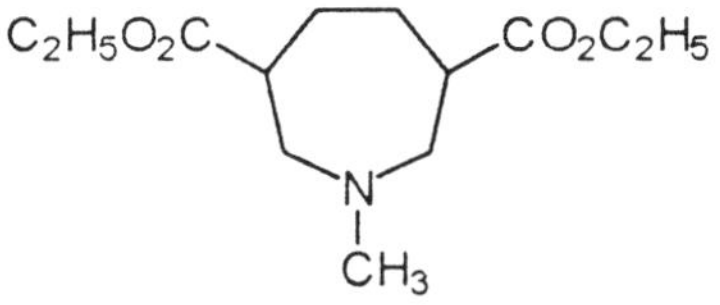

(82)

6. C–N Bond Cyclisations

In this section we consider processes which are effectively described as intramolecular hydroamination:

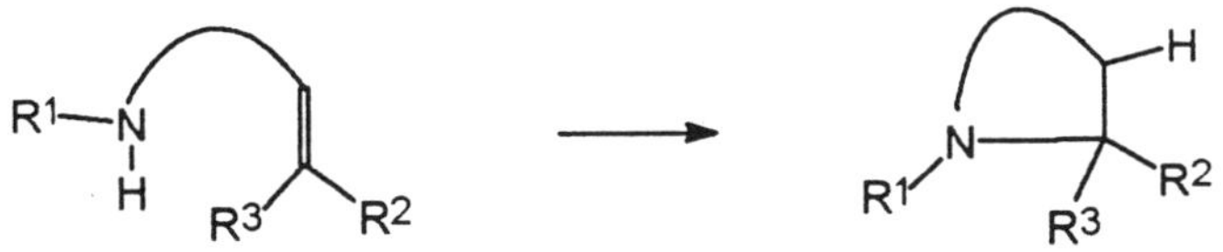

Photochemically induced versions are known: styrylamines (**83**; $n = 4, 5$) give azepanes but not as major products. Thus **83** ($n = 4$) did give **84** especially in

(**83**)   (**84**)   (**85**)   (**86**)

acetonitrile solution but the major product was the piperidine **85**.[70] Perversely **83** ($n = 5$) showed a reversal of these ratios, the azepane **86** being a minor product detected with the azocine **87**. One of the products formed by irradiating **88** was the azepane **89** (35%).[71] However these processes do not yet have the status of useful syntheses.

(**87**)   (**88**)   (**89**)   (**90**)

Allenic amines are showing some promise. Treatment of the allenic salt **90** with I_2 in CH_2Cl_2 followed by triethylamine gave the product **91** in 20% yield.[72] Cyclisation of the base (from **90**) was effected using $PdCl_2(PhCN)_2/CO/MeOH/Et_3N$, the product (**92**) being obtained in 23% yield.[73,74]. The *N*-tosyl analogues

(**91**)   (**92**)

(**93**)   (**94**)

could not be cyclised. An organolanthanide-induced cyclisation of the alkenamine **93** to the azepane **94** has been reported;[75] hopefully this type of process could be adapted for working on a moderate scale.

7. Ring Contractions and Fragmentations

Some interesting azepane structures have been reported arising from the above approaches. The 1-aza[3.2.1]bicyclooctanone **95** (originating from the diester **96**) was reduced (Wolff–Kishner) and subjected to Hofmann degradation

(95) (96) (97) (98)

(MeI/Ag$_2$O) to give the 4-vinylazepane **97**, which was converted into several other azepanes.[76] The tosyloxy quinuclidene **98** underwent fragmentation with alkali yielding the methylene azepanes **99** (R = H, CH$_3$).[77]

(99) (100) (101) (102)

Favorskii rearrangement of α-halolactams allowed access to several cyclic α-amino acids including the azepane acid **100**.[78,79] Certain 2-arylpiperidinium methylides, generated from silanes **101** by CsF treatment, yielded interesting structures (**102**, R^1, R^2, various: H, CH$_3$, OCH$_3$), thermolysis of the latter generating mainly styrylamines (**103**) with azepanes (**104**) as the minor products, produced presumably by a Stevens type of rearrangement.[80]

(103) (104) (105) (106)

8. Ring Expansion Processes

The earliest example of pyrrolidine ring expansion dates from 1964 when it was shown[81] that the haloamine **105** (R = CH$_3$, X = Cl) reacted with nucleophiles

(OH⁻, NH$_2$CH$_2$Ph) to give some of the normal displacement products **105** (R = CH$_3$, X = OH or NHCH$_2$Ph) along with azepanes **106** (R = CH$_3$, Y = OH or NHCH$_2$Ph). It was considered that these reactions involved the azetidinium ion (**107**, R = CH$_3$).

(**107**)   (**108**)   (**109**)   (**110**)

More recently, a thermal version has been reported in which radicals are implicated. Thus, conversion of the alcohol **105** (R = CH$_2$Ph, X = OH) to certain thiono- and dithionocarbonates (e.g. **105**, R = CH$_2$Ph, X = OCS.OPh, OCSSPh) allowed thermal reactions to proceed which relied on the ease of carbonyl sulphide (COS) elimination. This led to the production of azepanes (**106**, R = CH$_2$Ph, Y = OPh or SPh) probably via the ion **107** (R = CH$_2$Ph): again pyrrolidine derivatives (**105**, R = CH$_2$Ph, Y = OPh or SPh) intruded.[82]

Expansion of the tetrahydropyridinium salt **108** through the aziridinium salt **109** gave the 3-substituted azepane **110**.[83] The extensive researches of Yamamoto and co-workers into organoaluminium reagents uncovered a ring expansion/nitrogen insertion reaction involving hydroxylamine carbonates.[84] For example, the carbonate **111** reacted with the appropriate organoaluminium

(**111**)   (**112**)   (**113**)

reagent (R = CH$_3$ or ^{n}BuC≡C) giving azepanes (**112**, R = CH$_3$ or ^{n}BuC≡C) in good yields.

As discussed in Section 3, derivatives of cyclohexanone oximes undergo Beckmann rearrangement/reduction in 'one-pot' processes which overall could be regarded as ring expansions. In addition to those already mentioned,[45-60] we should comment here on the fact that azepanes have been found, along with several other products, when a number of cyclohexenone oximes were treated with LiAlH$_4$.[85] Indeed, oximes themselves have been subjected to the trialkylaluminium/diisobutylaluminium hydride protocol previously (Section 3) described for oxime sulphonates. In this way oxime **42** gave the azepane **45**, although no figure was given for the optical rotation.[46] A variation is known in which the mesyl derivative of cyclohexanone oxime can be treated sequentially with diethylaluminium iodide, phenylmagnesium bromide and diisobutylaluminium hydride to give an 81% yield of 2-phenylazepane (113).[86] Aluminium hydride has also been shown to induce ring expansion/reduction of the O-methyl ethers of cyclohexenone oximes.[87,88]

9. Alkenyl-Substituted Azepanes

2-Alkenyl-substituted azepanes (114) have received a good deal of attention, partly because of their accessibility and partly for their synthetic utility and structural relation to biologically active molecules.[89]

(114) (115) (116) (117)

Caprolactam acetals (115, $R^1 = CH_3, C_2H_5$) react with various activated methylene compounds such as malonates, β-keto esters and 1,3-diones, the products (114) being generally obtained in fair to good yield.[89-96] Enol ethers (116) have also been used;[96] this reaction may be promoted by piperidinium acetate,[97] although catalysis by nickel salts seems now to be favoured.[98-100] Thio analogues (117) appear to be more reactive; they give products (114) with activated methylene compounds in presence of triethylamine[101] or pyridine.[102] In some cases, heating the components together brings about the formation of products 114; in this respect malononitrile is particularly reactive.[103] Iminium salts (118, R = CH_3, CH_2Ph; X = OCH_3, SCH_3, Cl) have been shown to react with Meldrum's acid in the presence of triethylamine giving the useful products 119.[104,105]

(118) (119) (120) (121)

It is possible to deprotonate (LDA) the tetrahydroazepine 120 and to react the anion so formed with nitriles to yield azadienes (121, R = Ph or cyclohexyl).[106] In the presence of diethylaluminium chloride, silyl enol ethers react with cyclohexanone oxime mesylates giving 2-alkenylazepanes. For example, silyl enol ether 122 reacted with cyclohexanone oxime mesylate 123 providing the product 124 in 82% yield.[107]

(122) (123) (124)

(125) (126)

An interesting intramolecular aza-Wittig Reaction is recorded;[108] for instance triphenylphosphine treatment of the diketoazide 125 gave the 2-alkenylazepane

(127) Ph$_3$SiC≡CLi (128) (129)

126. When piperidone 127 reacted with lithium triphenylsilylacetylide (128), a ring expansion reaction ensued;[109] the structure of the product was later corrected[110] to 129.

PROPERTIES AND REACTIONS

Azepanes, as one would expect, are moderately strong bases, forming stable salts such as hydrochlorides, picrates, fumarates and hydrobromides. The tertiary bases (N-alkylazepanes) also form methiodides. pK_a values have been recorded for azepane (10.30),[156] N-methylazepane (10.19) and N-propylazepane (10.45).[111] The azepane bases are, in general, stable distillable liquids, oils or low-melting solids (see tables). Bifunctional bases have been shown to form metal complexes; for example, 3-aminoazepane (38, R^1 = H, R^2 = NH$_2$; and 44) forms Cu^{2+} complexes[39] and 1-azepanylacetic acid (9, R = CH$_2$CO$_2$H) forms Pd^{2+} and Pt^{2+} complexes.[112] N-Alkylation of N-unsubstituted azepanes is easily brought about by reaction with alkyl halides in the presence of Et$_3$N[113] or sodium hydroxide solution.[114] Acylation is also carried out using Et$_3$N[113] and sulphonation using pyridine.[61] N-Unsubstituted azepanes take part in the usual Michael-type additions with acrylonitrile[115] or ethylene sulphide[116] giving 9 (R = CH$_2$CH$_2$CN and CH$_2$CH$_2$SH, respectively). Reaction of azepane with

Azepanes (Hexahydroazepines)

(9)   (130)   (131)   (132)

hydrogen peroxide gave the hydroxylamine (9, R = OH).[117] N-Arylation has been achieved by use of benzynes[118] or with o- or p-nitrohalobenzenes.[119]

1. Substitution at 2-Position

This area has been extensively investigated. Seebach et al.[120] showed that N-nitrosoazepane (9, R = NO) could be deprotonated at C-2 (LDA/THF/ −80 °C) and reacted with electrophiles; for example benzophenone gave 130 [R^1 = NO, R^2 = CH(OH)Ph$_2$]. In a similar vein, Meyers and co-workers demonstrated that t-butylformamidines (9, R = CH=N^{t}Bu) could be deprotonated (^{n}BuLi) and reacted with electrophiles (PhSeSePh, ClCO$_2$CH$_3$, n-heptyl and n-propyl) to yield the corresponding 1,2-disubstituted azepanes (130, R^1 = CH=N^{t}Bu, R^2 = PhSe, CO$_2$CH$_3$, C$_7$H$_{15}$, C$_3$H$_7$). Removal of the N-protecting groups (KOH/CH$_3$) gave 2-substituted azepanes in acceptable yields.[121,122] The N-t-butoxycarbonyl group has also been shown to fulfil a similar role: in this case 9 (R = CO$_2^t$Bu) reacted in diethyl ether with sec-butyllithium/TMEDA and then treatment with dimethyl sulphate gave the product (130, R^1 = CO$_2^t$Bu, R^2 = CH$_3$); alternatively, addition to the lithio salt of DMF led to isolation of 130 (R^1 = CO$_2^t$Bu, R^2 = CHO).[123] Interestingly, repetition of this protocol can provide 2,7-disubstituted azepanes in which the substituents are mutually trans [131, E = CH$_3$, Si(CH$_3$)$_3$ and CHO].[123] It had previously been shown[124] that tributyltin chloride participated in these reactions leading to the isolation of 130 (R^1 = CO$_2^t$Bu, R^2 = Bu$_3$Sn). It has been pointed out that the successful lithiation of all of these N-substituted azepanes depends upon the formation of dipole-stabilised carbanions.[120–125] The o-iodobenzyl group [in 9, R = CH$_2$C$_6$H$_4$I(o)] has been used in samarium iodide-mediated C−C bond formation: thus, reaction with pentan-3-one led to 130 [R^1 = o-iodobenzyl, R^2 = C(OH)(C$_2$H$_5$)$_2$].[125]

Anodic electrolytic oxidation of N-formyl- and N-acetylazepanes has proved to be a very fruitful operation.[126,127] The products are C$_2$-alkoxy-N-formyl- or -N-acetylazepanes (130, R^1 = CHO, COCH$_3$, R^2 = OCH$_3$ or OC$_2$H$_5$) depending on which alcohol was employed. These methoxy- or ethoxyazepanes can be perceived as precursors to N-acyl iminium salts (132) and, as such, react with nucleophiles in situ giving C-2-substituted products; usually AlCl$_3$ has been used as promoter. In this way arenes and β-keto esters have been made to react giving a variety of products [130, R^1 = CHO,

R^2 = aryl or XC(CO$_2$C$_2$H$_5$)COY] which are listed in the tables.[128-131] This type of protocol has been extended to the production of 2-alkynylazepanes using bis-trimethylsilylacetylene.[132-134]

Intramolecular versions of *N*-acyliminium ion reactions have allowed cyclisations to be developed. Thus several 3-alkenyl- and alkynylalkyl-substituted 2-ethoxyazepanes have been cyclised under the influence of SnCl$_4$ at −23 °C (e.g. 133 → 134).[135] The required iminium ions have also been reached by a decarboxylative process:[136] acid 135 was treated with POCl$_3$ at 100 °C to yield the 1-azabicyclo[5.4.0]undecane 136.

(133)   (134)

(135)   (136)

2. Annulation to the Azepane Ring

In addition to the intramolecular iminium ion cyclisations referred to above, there are only a few other examples of ring annulation to azepanes. Thiazetidine dioxide (138) was isolated when sulphonyl chloride hydrochloride 137 was treated in chloroform with ammonia at 0 °C.[137] A Dieckmann cyclisation

(137)   (138)   (139)   (140)

applied to the diester 139 was reported to give (after hydrolysis) the expected ketone (140); unfortunately, no experimental details were provided.[138]

3. Ring-Expansion, Contraction or Fission

Certain ylids (**141**, R = H or CH$_3$, X = −), obtained by desilylation of **141** [R = H or CH$_3$, X = Si(CH$_3$)$_3$] by CsF in DMF, led to a mixture of products (**142** and **143**) which reacted with KOH or HCl giving **144** (R = H or CH$_3$).[139]

(141) (142)

(143) (144)

On the other hand, a contraction was seen when compound **76a** was treated with DBU in benzene at room temperature. The product (**145**) was also

(76a) (76b) (145) (146)

obtained from **76b** but more slowly and in poorer yield. The *cis*-isomer of **77** also provided **145** on treatment with DBU or KOtBu.[67]

It has been reported that the *trans*-diol **146** (R = H) or its acetate (R = CH$_3$CO) undergo ring contraction to **147** and **148** on treatment with various acetic acid/acetic anhydride combinations,[140] depending upon the temperature used. Compound **148** was shown to lie on the path between **146** and **147** and a plausible mechanism (involving **149**) was suggested.[140]

The N−C-2 bond of some azepanes has been shown to be vulnerable. Thus, when compound **150** was treated with methyl iodide, one of the products was

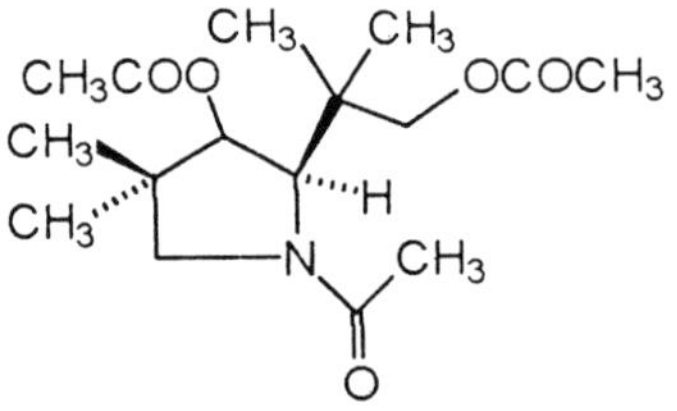

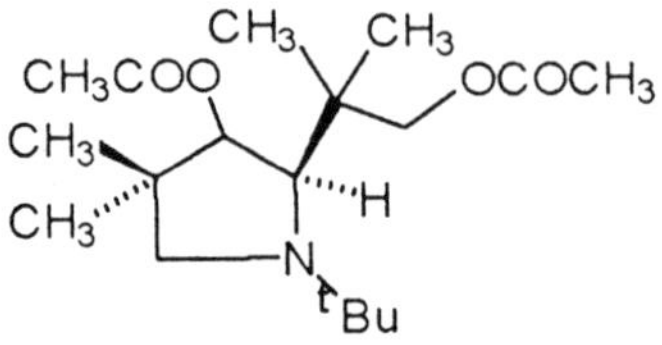

(147) (148) 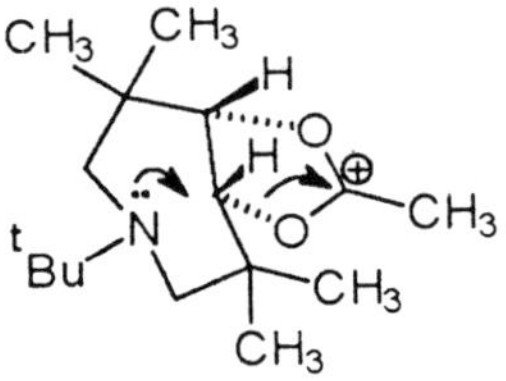 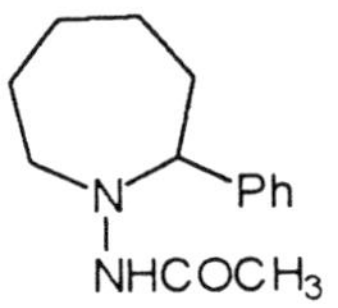 (149) (150)

a ring-opened salt (151).[141] Not surprisingly, N-phenylazepane[142] reacted with cyanogen bromide to give the bromocyanide 152.

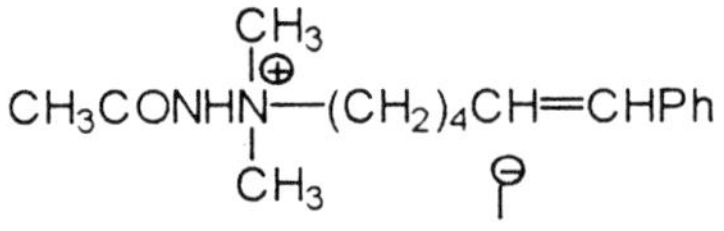

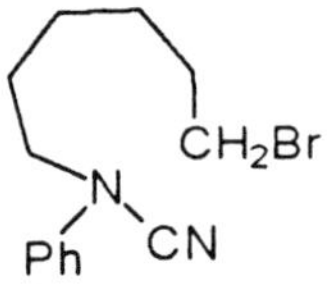

(151) (152)

4. Oxidation

The azepane ring appears to be stable to oxidation by permanganate, 97 being oxidised at the side-chain alkenyl group to yield the corresponding N-methylazepane-4-carboxylic acid (106, R = CH_3, Y = CO_2H).[76] Biological oxidation

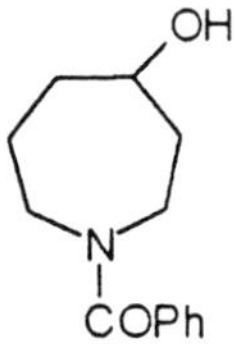

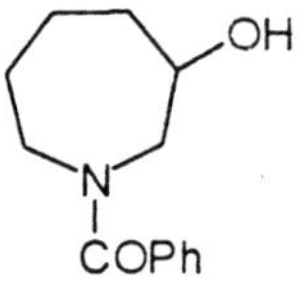

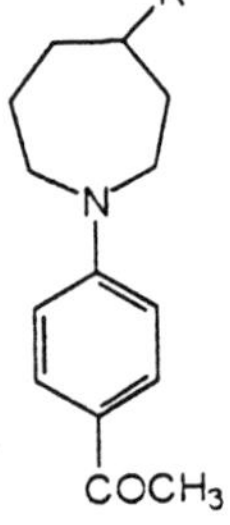

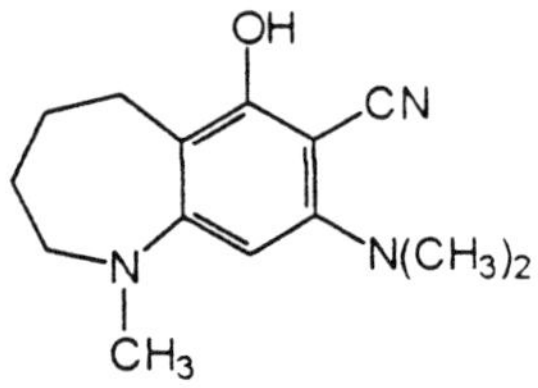

(153) (154) (155) (156)

using a strain of *Beauveria sulfurescens* has been reported by two groups of workers.[143,144] N-Benzoylazepane (9, R = PhCO) yielded a mixture of 4- and 3-hydroxy compounds (153 and 154),[143] while 155 (R = H) gave the alcohol (155, R = OH) in good yield.

5. 2-Alkenylazepanes

The ready availability of these compounds has allowed their chemistry to be thoroughly investigated. Protonation occurs on the exocyclic carbon atom (β-atom of an enamine) as one might expect.[93] In a few cases, the precise stereochemistry of the compounds **114** has been established:[93,100,148] however, it is well known that isomerisation is facile, particularly with heating, especially in cases where neither X nor Y (in **114**) is hydrogen.[93] In Table 4, we present the structures of 2-alkenylazepanes as recorded in the literature, but their accuracy cannot be guaranteed.

2-Alkenylazepanes have proved useful for annulation of other rings on to the azepane. A 1-benzazepine (**156**) arose when **157** was heated[95] and treatment of

(**157**)   (**158**)

(**159**)   (**160**)

158 with ammonia in a bomb yielded the pyrimido[3,4-b]azepine structure (**159**);[96] a variant of the latter[145] gave **160**.

Pyrimido[1,6-a]azepines (**161**, R^1 = Ph, cC_6H_{11}; R^2 = p-tolyl, ^iPr, ^iBu) were prepared from compounds **121** and aldehydes under the influence of

(**161**)   (**162**)   (**163**)   (**164**)

$BF_3(C_2H_5)_2O$;[106] alternatively a pyrimidine structure (**162**) can be obtained from **121** using 'triphosgene' ($Cl_3COCOOCCl_3$).[106] Meldrum's acid derivatives (**119**) have proved very useful. Thus, flash vacuum pyrolysis of **119** (R = CH_3, CH_2Ph) gave the pyrroloazepine **163** (R = H, Ph),[105] whilst compound **119** [R = $(CH_2)_3Cl$] produced the acid chloride **164**, which could be converted further.[146] Furthermore, pyrolysis of Meldrum's acid derivative **119** (R = H) in the presence of alcohols, ethanethiol and diethylamine yielded other alkenylazepanes [**114**, R = X = H, Y = CO_2R, $COSC_2H_5$, $CON(C_2H_5)_2$].[147] N-Methyl analogue **119** (R = CH_3) gave similar results.[104]

(114)   (119)   (121)

Alkylation (CH_3I/K_2CO_3) of **114** (R = X = H, Y = $CO_2C_2H_5$) yielded exclusively **114** (R = H, X = CH_3, Y = $CO_2C_2H_5$)[148] and acylation with both aliphatic and aromatic acid chlorides gave products **114** (R = H, X = $CO_2C_2H_5$, Y = CH_3CO, $CH_3CH_2CH_2CO$, PhCO).[149] Decarboxylation of the latter proceeds in fair to good yields using boric acid.[99]

Isothiazolo[2,3-a]azepines (**165**, R = Ph, OC_2H_5) were shown[150,151] to arise when isothiocyanate adducts (**114**, R = H, X = NO_2, Y = CSNHCOR) were

(165)   (166)   (167)

treated with bromine in chloroform or acetic acid. Pyrazoles or isoxazoles arose by reaction of certain alkenylazepanes (**114**, R = CH_3, X = COAryl, Y = H) with hydrazines or hydroxylamine. Thus **166** and **167** were obtained.[90]

Long-chain ω-amino acids were obtained by treating ketoenamines with aqueous sodium hydroxide.[152] When this useful protocol was applied to the

(168) $H_2N(CH_2)_5CO(CH_2)_5CO_2H$
         (169)

alkenylazepane ketone **168**, the product **169** was obtained in 77% yield; reduction (Wolff–Kishner) of the carbonyl group allowed the isolation of the ω-amino acid [$NH_2(CH_2)_{11}CO_2H$] in 93% yield.

Azepanes (Hexahydroazepines) 31

Table 1. Mono-N-substituted azepanes

Structure	M.p. (°C)/b.p. (°C/mmHg)	Derivatives and m.p. (°C)	Spectroscopic evidence/data	Ref.
N–H	138–138.2/~760 136/~760 136–137	HCl 235–236(d) Picrate 144.5	n_D^{23} 1.4654 n_D^{20} 1.4658	153 154 155 33
N–CH$_3$	146/~760 138–140/748	Picrate 202 Picrate 217 HCl 180.5–182.5 CH$_3$I 265		156 32a 157 2
N–C$_2$H$_5$	90.5–91.5/90 153.5/741 107	Picrate 173–173.5 CH$_3$I 244.5–245.5(d) Picrate 173	n_D^{20} 1.4571 n_D^{23} 1.4541	158 19
N–CH$_2$CH$_2$CH$_3$	176–177 102.5/77.5	Picrate 92.5	n_D^{20} 1.4555; d^{20} 0.8417 n_D^{23} 1.4538	159 19
N–CH(CH$_3$)$_2$			No data	160
N–(CH$_2$)$_3$CH$_3$	196–197		n_D^{20} 1.4573; d^{20} 0.8432	159
N–(CH$_2$)$_4$CH$_3$	94–95/13	HCl 217–218	n_D^{25} 1.4551	4
N–CH$_2$Ph	130–132/12 75/0.25	HCl 158.5–159.5 HCl 163–165	n_D^{25} 1.5243 ^{1}H, ^{13}C NMR, MS	4 115 9

Table 1 (continued)

Structure	M.p. (°C)/b.p. (°C/mmHg)	Derivatives and m.p. (°C)	Spectroscopic evidence/data	Ref.
N-SO$_2$-C$_6$H$_4$-CH$_3$ (azepane)	75 164/0.3			14
N-CHO (azepane)	102–103/5 127–130/23		n_D^{25} 1.4886 n_D^{20} 1.4903	117 162
N-CO$_2$CH$_3$ (azepane)	120–121/30 96–97/13		^{1}H NMR, IR	59 163
N-CO$_2$C$_2$H$_5$ (azepane)	46–49/0.1 118–120/20		n_D^{25} 1.4622 n_D^{21} 1.4635, IR	55 53, 54
N-CO$_2$Ph (azepane)	128–130/0.6		n_D^{25} 1.5328, IR	55
N-CO$_2$CH$_2$Ph (azepane)	Vac. dist.		^{1}H, ^{13}C NMR, MS	123
N-CON⊲ (azepane)	114/3	HCl 104		113
N-COSC$_2$H$_5$ (azepane)	136.5–137		n_D^{25} 1.5156	164

Azepanes (Hexahydroazepines) 33

Table 1 (continued)

Structure	M.p. (°C)/b.p. (°C/mmHg)	Derivatives and m.p. (°C)	Spectroscopic evidence/data	Ref.
N-COCO$_2$CH$_3$	180–183/35		n_D^{25} 1.4839; d^{27} 1.1062	165
N-CH$_2$CO$_2$H		HCl 187–188 HCl 185–187		166 112
N-CH$_2$CO$_2$CH$_3$	89–90/8	Picrate 111.5–113	n_D^{20} 1.4660; d^{20} 1.0066	159
N-CH$_2$CO$_2$C$_2$H$_5$	96–97/8	Picrate 87–88	n_D^{20} 1.4628; d^{20} 0.9852	159
N-CH$_2$CO$_2^n$Pr	119–120/5	Picrate 104–105	n_D^{20} 1.4624	159
N-CH$_2$CO$_2^i$Pr	115–115.5/11	Picrate 120–121	n_D^{20} 1.4590	159
N-CH$_2$CO$_2^n$Bu	130–132/12	Picrate 77–77.5	n_D^{20} 1.4743	159
N-CH$_2$CN	102–103/14 110–112/15		n_D^{27} 1.4712	115 167

Table 1 (continued)

Structure	M.p. (°C)/b.p. (°C/mmHg)	Derivatives and m.p. (°C)	Spectroscopic evidence/data	Ref.
azepine-CH₂CHO	69–71/0.2	HCl 92–94	n_D^{27} 1.4821	115
azepine-CH₂CH₂OH	99–101/15 62–63/0.34	Picrate 105–109	^{1}H NMR, MS	168 9
azepine-CH₂CH₂Cl		HCl 208–209		169
azepine-CH₂CH₂CN	121–123/14		n_D^{30} 1.4710	115
azepine-CH₂CH₂CO₂CH₃	125–132/14		n_D^{27} 1.4580	115
azepine-(CH₂)₃OH	112–114/9	HCl 153–154	n_D^{22} 1.4825	33
azepine-CH₂CHOH-CH₃	97–100/11	HCl 131–132	n_D^{22} 1.4680	33
azepine-CHCH₂OH-CH₃	110–112/17	HCl 111–112	n_D^{22} 1.4748	33

Azepanes (Hexahydroazepines)

Table 1 (continued)

Structure	M.p. (°C)/b.p. (°C/mmHg)	Derivatives and m.p. (°C)	Spectroscopic evidence/data	Ref.
Azepane-N-(CH$_2$)$_4$OH	110–112/4 111–115/2.8	CH$_3$I 147.5–148.5	n_D^{20} 1.4813 d^{20} 0.9487, IR	171 170
Azepane-N-CH$_2$CH(Ph)C(CH$_3$)(Ph)CN	101.5–102.5	HBr 230–231.5 HI 228–229 HCl 218.5–220		157
Azepane-N-CH(CH$_3$)-C≡CH	82–83/15			172
Azepane-N-CH$_2$CH$_2$N(C$_2$H$_5$)$_2$	122–126/14	2HCl 222–224 (d)		114
Azepane-N-CH$_2$CH$_2$N(piperidine)	151–153/17	2HCl 308–311 (d) 2HBr 315–317 (d)		114
Azepane-N-CH$_2$CH$_2$NH$_2$	212/~760 71–73/5 66–67/1	Dipicrate 207 2HCl 114–115		173 174 167 115
Azepane-N-(CH$_2$)$_3$N(CH$_3$)$_2$		2HCl 272–275		115
Azepane-N-CH$_2$CH(NH$_2$)CO$_2$H	167–168			166

Table 1 (continued)

Structure	M.p. (°C)/b.p. (°C/mmHg)	Derivatives and m.p. (°C)	Spectroscopic evidence/data	Ref.
Azepine-CH(Ph)CO₂C₂H₅	165/3		n_D^{26} 1.5360	175
Azepine-C(CH₃)(CH₃)CH₂CH₃	90–92/12	HCl 244–245		142
N-Ph azepine	66/0.18 146–148/12	HCl 194 HBr 219	^{1}H NMR, MS	9 142
N-(2,6-dinitrophenyl) azepine	73–74			119
N-(2-amino-4-nitrophenyl) azepine	67–68.5 172.8–173.5/0.15			119
N-(4-amino-2-nitrophenyl) azepine	76–77			119

Azepanes (Hexahydroazepines)

Table 1 (continued)

Structure	M.p. (°C)/b.p. (°C/mmHg)	Derivatives and m.p. (°C)	Spectroscopic evidence/data	Ref.
N-(2,4-dinitrophenyl)azepane	108–109			119
N-(2-amino-4-chlorophenyl)azepane	198–200/20	HCl 200–202	n_D^{23} 1.5800	119
N-(2-aminophenyl)azepane	160/10	HCl 165–166	$n_D^{23.5}$ 1.5640	119
N-(2-nitrophenyl)azepane	Yellow oil		$n_D^{22.3}$ 1.6010	119
N-(4-methylphenyl)azepane	62/0.12		^{1}H, ^{13}C NMR, MS	9

Table 1 (continued)

Structure	M.p. (°C)/b.p. (°C/mmHg)	Derivatives and m.p. (°C)	Spectroscopic evidence/data	Ref.
1-(3-methylphenyl)azepane	71/0.1		^{1}H NMR, MS	9
1-(4-methoxyphenyl)azepane	87/0.24		^{1}H, ^{13}C NMR, MS	9
1-(4-chlorophenyl)azepane	75/0.16		^{1}H NMR, MS	9
1-(4-methylbenzyl)azepane	71–72/0.12		^{1}H, ^{13}C NMR, MS	9
1-(4-chloro-2-nitrophenyl)azepane	56			119

Table 1 (continued)

Structure	M.p. (°C)/b.p. (°C/mmHg)	Derivatives and m.p. (°C)	Spectroscopic evidence/data	Ref.
azepane-N-(2,4-dimethoxyphenyl)	109–110/0.01	Picrate 141.5–142.5		118
azepane-N-CH(Ph)CH$_2$OH	165/3		n_D^{26} 1.5360	175
azepane-N-CH(Ph)CO$_2$C$_2$H$_5$	163/5		n_D^{31} 1.5300	175
azepane-N-OH	94–95/13 64–65/1.4		n_D^{26} 1.4852	176 177
azepane-N-CN	110/0.1			177
azepane-N-NH$_2$	86–86.5/25			178
azepane-N-NO	83–86/2			178

Table 1 (continued)

Structure	M.p. (°C)/b.p. (°C/mmHg)	Derivatives and m.p. (°C)	Spectroscopic evidence/data	Ref.
azepane-N-CH₂CH₂SH	86/5		n_D^{20} 1.5076	116
azepane-N-(CH₂)₃Cl		HCl 208–209		33
azepane-N-CH₂CH₂OCHPh₂	158–160/0.01	HCl 144–146		179
azepane-N⁺HPPh₃ X⁻	X = Br 178–180 X = Cl: 168–170		¹H NMR	180
azepane-N-COCl	116–118/11			163
azepane-N-COCH₂CH₃	123/9.5		n_D^{23} 1.4821	19
azepane-N-CH(CH₃)CH₂C₂H₅	65.5/12	Picrate 130–131	n_D^{20} 1.4523	161

Azepanes (Hexahydroazepines)

Table 1 (continued)

Structure	M.p. (°C)/b.p. (°C/mmHg)	Derivatives and m.p. (°C)	Spectroscopic evidence/data	Ref.
N–CH₂CH₂Ph azepane	71.2/0.18		^{1}H, ^{13}C NMR, MS	9
N–CH(CH₃)Ph azepane	64/0.07		^{1}H, ^{13}C NMR, MS	9
N–cyclohexyl azepane	47/0.08		^{1}H NMR, MS	9
N–CH(CH₃)₂ azepane (isopropyl)	85–86/12		n_D^{20} 1.4845	161

Table 2. Mono-2-substituted azepanes

Structure	M.p. (°C)/b.p. (°C/mmHg)	Derivatives and m.p. (°C)	Spectroscopic evidence/data	Ref.
2-CH$_3$ azepane	59–61/22 148–150/748	HCl 199–201 NSO$_2$Ph 75–77 HCl 156	n_D^{27} 1.4569 n_D^{20} 1.4588 ^{1}H NMR	60 33 85 31
2-C$_2$H$_5$ azepane			^{1}H NMR, IR claimed	48
2-C$_3$H$_7$ azepane	189–191 79–81/18	HCl 160–162	n_D^{20} 1.4580 d^{20} 0.8574 IR, ^{1}H NMR	118 21 49
2-C$_4$H$_9$ azepane	Chromatography			50
2-CH$_2$OH azepane	90–95/1			34
2-Ph azepane	142–145/14	N-Acetyl 130–5/0.05 N-Tosyl 101–103		61 182
2-CO$_2$H azepane (d/l)	208–209 205–208	HCl 214–215 Phenylthiohy-dantoin 165–166 + Tosyl acid		16 78
2-CO$_2$H azepane (S)	183.3–184.4 (d)		α_D − 21 ($c = 1.02$/H$_2$O) IR, MS, ^{1}H NMR	18

Table 2 (continued)

Structure	M.p. (°C)/b.p. (°C/mmHg)	Derivatives and m.p. (°C)	Spectroscopic evidence/data	Ref.
azepane-2-CO₂C₂H₅	55–57/1.5	HCl 166–167		53, 16
azepane-2-CH₂SH	60 (subl.)			137
azepane-2-(3,4-dimethoxyphenyl)	135–140/12	HCl 176–180 Picrate 178–182		61
azepane-2-(2-methylphenyl)	155/10	N-CH₂Si(CH₃)₃ 140/1.5	¹H NMR	139
azepane-2-C(Ph₂)OH		HCl 253–254		183
azepane-2-C≡C-nBu			Data in Registry	46, 49

Table 2 (continued)

Structure	M.p. (°C)/b.p. (°C/mmHg)	Derivatives and m.p. (°C)	Spectroscopic evidence/data	Ref.
azepane with CSi(CH₃)₃ substituent (**R**)		Oxalate 145–146 N-Acetyl 72–74/0.02	$[\alpha]_D^{22} - 29.8$ ($c = 1.4$, CH_3OH) $[\alpha]_D^{22} + 18.3$ ($c = 1.4$, CH_3OH) $[\alpha]_D^{22} + 190$ ($c = 1.4$, CH_3OH)	132, 134
azepane with C≡CH substituent (**R**)		Oxalate 157–158 N-Acetyl 101–102	$[\alpha]_D^{22} + 12.8$ ($c = 1.4$, CH_3OH) $[\alpha]_D^{22} + 197$ ($c = 1.4$, CH_3OH)	132, 134
azepane with C≡CSi(CH₃)₃ substituent (**S**)		Oxalate 145–146 N-Acetyl 70–72/0.01	$[\alpha]_D^{22} + 29.9$ ($c = 1.4$, CH_3OH) $[\alpha]_D^{22} - 18.7$ ($c = 1.4$, CH_3OH) $[\alpha]_D^{22} - 190$ ($c = 1.4$, CH_3OH)	132, 134
azepane with C≡CH substituent (**S**)		Oxalate 157–158 N-Acetyl 101–102	$[\alpha]_D^{22} - 12.7$ ($c = 1.4$, CH_3OH) $[\alpha]_D^{22} - 194$ ($c = 1.4$, CH_3OH)	132, 134
azepane with cyclohexenyl substituent	90–90.5/3	N-COCHBrCH₃ 230	IR, ¹H NMR	233
azepane with indol-3-yl substituent	202		IR	181

Azepanes (Hexahydroazepines) 45

Table 3. Mono-3-substituted azepanes

Structure	M.p. (°C)/b.p. (°C/mmHg)	Derivatives and m.p. (°C)	Spectroscopic evidence/data	Ref.
3-CH₃ azepane			^{13}C NMR	85, 87
3-^{n}Pr azepane	194–196/760		n_D^{20} 1.4620, d^{20} 0.8649	184
3-CH₂OH azepane	97–98/1	HCl 122–124		154
3-CO₂H azepane		HBr 145–146.5	IR, ^{1}H NMR	185
3-(4-OCH₃-C₆H₄) azepane			^{1}H NMR, MS	80
3-(4-CH₃-C₆H₄) azepane			^{1}H NMR, MS	80

Table 3 (continued)

Structure	M.p. (°C)/b.p. (°C/mmHg)	Derivatives and m.p. (°C)	Spectroscopic evidence/data	Ref.
3-pyrrolidinyl-azepane (NH)	Pale yellow oil	Fumarate 155–156 N-(3,4-Dichlorophenyl)acetyl 156–158	^{1}H NMR, MS	186
3-hydroxy-azepane (NH)	GLC pure		IR	187

Azepanes (Hexahydroazepines)

Table 4. Mono-4-substituted azepanes

Structure	M.p. (°C)/b.p. (°C/mmHg)	Derivatives and m.p. (°C)	Spectroscopic evidence/data	Ref.
4-NH$_2$ azepane	153–156/748	HCl 174–175 CH$_3$I 192 3,5-Dinitro-benzoate 125–126		33 77
4-CO$_2$H azepane		HBr 111.5–112.5	IR, pK_a(H$_2$O) 3.87, 10.46	51
4-OCH$_3$ azepane	75–75.6/15		n_D^{25} 1.4673	188
4-OH azepane	82/0.5	Picrolonate 225		189, 190
4-NH$_2$ azepane	79–80/10	Chloroaurate 209 (d) Picrate 224 (d) Ring N-benzyl 105–106/0.4		189
4-N(CH$_3$)$_2$ azepane	75/8		n_D^{25} 1.4750	188

Table 4 (continued)

Structure	M.p. (°C)/b.p. (°C/mmHg)	Derivatives and m.p. (°C)	Spectroscopic evidence/data	Ref.
(azepane with NHPh at 4-position, NH)	130–133/0.5 52–54	N^1-CH$_3$, N^2-acetyl 137–139 N^1-CH$_2$Ph, N^2-acetyl 169–171 N^1-CH$_2$–◁, N^2-acetyl 84–86	^{1}H NMR	191

Azepanes (Hexahydroazepines)

Table 5. 1,2-Disubstituted azepanes

Structure	M.p. (°C)/b.p. (°C/mmHg)	Derivatives and m.p. (°C)	Spectroscopic evidence/data	Ref.
N-CH₃ substituent on ring N: CH₃; C2: CH₃	80–90/5	Picrate 233 Picrolonate 162–164 Picrate 237–238		31 192
N-substituent: SO₂Ph; C2: CH₃	75–77			60
N-substituent: CO₂CH₂Ph; C2: CH₃	Kugelrohr distilled		^{1}H NMR	123
N-substituent: (CH₂)₄CH₃; C2: CH₃	117–118/22	Picrate 79–80	n_D^{25} 1.4530	5
N-substituent: (CH₂)₂N(piperidine); C2: CH₃	177–181/25	2HCl 250–252 (d)		114
N-substituent: (CH₂)₂Cl; C2: CH₃		Picrate 127–128		33
N-substituent: (CH₂)₃Cl; C2: CH₃		HCl 113–114		33

Table 5 (continued)

Structure	M.p. (°C)/b.p. (°C/mmHg)	Derivatives and m.p. (°C)	Spectroscopic evidence/data	Ref.
azepane, N-(CH$_2$)$_3$Ph, 2-CH$_3$	106–109/1	Picrate 105–106		5
azepane, N-CH$_3$, 2-C$_2$H$_5$	174–175/755	Picrate 166.5–167.5	n_D^{20} 1.4472	193
azepane, N-CH$_3$, 2-CO$_2$H	183–184	Ethyl ester 44–50/1.5	n_D^{20} 1.4584	16
azepane, N-COCH$_3$, 2-CO$_2$C$_2$H$_5$	107–109/1		n_D^{20} 1.4779	8
azepane, N-CH$_3$, 2-CONH$_2$	74–75 119–121/1.5	N-CH$_3$ 96/1.5	n_D^{20} 1.4888	8
azepane, N-C$_2$H$_5$, 2-CO$_2$C$_2$H$_5$	70–72/2		n_D^{24} 1.5420	16
azepane, N-CH$_2$Ph, 2-CO$_2$CH$_3$			^{1}H, ^{13}C NMR, IR	194

Table 5 (continued)

Structure	M.p. (°C)/b.p. (°C/mmHg)	Derivatives and m.p. (°C)	Spectroscopic evidence/data	Ref.
azepane-2-CO$_2$H, N-nBu	140–141	Ethyl ester 78–81/1.5	n_D^{24} 1.4544	16
azepane-2-CO$_2$C$_2$H$_5$, N-(CH$_2$)$_2$CO$_2$C$_2$H$_5$	118–119/1.5			16
azepane-2-CH(CO$_2$CH$_3$)$_2$, N-CHO	79–81			195
azepane-2-OCH$_3$, N-CHO	102–105/4			126
azepane-2-CH$_3$, N-CH$_2$CH$_2$OH	94–97/7	CH$_3$I 213–215	n_D^{22} 1.4790	33
azepane-2-CH$_3$, N-CH(CH$_3$)CH$_2$OH	113–116/11	HCl 121–122	n_D^{22} 1.4745	33

Table 5 (continued)

Structure	M.p. (°C)/b.p. (°C/mmHg)	Derivatives and m.p. (°C)	Spectroscopic evidence/data	Ref.
azepine-CH₃, N-CH₂CHOH-CH₃	96–98/7	CH_3I 180–181	n_D^{22} 1.4670	33
azepine-CH₃, N-(CH₂)₃OH	119–122/7	HCl 114–115	n_D^{22} 1.4809	33
azepine-OC₂H₅, N-CO₂CH₂Ph	Chromatography			41
azepine-OC₂H₅, N-CO₂C₂H₅	83/2 90–92/3		IR, MS, ^{1}H NMR	41 44
azepine-OC₂H₅, N-CO₂^{t}Bu	78/2			41
azepine-C(=CH₂)CO₂CH₃, N-CH₂Ph	Chromatography		IR, MS, ^{1}H NMR	74
azepine-CH₂OH, N-CH₂CH₂OH	140–160/1.0–1.5		IR	34

Azepanes (Hexahydroazepines)

Table 5 (continued)

Structure	M.p. (°C)/b.p. (°C/mmHg)	Derivatives and m.p. (°C)	Spectroscopic evidence/data	Ref.
N-CH₂CH₂Cl, 2-CH₂Cl azepane	132–133			34
N-CH₃, 2-C(CH₃)₂OH azepane	104–106/14			183
N-CH₃, 2-CH₂OH azepane	52–53/2		n_D^{20} 1.4557	8
N-CH₃, 2-CO₂(CH₂)₂N(CH₃)₂ azepane	78–79/1		n_D^{20} 1.4681	8
N-CH₃, 2-CH₂NH₂ azepane	68–69/6		n_D^{20} 1.4834	8
N-NO, 2-CHPh₂ azepane	158.5	HCl 271 (d)	^{1}H NMR	120
N-COCH₃, 2-OCH₃ azepane	53/0.07		n_D^{25} 1.4775, ^{1}H NMR	127

Table 5 (continued)

Structure	M.p. (°C)/b.p. (°C/mmHg)	Derivatives and m.p. (°C)	Spectroscopic evidence/data	Ref.
azepane-N-NO, 2-OC₂H₅	60/0.1		¹H NMR, IR, MS	180
azepane-N-CHO, 2-(2,4,6-trimethoxyphenyl)		-OCH₃ 99–105	MS, ¹H NMR	129
azepane, N-(2,4-dimethoxyphenyl), 2-nPr	128–130/0.01			118
azepane-N-CHO, 2-(4-methylphenyl)	146–150/0.5		MS, ¹H NMR	130
azepane-N-CHO, 2-Ph	134/1.2 143–148/0.8		¹H NMR	128, 130
azepane-N-NO, 2-Ph	138/0.02		n_D^{20} 1.5602, IR, MS, ¹H NMR	141

Table 5 (continued)

Structure	M.p. (°C)/b.p. (°C/mmHg)	Derivatives and m.p. (°C)	Spectroscopic evidence/data	Ref.
azepane-2-Ph, N-NH₂	102–105/0.7	N-Acetyl 81.5–83 CH₃I 165–167	n_D^{20} 1.5474, IR, MS, ^{1}H NMR	141
azepane-2-Ph, N-(CH₂)₂CO₂CH₃	130–132/0.25			61
azepane-2-(3,4-di-OCH₃-C₆H₃), N-(CH₂)₃OH	165–170/0.1			61
azepane-2-(3,4-di-OCH₃-C₆H₃), N-(CH₂)₂CO₂CH₃	171–175/0.3			61
azepane-2-Ph, N-CH₂Si(CH₃)₃	150/6		^{1}H NMR	139
azepane-2-CO₂CH₂Ph, N-(CH₂)₃CH(CO₂C₂H₅)₂	205–210/0.1			136
azepane-2-C≡C-^{n}Pent, N-CH₃	68–70/0.5		IR, ^{1}H NMR	195

Table 5 (continued)

Structure	M.p. (°C)/b.p. (°C/mmHg)	Derivatives and m.p. (°C)	Spectroscopic evidence/data	Ref.
azepine-N(CH₂Ph)-C≡C-C₄H₉				84
azepine-N(C₂H₅)-C≡C-Ph			^{1}H NMR, IR, MS	196
azepine-N(CH₃)-C≡C-Ph	90–95/0.5		IR, ^{1}H NMR	195
azepine-N(CH₂Ph)-CH₂COCH₃	Chromatography		^{1}H, ^{13}C NMR, MS	71
azepine-N(CH₂Ph)-C(OH)(C₂H₅)₂			Supplement	125
azepine-N(CHO)-C(CH₃)(CO₂C₂H₅)₂	163–170/0.7–0.9		MS, ^{1}H NMR	131
azepine-N(CHO)-C(Ph)(CO₂C₂H₅)₂	72–76		MS, ^{1}H NMR	131

Azepanes (Hexahydroazepines)

Table 5 (continued)

Structure	M.p. (°C)/b.p. (°C/mmHg)	Derivatives and m.p. (°C)	Spectroscopic evidence/data	Ref.
azepane-CH(barbiturate), N-CHO	124–127		^{1}H NMR, MS (no M$^+$)	131
azepane-CH(CH$_2$CO$_2$CH$_3$), N-CO$_2$C$_2$H$_5$	125/3		IR, MS, ^{1}H NMR	44
azepane-CH(indol-3-yl), N-CO$_2$C$_2$H$_5$		Chromatography	IR, MS, ^{1}H NMR	44
azepane-CH(CN), N-CO$_2$C$_2$H$_5$	100–102/2		IR, MS, ^{1}H NMR	44
azepane-CH(N$_3$), N-CO$_2$C$_2$H$_5$		Chromatography	IR, MS, ^{1}H NMR	44
azepane-CH(S(CH$_2$)$_3$SH), N-CO$_2$C$_2$H$_5$		Chromatography	IR, MS, ^{1}H NMR	44

Table 5 (continued)

Structure	M.p. (°C)/b.p. (°C/mmHg)	Derivatives and m.p. (°C)	Spectroscopic evidence/data	Ref.
azepane, N-CHO, 2-CHPh(CO$_2$H)	214–215		MS, ^{1}H NMR	131
azepane, N-CHO, 2-CH(COCH$_3$)(CO$_2$CH$_3$)	87–91		MS, ^{1}H NMR	131
azepane, N-CHO, 2-(1-CO$_2$C$_2$H$_5$-2-oxocyclopentyl)	187–191/1.3 119–122		MS, ^{1}H NMR	131

Azepanes (Hexahydroazepines)

Table 6. 1,3-Disubstituted azepanes

Structure	M.p. (°C)/b.p. (°C/mmHg)	Derivatives and m.p. (°C)	Spectroscopic evidence	Ref.
1-CH₃, 3-CO₂CH₃ azepane			IR, ^{1}H NMR	37
1-CH₃, 3-CH₂OH azepane			IR, ^{1}H NMR	37
1-CH₃, 3-OH azepane	78–79/15	O-Acetyl 81–82/10 O-Benzoyl 125/2–3 CH₃I 141–142	IR	197
1-COPh, 3-OH azepane	Chromatography		IR, ^{1}H, ^{13}C NMR	143
1-CH₂Ph, 3-CH₃ azepane			^{1}H NMR, MS, IR	36
1-N⁺–O⁻, 3-OH azepane	Wax		^{13}C NMR, MS, IR	198

Table 6 (continued)

Structure	M.p. (°C)/b.p. (°C/mmHg)	Derivatives and m.p. (°C)	Spectroscopic evidence	Ref.
azepane N-oxide with 3-OH	51.5–52.5		^{13}C NMR, MS, IR	198
1-methyl-3-phenylazepane			^{1}H NMR, MS	80
1-n-propyl-3-(3-hydroxyphenyl)azepane		HCl 148–150		199
1-phenyl-3-hydroxyazepane			^{1}H, ^{13}C NMR, MS, IR	12
1-(2-methylphenyl)-3-hydroxyazepane			^{1}H, ^{13}C NMR, MS, IR	12
1-(3-methylphenyl)-3-hydroxyazepane			^{1}H, ^{13}C NMR, MS, IR	12

Azepanes (Hexahydroazepines) 61

Table 6 (continued)

Structure	M.p. (°C)/b.p. (°C/mmHg)	Derivatives and m.p. (°C)	Spectroscopic evidence	Ref.
1-(4-methylphenyl)azepan-3-ol			^{1}H, ^{13}C NMR, MS, IR	12
1-(2-methoxyphenyl)azepan-3-ol			^{1}H, ^{13}C NMR, MS, IR	12
1-(3-methoxyphenyl)azepan-3-ol			^{1}H, ^{13}C NMR, MS, IR	12
1-(4-methoxyphenyl)azepan-3-ol			^{1}H, ^{13}C NMR, MS, IR	12

Table 6 (continued)

Structure	M.p. (°C)/b.p. (°C/mmHg)	Derivatives and m.p. (°C)	Spectroscopic evidence	Ref.
[azepane-3-ol N-(2-chlorophenyl)]			^{1}H, ^{13}C NMR, MS, IR	12
[azepane-3-ol N-(3-chlorophenyl)]			^{1}H, ^{13}C NMR, MS, IR	12
[azepane-3-ol N-(4-chlorophenyl)]			^{1}H, ^{13}C NMR, MS, IR	12

Azepanes (Hexahydroazepines)	63

Table 7. 1,4-Disubstituted azepanes

Structure	M.p. (°C)/b.p. (°C/mmHg)	Derivatives and m.p. (°C)	Spectroscopic evidence/data	Ref.
4-CH₃, N-CH₃		Picrate 188–191		77
4-C₂H₅, N-CH₃		Picrate 146–147		76
4-CH₃, N-COPh	120–128/ 0.08–0.09		IR, ^{1}H NMR	200
4-OH, N-COPh			IR IR, ^{1}H NMR	201 143
4-Ph, N-CH₃	88–90/0.25	Picrate 149–150 HCl 78–79 CH₃I 146–147 CH₃CH₂CH₂I 166–169		28 52

Table 7 (continued)

Structure	M.p. (°C)/b.p. (°C/mmHg)	Derivatives and m.p. (°C)	Spectroscopic evidence/data	Ref.
4-Cl, N-CH₃ azepane	30–35/2	HCl 125–126 Picrate 173–174 Picrate 138–139		81 232
4-Cl, N-CH₂Ph azepane	104–105/0.3			189
4-OH, N-CH₂Ph azepane	129/0.5	Picrate 136–138	IR, ^{1}H NMR, MS	189 82, 232
4-vinyl, N-CH₃ azepane	173–176	Picrate 113–115		76
4-CO₂C₂H₅, N-CH₃ azepane	102–104/10	Picrate 128–130		76

Table 7 (continued)

Structure	M.p. (°C)/b.p. (°C/mmHg)	Derivatives and m.p. (°C)	Spectroscopic evidence/data	Ref.
4-OH, N-CH₃ azepane	95–98/12 96/11 105–107/2	Picrate 180–181 Benzoate 104–105/0.07 Picrate 141–143	n_D^{23} 1.4863 IR	81 232
4-OH, N-C(O)OCH=CH₂ azepane			^{1}H NMR, IR, MS	82
4-Cl, N-C(O)OCH=CH₂ azepane		Chromatography	^{1}H NMR, IR, MS	82
4-OPh, N-CH₂Ph azepane		Chromatography	^{1}H NMR, IR, MS	82
4-SPh, N-CH₂Ph azepane		Chromatography	^{1}H NMR, MS	82

Table 7 (continued)

Structure	M.p. (°C)/b.p. (°C/mmHg)	Derivatives and m.p. (°C)	Spectroscopic evidence/data	Ref.
azepine with O-pyridyl, N-CH₂Ph		Chromatography	¹H NMR, MS	82
azepine with NHPh, N-CH₂CHOH(Ph)			¹H NMR, MS	202
azepine with N(CH₃)₂, N-C₆H₄NO₂	242–244			188
azepine with OCH₃, N-C₆H₄NH₂		HCl 184–187		188

Table 7 (continued)

Structure	M.p. (°C)/b.p. (°C/mmHg)	Derivatives and m.p. (°C)	Spectroscopic evidence/data	Ref.
(4-hydroxyazepane N-oxide, N=N form)	43–45		^{13}C NMR, MS, IR	198
(4-hydroxyazepane N-oxide isomer)	80.5–81.5		^{13}C NMR, IR	198
(4-(indol-3-yl)-1-methylazepane)				203
(4-benzylamino-1-methylazepane)	96–98/0.02	Dipicrate 177–179		81
(4-methyl-1-(2-diethylaminoethyl)azepane)	136–141/15	2HCl 201–203		114

Table 7 (continued)

Structure	M.p. (°C)/b.p. (°C/mmHg)	Derivatives and m.p. (°C)	Spectroscopic evidence/data	Ref.
4-methyl-1-(1-hydroxypropan-2-yl)azepane	111–113/11	CH$_3$I 77–78	n_D^{22} 1.4705	33
4-methyl-1-(2-hydroxypropyl)azepane	101–104/12	HCl 105–106	n_D^{22} 1.4643	33
4-methyl-1-(3-hydroxypropyl)azepane	123–127/12	HCl 115–116	n_D^{22} 1.4780	33
4-methyl-1-(2-chloropropyl)azepane		HCl 198–199 Picrate 127–128		33
4-methyl-1-(3-chloropropyl)azepane		HCl 158–159		33

Azepanes (Hexahydroazepines) 69

Table 7 (continued)

Structure	M.p. (°C)/b.p. (°C/mmHg)	Derivatives and m.p. (°C)	Spectroscopic evidence/data	Ref.
4-NHPh, N-CO₂C₂H₅ azepane	170–175/0.75		^{1}H NMR	191
4-CH₃, N-(CH₂)₂CN azepane	129–130/15		n_D^{30} 1.4692	115
4-CO₂C₂H₅, N-CO₂Ph azepane			IR, ^{1}H NMR, MS	68
4-(4-OCH₃-C₆H₄), N-CH₃ azepane		CH₃I 142–143.5		62
4-OH, N-CO₂C₂H₅ azepane	143–144/0.5	OCO₂C₂H₅ 155–7/0.5	IR, ^{1}H NMR, MS	232

Table 8. Miscellaneous disubstituted azepanes

Structure	M.p. (°C)/b.p. (°C/mmHg)	Derivatives and m.p. (°C)	Spectroscopic evidence/data	Ref.
	169–170.5/760			188
	53–55/4	N-Benzoyl 200–202/13 HCl 197–200 Chloroplat. 195	n_D^{20} 1.4550, d_4^{20} 0.8461	21
	125–140/0.01		IR, ^{1}H NMR, MS	22
	130–150/0.01		IR, ^{1}H NMR, MS	22
	2 isomers		IR, ^{1}H NMR	85
	81–83		IR, ^{1}H NMR	58

Table 8 (continued)

Structure	M.p. (°C)/b.p. (°C/mmHg)	Derivatives and m.p. (°C)	Spectroscopic evidence/data	Ref.
			X-ray	204
	95/10		n_D^{20} 1.4600, d^{20} 0.8585	184
	71/8		n_D^{20} 1.4590, d^{20} 0.8518	184
			^{13}C, ^{1}H, ^{19}F NMR, MS, IR	205
	107–118/2		^{1}H NMR, IR, MS	206

Table 8 (continued)

Structure	M.p. (°C)/b.p. (°C/mmHg)	Derivatives and m.p. (°C)	Spectroscopic evidence/data	Ref.
2,7-diphenyl azepane (Ph, N-H, Ph)	96–98	Picrate 175–180		207
4-Ph, 2-CH₂OH azepane	120–133/0.08	N-Benzoyl 170–172.5	IR	35
4-CH₃, 2-iPr azepane	170–176/760 84/18	HCl 149, $[\alpha]_D^{23} - 0.8$ ($c = 4$, CH_3OH)	IR, ^{1}H NMR $[\alpha]_D^{22} + 7.4$ ($c = 0.55$, C_2H_5OH)	40

Azepanes (Hexahydroazepines)

Table 9. Miscellaneous trisubstituted azepanes

Structure	M.p. (°C)/b.p. (°C/mmHg)	Derivatives and m.p. (°C)	Spectroscopic evidence/data	Ref.
4,5-(OH)$_2$, N-SO$_2$CH$_3$ azepane	170–173	Bis-*O*-methylsulphonyl 137–138	^{1}H NMR, IR	206
(±) 4-OH, 3-CO$_2$CH$_3$, N-CO$_2$CH$_3$ azepane	180/60		IR	185
(±) 4-OH, 3-CO$_2$CH$_3$, N-CO$_2$CH$_3$ azepane (diastereomer)	55.5–57.5		IR	185
4-Ph, 2-CO$_2$H, N-COPh azepane	Impure		IR	35
4-Ph, 2-CHO, N-COPh azepane	Impure		IR	35
2,7-(CN)$_2$, N-CH$_2$Ph azepane	R^*R^* oil R^*S^* 75–76		IR, MS, ^{1}H NMR IR, MS, ^{1}H NMR	208

Table 9 (continued)

Structure	M.p. (°C)/b.p. (°C/mmHg)	Derivatives and m.p. (°C)	Spectroscopic evidence/data	Ref.
(azepane with CH₃-substituted propynyl, OC₂H₅, N-CO₂ᵗBu)	Oil		IR, MS, ¹H NMR	135
(5-methyl-2-isopropyl azepane, N-NH₂)	106/0.9	HBF₄ 138	$[\alpha]_D^{23} + 21.4$ ($c = 5.7$, C_6H_6) IR, MS, ¹H NMR	40
(5-methyl-2-isopropyl azepane, N-NO)	100/0.7		$[\alpha]_D^{23} - 120$ ($c = 0.5$, C_2H_5OH) IR, MS, ¹H NMR	40
(3-Br, 2-OCH₃, N-CO₂CH₃ azepane)			IR, MS, ¹H NMR	209
(3-I, 2-OCH₃, N-CO₂CH₃ azepane)			IR, MS, ¹H NMR	209
(4-OH, 2-CH(CH₃)₂, N-CHO azepane)			IR, MS, ¹H NMR	25

Table 9 (continued)

Structure	M.p. (°C)/b.p. (°C/mmHg)	Derivatives and m.p. (°C)	Spectroscopic evidence/data	Ref.
2,7-di(OCH₃)-N-COCH₃ azepane	74–75/0.9		n_D^{25} 1.4673, ^{1}H NMR	127
4-Ph-3-CH₃-N-CH₃ azepane	98–100/0.25	CH₃I 184–190	n_D^{31} 1.5251	210
4-Ph-2-CH₃-N-CH₃ azepane	106–108/0.2	Picrate 128–130	n_D^{27} 1.5255	52
3-CN-4-NH-N-CH₃ azepane	140–142 or 108–109			211
3-CN-4-NH-N-CH₃ azepane (isomer)	108–109 or 140–142			211
4,5-di(C₂H₅)-N-(n-C₅H₁₁) azepane	98–100/2		n_D^{25} 1.4570	5

Table 9 (continued)

Structure	M.p. (°C)/b.p. (°C/mmHg)	Derivatives and m.p. (°C)	Spectroscopic evidence/data	Ref.
4,5-bis(CO$_2$C$_2$H$_5$)-1-CH$_3$ azepane			No data	212, 69
2,7-bis(CH$_2$OH)-1-CH$_2$Ph azepane	196–199/0.4			7
2,7-bis(CH$_2$Cl)-1-CH$_2$Ph azepane	145–150/0.5, 83–85			7
2-CHO, 7-CH$_3$, 1-CO$_2$CH$_2$Ph azepane	85–89		^{13}C NMR cis and trans	123
2,7-bis(CO$_2$C$_2$H$_5$)-1-CH$_2$Ph azepane	135–150/0.05 'impure'			7
3,4,5-triOH azepane		Tetraacetate 112–114	$[\alpha]_D^{20}$ + 4.8 (c = 1, CH$_3$OH) IR, MS	20

Table 9 (continued)

Structure	M.p. (°C)/b.p. (°C/mmHg)	Derivatives and m.p. (°C)	Spectroscopic evidence/data	Ref.
2-methyl-5-ethyl-7-n-propyl azepane	85–88/5	N-Benzoyl 205–8/14 HCl 198–200 Chloroplat. 185	d_4^{20} 0.8463, n_D^{20} 1.4570	21
2,7-dimethyl-4-methyl-n-propyl azepane	61–63/2	N-Benzoyl 192–5/14 HCl 168 Chloroplat. 190	d_4^{20} 0.8398, n_D^{20} 1.4550	21
2-isopropyl-4-methyl-7-methyl azepane, cis and trans	170–176		IR, ^{1}H NMR	46
4-Ph-5-CO$_2$H, N-CO$_2$C$_2$H$_5$ azepane	146–147		IR, ^{1}H NMR	226
4-Ph-5-CHO, N-CO$_2$C$_2$H$_5$ azepane	180–183/0.1		IR, ^{1}H NMR	226

Table 10. Miscellaneous tetra- and higher substituted azepanes

Structure	M.p. (°C)/b.p. (°C/mmHg)	Derivatives and m.p. (°C)	Spectroscopic evidence/data	Ref.
CH₃O₂C-, -CO₂CH₃, CH₃, N-H, CH₃	92–94/0.05		$n_D^{21.5}$ 1.4792, n_D^{25} 1.4698, IR	56
CH₃O₂C-, -CO₂CH₃, CH₃, N-CH₃, CH₃	130/0.05		n_D^{25} 1.4596, ^{1}H NMR	57
C₂H₅O₂C-, -CO₂C₂H₅, CH₃, N-H, CH₃	109–110/0.3		n_D^{25} 1.4716, ^{1}H NMR, IR	56
CH₃OCO, OCOCH₃, CH₃OCO, OCOCH₃, N-COCH₃	Syrup		$[\alpha]_D^{20} - 7.3$ ($c = 1$, CH₃OH) IR, MS	20
CH₃OCO, OCOCH₃, CH₃OCO, OCOCH₃, N-COCH₃	149.5–152		IR, MS	20
HO, OH, HO, OH, N-H	Yellow syrup		$[\alpha]_D^{23} + 19.9$ ($c = 2$, H₂O)	20

Azepanes (Hexahydroazepines)

Table 10 (continued)

Structure	M.p. (°C)/b.p. (°C/mmHg)	Derivatives and m.p. (°C)	Spectroscopic evidence/data	Ref.
(HO, OH, HO, OH azepane with N-COCH₃)	113–114.5		$[\alpha]_D^{20} + 44.5$ $(c = 1, H_2O)$	20
(HO, OH, HO, OH azepane with N-NO)	162–163.5		$[\alpha]_D^{23} + 9.7$ $(c = 1, H_2O)$	20
(tetramethyl dihydroxy azepane with N-C(CH₃)₂CH₂OH)	136–137		IR, MS, ^{1}H, ^{13}C NMR	213
(tetramethyl dihydroxy azepane with N-C(CH₃)₂CH₂OH, other stereo)	161–162 (subl.)		MS, ^{1}H, ^{13}C NMR	213
(tetramethyl diacetoxy azepane with N-^{t}Bu)	65–66			140

Table 10 (continued)

Structure	M.p. (°C)/b.p. (°C/mmHg)	Derivatives and m.p. (°C)	Spectroscopic evidence/data	Ref.
	140–141	Di-O-acetate 91.5–92.5	IR, ^{1}H NMR, MS	214
	140–142 (subl.)	Di-O-acetate 79–82	IR, ^{1}H NMR, MS	214
	91–94		IR, ^{1}H NMR, MS	214
	122–123		IR, ^{1}H NMR, MS	214
	81/760		IR, NMR, MS	215

Table 10 (continued)

Structure	M.p. (°C)/b.p. (°C/mmHg)	Derivatives and m.p. (°C)	Spectroscopic evidence/data	Ref.
Perfluoroazepane N-(CH$_2$)$_3$Cl	150–151.5/760		n_D^{20} 1.3178	216
Perfluoroazepane N-(CF$_2$)$_2$Cl	124–125.5/760		n_D^{20} 1.3156	216

Table 11. 2,2-Disubstituted azepanes

Structure	M.p. (°C)/b.p. (°C/mmHg)	Derivatives and m.p. (°C)	Spectroscopic evidence/data	Ref.
2,2-(CH₃)₂, N-CH₃	163–165/760 171/~760	Picrate 248–249 CH₃I 211	d^{20} 0.8564, n_D^{20} 1.4566	32b 217
2,2-(C₂H₅)₂, N-CH₃	99/17 95.5/15	Picrate 173–174 CH₃I 131	d^{20} 0.8785, n_D^{20} 1.4705	32b 217
2,2-(nPr)₂, N-CH₃	108–109/9	Picrate 149.5–150		32b
2,2-(nBu)₂, N-CH₃	135.5–135.5/10	HCl 107–110 Picrate 93.5–94	n_D^{20} 1.473	32b
2,2-(Ph)₂, N-H	Details in supplement		Spectra in supplement	218
2,2-(OC₂H₅)₂, N-CH₃	88/22		n_D^{20} 1.4758	219
2-CH₃, 2-allyl, N-H	Chromatography		¹H NMR	46

Azepanes (Hexahydroazepines)

Table 11 (continued)

Structure	M.p. (°C)/b.p. (°C/mmHg)	Derivatives and m.p. (°C)	Spectroscopic evidence/data	Ref.
azepane, N-H, 2-CH₃, 2-allyl	Chromatography			50
azepane, N-H, 2-CO₂H, 2-CH₃	>255 (d)		^{1}H NMR, IR, $\alpha_D - 17$ ($c = 1$, H₂O)	18
azepane, N-H, 2-CH₂Ph, 2-CO₂CH₃		HCl >280	IR, ^{1}H NMR	15
azepane, N-H, 2-Ph, 2-CO₂CH₃		HCl 190–192	IR, ^{1}H NMR	15
azepane, N-H, 2-CH₃, 2-CO₂CH₃	151–152		IR, ^{1}H NMR	15
azepane, N-H, 2-CH₃, 2-CO₂H	>280	HCl 253–255	IR, ^{1}H NMR	15
azepane, N-H, 2-CO₂C₂H₅, 2-CH₂Ph			^{1}H NMR, $\alpha_D - 7.6$ ($c = 1$, CHCl₃), ee > 95%	45

Table 11 (continued)

Structure	M.p. (°C)/b.p. (°C/mmHg)	Derivatives and m.p. (°C)	Spectroscopic evidence/data	Ref.
azepane with $CO_2C_2H_5$ and CH_2-naphthyl substituents, NH			^{1}H NMR, $\alpha_D - 10.2$ ($c = 1$, $CHCl_3$), ee > 95%	45
azepane with $CO_2C_2H_5$ and CH_2-C$_6$H$_4$-Br substituents, NH			^{1}H NMR, $\alpha_D - 3$ ($c = 1$, $CHCl_3$), ee > 95%	45
azepane with $CO_2C_2H_5$ and CH_2-C$_6$H$_4$-Cl substituents, NH			^{1}H NMR, $\alpha_D - 10.5$ ($c = 1$, $CHCl_3$), ee > 95%	45
CH_3, CH_3, vinyl-substituted azepane, N–OH, $2R^*6S^*$	44–46		IR, MS, ^{1}H, ^{13}C NMR	235
CH_3, CH_3, vinyl-substituted azepane, N–Tos, $2R^*6S^*$	59–60		IR, MS ^{1}H, ^{13}C NMR, X-ray	235

Table 12. 3,3-Disubstituted azepanes

Structure	M.p. (°C)/b.p. (°C/mmHg)	Derivatives and m.p. (°C)	Spectroscopic evidence	Ref.
3-C₂H₅, 3-OH, N-CH₃ azepane		HClO₄ 85.5–86.5	IR, ¹H NMR	83
3-C₂H₅, 3-OCH₃, N-CH₃ azepane		HClO₄ 111–111.5	¹H NMR, IR	83
3-Ph, 3-OH, N-H azepane	203–204		IR	38
3-(3-hydroxyphenyl), 3-C₂H₅, N-H azepane	116–118	HBr 187–188		220
3-(3-methoxyphenyl), 3-C₂H₅, N-CH₃ azepane	108–112/0.1	HBr 140–142		220

Table 12 (continued)

Structure	M.p. (°C)/b.p. (°C/mmHg)	Derivatives and m.p. (°C)	Spectroscopic evidence	Ref.
3-(3-hydroxyphenyl)-3-ethyl-1-methylazepane	127.5–133	HCl 184–186		220
3-phenyl-3-ethylazepane (NH)	110–112/0.5	Picrate 118–120 N-Propionyl 145–150/0.2		230

Azepanes (Hexahydroazepines)

Table 13. 4,4-Disubstituted azepanes

Structure	M.p. (°C)/b.p. (°C/mmHg)	Derivatives and m.p. (°C)	Spectroscopic evidence/data	Ref.
4-OH, 4-CH₃ azepane (NH)		HCl 134–136		77
4-Ph, 4-OH azepane (NH)		HCl 98		190
4-Ph, 4-OH azepane (N-CH₂Ph)	53	HCl 204		190
4-Ph, 4-OH azepane (N-CH₃)	120–122/0.3		n_D^{26} 1.5515	211
4-Ph, 4-OCOCH₃ azepane (N-CH₃)	120–125/0.3 147–152/0.3	CH₃I 215–217 (d) CH₃I 218–219 Picrate 183–183.5	n_D^{29} 1.5375	211 52

Table 13 (continued)

Structure	M.p. (°C)/b.p. (°C/mmHg)	Derivatives and m.p. (°C)	Spectroscopic evidence/data	Ref.
4-OH, 4-C₂H₅, N-CH₃ azepane	102–105/20			76
4-Ph, 4-OH, N-(CH₂)₂Ph azepane		HCl 163		190
4-Ph, 4-CONH₂, N-CH₃ azepane + H₂O	95–96			28
4-Ph, 4-CN, N-CH₃ azepane	119–121/0.25 130–132/0.35 133–135/1	CH_3Cl 270 Picrate 173–175 CH_3Br 248–249 HCl 150–151 CH_3Br 241–241.5 CH_3Cl 245–246	n_D^{22} 1.5341, d^{22} 1.030	28, 52 26
4-Ph, 4-CN, N-C₂H₅ azepane	136–139/1	HCl 182–183		26

Azepanes (Hexahydroazepines)

Table 13 (continued)

Structure	M.p. (°C)/b.p. (°C/mmHg)	Derivatives and m.p. (°C)	Spectroscopic evidence/data	Ref.
4-Ph, 4-CO₂CH₃, 1-CH₃ azepane	124–125/0.3	HCl 174–176	n_D^{28} 1.5278, d_4^{28} 1.062	28
4-Ph, 4-CO₂H, 1-C₂H₅ azepane	171–173	Ethyl ester HCl 178–179		26
4-Ph, 4-CO₂H, 1-CH₃ azepane	222–223	Sulphate 250–251 (d)		28
4-CN, 4-(3-OCH₃-C₆H₄), 1-CH₃ azepane	150–154/0.3	CH₃Cl 212–213	n_D^{22} 1.5332, d^{22} 1.062	30
4-Ph, 4-CO₂C₂H₅, 1-CH₃ azepane	128–130/0.3	HCl 151–153 Picrate 169–170 CH₃Br 215–217	n_D^{28} 1.5220, d_4^{26} 1.038	28
4-Ph, 4-COC₂H₅, 1-CH₃ azepane	132–136/0.3	CH₃I 217–219 HCl 122–125	n_D^{26} 1.5302	30

Table 13 (continued)

Structure	M.p. (°C)/b.p. (°C/mmHg)	Derivatives and m.p. (°C)	Spectroscopic evidence/data	Ref.
4-(COOC$_2$H$_5$)-4-(3-hydroxyphenyl)-1-methylazepane	135–136			30
4-methyl-4-hydroxy-1-benzoylazepane	94–96		$\alpha_D - 4$ ($c = 0.747$, CHCl$_3$) IR, ^{1}H NMR	200
4-phenyl-4-(n-butyryl)-1-methylazepane	132–133/0.3	Picrate 138–140	n_D^{26} 1.5300	30, 221
5-hydroxy-4,4-diphenyl-1-methylazepane	128.2–129.8			222
4-phenyl-4-hydroxy-3-allyl-1-methylazepane	76–77	HCl 172–173 O-Propionyl HCl 142–143		223

Table 13 (continued)

Structure	M.p. (°C)/b.p. (°C/mmHg)	Derivatives and m.p. (°C)	Spectroscopic evidence/data	Ref.
Ph, OCOC$_2$H$_5$, CH$_3$ azepane N-CH$_3$	124–128/0.3	HCl 201–202 Picrate 114–115	n_D^{25} 1.5142	224 231
Ph, OCOC$_2$H$_5$, CH$_3$ azepane N-CH$_3$	126/0.3	HCl 207 HBr 201–202 Picrate 162–163	n_D^{28} 1.5182, n_D^{21} 1.5215 ^{1}H NMR	52, 224 231
Ph, OCOC$_2$H$_5$, CH$_3$ azepane N-CH$_3$	138–148/0.3	CH$_3$I 60–65		52
Ph, Ph, CH$_3$ azepane N-CH$_3$		CH$_3$Cl 265–267 CH$_3$I 234–236		27
CH$_3$, Ph, CN azepane N-CH$_3$	128–130/0.4	CH$_3$Cl 253–256	n_D^{29} 1.5332	29

Table 13 (continued)

Structure	M.p. (°C)/b.p. (°C/mmHg)	Derivatives and m.p. (°C)	Spectroscopic evidence/data	Ref.
(4-Ph, 4-CO₂CH₃, 2-CH₃, N-CH₃ azepane)	120–122/0.25	Picrate 159–161 HCl 206–207	n_D^{32} 1.5250, d^{32} 1.060	210
(4-Ph, 4-CO₂C₂H₅, 2-CH₃, N-CH₃ azepane)	122–124/0.25	Picrate 177–178 HCl 145–146.5		210
(4-Ph, 4-COC₂H₅, 2-CH₃, N-CH₃ azepane)	135–142/0.3	CH₃I 183–185	$n_D^{27.5}$ 1.5350, d^{26} 1.035	30
(4-Ph, 4-CO₂^{i}Pr, 2-CH₃, N-CH₃ azepane)	118–120/0.25	Picrate 193–194 (d)	n_D^{27} 1.5160, d^{27} 0.999	210
(4-Ph, 4-CO₂^{n}Pr, 2-CH₃, N-CH₃ azepane)	128–130/0.25	Picrate 169–171	$n_D^{27.5}$ 1.5160, $d^{27.5}$ 1.012	210

Azepanes (Hexahydroazepines)

Table 13 (continued)

Structure	M.p. (°C)/b.p. (°C/mmHg)	Derivatives and m.p. (°C)	Spectroscopic evidence/data	Ref.
(4-CH₃, 3-Ph, 3-COC₂H₅, N-CH₃ azepane)	130–145/0.3			29
(4-CH₃, 3-Ph, 3-CO₂C₂H₅, N-CH₃ azepane)	126–128/0.3	CH₃I 165–170	$n_D^{28.5}$ 1.5235	29
(4-CH₃, 3-Ph, 3-CO₂H, N-CH₃ azepane)	182–184			29
(7-CH₃, 4-Ph, 4-CO₂C₂H₅, N-CH₃ azepane)	134–140/0.3 126–130/0.2	CH₃I 222–223	n_D^{23} 1.5232, d^{23} 1.044	29
(5-CH₃, 4-Ph, 4-CO₂C₂H₅, N-CH₃ azepane)	122–124/0.3		n_D^{24} 1.5278	29

Table 13 (continued)

Structure	M.p. (°C)/b.p. (°C/mmHg)	Derivatives and m.p. (°C)	Spectroscopic evidence/data	Ref.
	122–126/0.3	HCl 179–181	n_D^{23} 1.5301	29
	117–120/0.2	Picrate 195–197 (d) HCl 258–259 CH$_3$I 259–261	n_D^{26} 1.5330, d^{26} 1.016	210
	123–126/0.2	Picrate 199–202 CH$_3$I 254–255 CH$_3$Cl 259	n_D^{27} 1.5341, d^{27} 1.019	210
	120–140/0.3		n_D^{26} 1.5351	29
	125–130/0.3		n_D^{25} 1.5249, d^{25} 0.907	29

Table 13 (continued)

Structure	M.p. (°C)/b.p. (°C/mmHg)	Derivatives and m.p. (°C)	Spectroscopic evidence/data	Ref.
azepane with propargyl, OH, CH₃, N-(CH₂)₂Ph	140–150/0.1	HCl 147–149	$n_D^{22.6}$ 1.5392	225
azepane with propargyl, OH, CH₃, N-CH₃	130–135/25	HCl 148–150	$n_D^{22.6}$ 1.4665	225
azepane with OH, CH₃, CH₃, N-CH₃	95–100/8	CH₃I 225–232		64
azepane with Ph, CN, CH₃, CH₃, N-CH₃	160–162.5			29
azepane with Ph, CN, CH₃, CH₃, (CH₃)₂CH, N-CH₃	172/0.5			29

Table 13 (continued)

Structure	M.p. (°C)/b.p. (°C/mmHg)	Derivatives and m.p. (°C)	Spectroscopic evidence/data	Ref.
(azepane: 4-Ph, 4-CN, 3-C₂H₅, 2-nPr, N-CH₃)	170–174/0.8–0.9	Ethyl ester 174–8/0.7	n_D^{30} 1.5048	29
(azepane: 4-OCOCH₃, 4-Ph, 3-CH₃, N-CH₃)	142–152/0.3	HCl 216 CH₃I 95–97	$n_D^{27.5}$ 1.5242, ^{1}H NMR	224, 231
(azepane: 4-Ph, 4-OCOCH₃, 3-CH₃, N-CH₃)	142–152/0.3	HCl 196	^{1}H NMR	224, 231
(azepane: 4-OH, 4-Ph, 3-CH₃, N-CH₃)	91–92	Picrate 157	^{1}H NMR	231
(azepane: 4-Ph, 4-OH, 3-CH₃, N-CH₃)		HCl 250–251	^{1}H NMR	231

Table 13 (continued)

Structure	M.p. (°C)/b.p. (°C/mmHg)	Derivatives and m.p. (°C)	Spectroscopic evidence/data	Ref.
[azepane with OH, CH₂OH, aryl, N-CO₂C₂H₅]	Chromatography		IR, ^{1}H NMR	226
[azepane with OH, CH₂OH, aryl, N-CH₂Ph]	Chromatography		IR, ^{1}H NMR	226

Table 14. Alkenyl-substituted azepanes

Structure	M.p. (°C)/b.p. (°C/mmHg)	Derivatives and m.p. (°C)	Spectroscopic evidence/data	Ref.
azepane=C(COSCH$_3$)(NHCOPh)	200–203 (d)		IR	102
azepane=C(oxazolone-2-Ph)	132–134		IR	102
azepane=C(CN)$_2$	108	N-Acetyl 247–248.5 N-Methyl 81 N-Heptyl 200	IR	103
azepane=CH(NO$_2$)	89–91			92
azepane=C(CN)(NO$_2$)	137–139		IR	151
azepane=CH(CON(C$_2$H$_5$)$_2$)	145/0.05			147
azepane=CH(COSC$_2$H$_5$)	134/0.05			147

Azepanes (Hexahydroazepines)

Table 14 (continued)

Structure	M.p. (°C)/b.p. (°C/mmHg)	Derivatives and m.p. (°C)	Spectroscopic evidence/data	Ref.
	104/0.07			147
			¹H NMR, IR (not detailed)	107
			Data not given but compound used	106
	58 129–132/0.01		UV, IR	152
	164–165			89
	142–143 146			89 101

Table 14 (continued)

Structure	M.p. (°C)/b.p. (°C/mmHg)	Derivatives and m.p. (°C)	Spectroscopic evidence/data	Ref.
azepine=C(C₂H₅)CHO			¹H NMR, IR	107
azepine=cyclohexanone-CH₃			¹H NMR, IR	107
azepine=cyclohexane-1,3-dione	84			101
azepine=C(COCH₃)CO₂CH₃	150/0.05			101
azepine=C(CO₂CH₃)CO₂CH₃	150/0.05			101
azepine=C(C₂H₅)CO₂C₂H₅	100/0.01		IR, ¹H NMR	148

Azepanes (Hexahydroazepines)

Table 14 (continued)

Structure	M.p. (°C)/b.p. (°C/mmHg)	Derivatives and m.p. (°C)	Spectroscopic evidence/data	Ref.
azepane =C(CH$_3$)CO$_2$C$_2$H$_5$	93/0.07		IR, ^{1}H NMR	148
azepane =CH-CO$_2$C$_2$H$_5$ (Z)	55–56		^{1}H NMR	100
azepane =CH-COPh	76		IR, ^{1}H, ^{13}C NMR	99
azepane =CH-CO-^{n}Pr	Chromatography		IR, ^{1}H, ^{13}C NMR	99
azepane =CH-COCH$_3$	110/0.05		IR, ^{1}H, ^{13}C NMR	99
azepane =C(CO$_2$C$_2$H$_5$)$_2$	128/0.05		IR, ^{1}H NMR	149

Table 14 (continued)

Structure	M.p. (°C)/b.p. (°C/mmHg)	Derivatives and m.p. (°C)	Spectroscopic evidence/data	Ref.
azepine=C(CO$_2$C$_2$H$_5$)(CO—^{n}Pr)	124/0.01		IR, ^{1}H NMR	149
azepine=C(CO$_2$C$_2$H$_5$)(COPh)	44		IR, ^{1}H NMR	149
azepine=C(COCH$_3$)(COCH$_3$)	55		IR, ^{1}H NMR	149
azepine=C(CO$_2$C$_2$H$_5$)(COCH(CH$_3$)$_2$)	126/0.01		IR, ^{1}H NMR	149
azepine=C(CO$_2$C$_2$H$_5$)(COCH$_3$)	115/0.01		IR, ^{1}H NMR	149
azepine=C(CONHCH$_2$Ph)(CN)	138–140		MS	145
azepine=C(CONHPh)(CN)	162–164		MS	145

Table 14 (continued)

Structure	M.p. (°C)/b.p. (°C/mmHg)	Derivatives and m.p. (°C)	Spectroscopic evidence/data	Ref.
N-CH₃ azepane =C(CN)(CO₂C₂H₅)	73–76		IR, UV, ¹H NMR	92
N-CH₃ azepane =CH(NO₂)	84–85		IR, UV, ¹H NMR	92
N-CH₃ azepane =C(Ph)(CO₂CH₃)	76–79		IR, UV, ¹H NMR	92
N-CH₃ azepane =C(Ph)(CN)	79–83		IR, UV, ¹H NMR	92
N-CH₃ azepane =C(CONH₂)(CN)	189–190		IR, UV, ¹H NMR	92
N-CH₃ azepane =C(CO₂C₂H₅)(CO₂C₂H₅)	61–63		IR, UV, ¹H NMR	92, 227
N-CH₃ azepane -CH(H)-CO-O-C₆H₄-NO₂	126–127		¹H NMR	93

Table 14 (continued)

Structure	M.p. (°C)/b.p. (°C/mmHg)	Derivatives and m.p. (°C)	Spectroscopic evidence/data	Ref.
(azepine-N-CH₃, =C(H)-CO-O-C₆H₄-OCH₃)	92–93 80		^{1}H NMR	93 90
(azepine-N-CH₃, =C(COCH₃)₂)	152–155/2		^{1}H NMR	93
(azepine-N-CH₃, =C(Ph)(CO₂C₂H₅))	182–183/2		IR, UV, ^{1}H NMR	92, 93
(azepine-N-CH₃, =C(H)(COPh))	85–87 65		IR, UV, ^{1}H NMR	92, 93 90
(azepine-N-(CH₂)₃Cl, =C(H)(CO₂C₂H₅))	51–52		IR, ^{1}H NMR	146
(azepine-N-(CH₂)₃Cl, =C(dimedone))	169		IR, ^{1}H NMR	146

Azepanes (Hexahydroazepines)   105

Table 14 (continued)

Structure	M.p. (°C)/b.p. (°C/mmHg)	Derivatives and m.p. (°C)	Spectroscopic evidence/data	Ref.
(azepane=C(CH₂Ph)–dioxane-dione with CH₃,CH₃)	198		IR, ^{1}H NMR, MS	105
(N-CH₃ azepane=C(CN)–CO–N=CHN(CH₃)₂)	132–135		^{1}H NMR	94
(N-CH₃ azepane=CH–C(CO₂C₂H₅)(CN)=CH–N(CH₃)₂)	100–102		^{1}H NMR, IR, UV	95
(NH azepane=C(NO₂)–C(=S)–NHCO₂C₂H₅)	210–214			150
(N-CH₃ azepane=CH–CO–2-pyridyl)	103			90
(N-CH₃ azepane=CH–CO–2-furyl)	100			90

Table 14 (continued)

Structure	M.p. (°C)/b.p. (°C/mmHg)	Derivatives and m.p. (°C)	Spectroscopic evidence/data	Ref.
(2,4-dimethylphenoxy enamine of N-methylazepane)	90			90
(4-fluorophenoxy enamine)	50			90
(4-bromophenoxy enamine)	94			90
(4-chlorophenoxy enamine)	75			90
(3-nitrophenoxy enamine)	90			90

Azepanes (Hexahydroazepines)

Table 14 (continued)

Structure	M.p. (°C)/b.p. (°C/mmHg)	Derivatives and m.p. (°C)	Spectroscopic evidence/data	Ref.
(azepane=CH–CO–O–C₆H₄–NO₂, N–CH₃)	65			90
(azepane=CH–CO–O–C₆H₄–CH₃ ortho, N–CH₃)	90			90
(azepane=CH–CO–O–C₆H₄–CH₃ para, N–CH₃)	8			90
(azepane=C(CO₂CH₃)–CO–(CH₂)₂CO₂CH₃, NH)		Chromatography	IR, ^{1}H, ^{13}C NMR	99
(azepane=C(CO₂CH₃)–CO–(CH₂)₃CO₂CH₃, NH)		Chromatography	IR, ^{1}H, ^{13}C NMR	99
(azepane=C(dioxinone dimethyl), N–CH₃)	131		IR, ^{1}H NMR	105

Table 14 (continued)

Structure	M.p. (°C)/b.p. (°C/mmHg)	Derivatives and m.p. (°C)	Spectroscopic evidence/data	Ref.
azepane, N-CH₃, =CH-CO₂C₂H₅	106–108/0.05		IR, ¹H NMR	104
azepane, N-CH₃, =CH-C(OC₂H₅)=W(CO)₅	107		IR, MS, ¹H, ¹³C NMR	228
azepane, N-CH₃, =CH-C(OC₂H₅)=Cr(CO)₅	90		IR, MS, ¹H, ¹³C NMR	228
azepane, N-CH₃, =CH-CHO	Oil		IR, MS, ¹H, ¹³C NMR	109
azepane, N-CH₂Ph, =CH-CO₂ᵗBu	104–105		IR, ¹H NMR, MS	229
azepane, N-H, =CH₂		HCl 165		77

Table 14 (continued)

Structure	M.p. (°C)/b.p. (°C/mmHg)	Derivatives and m.p. (°C)	Spectroscopic evidence/data	Ref.
(4-methylene-1-methylazepane)		HCl 119–121 Picrate 176–177 CH$_3$I 206–209		77

REFERENCES

1. J.A. Moore and E. Mitchell, in R.C. Elderfield (Ed.), *Heterocyclic Compounds*, Vol. 9, Wiley, Chichester, 1967, p. 224.
2. A. Müller and E. Rölz, *Chem. Ber.*, 1928, **61**, 570.
3. R.G. Rice, E.J. Kohn and L.W. Daasch, *J. Org. Chem.*, 1958, **23**, 1352.
4. J.H. Paden and H. Adkins, *J. Am. Chem. Soc.*, 1936, **58**, 2491.
5. R.M. Hill and H. Adkins, *J. Am. Chem. Soc.*, 1938, **60**, 1033.
6. R.C. Fuson and W. Cole, *J. Am. Chem. Soc.*, 1938, **60**, 1237.
7. G. Cignarella, G. Nathansohn, G. Bianchi and E. Testa, *Gazz. Chim. Ital.*, 1962, **92**, 3.
8. A.M. Kritsyn, A.M. Likhosherstov and A.P. Skoldinov, *Dokl. Akad. Nauk SSSR*, 1968, **179**, 345; *Chem. Abs.*, 1968, **69**, 27228s.
9. S.C. Shim, C.H. Doh, T.J. Kim, H.K. Lee and K.D. Kim, *J. Heterocycl. Chem.*, 1988, **25**, 1383.
10. S.C. Shim, C.H. Doh, S.Y. Lee, W.H. Cho and K.T. Huh, *Taehan Hwahakhoe Chi*, 1990, **34**, 652; *Chem. Abs.*, 1991, **114**, 122027g.
11. S.C. Shim, C.H. Doh, B.W. Woo, H.S. Kim, K.T. Huh, W.H. Park and H. Lee, *J. Mol. Catal.*, 1990, **62**, L11; *Chem. Abs.*, 1991, **114**, 81549h.
12. Y.Z. Youn, D.Y. Lee, B.W. Woo, J.G. Shim, S.A. Chae and S.C. Shim, *J. Mol. Catal.*, 1993, **79**, 39; *Chem. Abs.*, 1993, **119**, 49212k.
13. J. Von Braun and A. Steindorff, *Chem. Ber.*, 1905, **38**, 3083.
14. K. Ziegler and Ph. Orth, *Chem. Ber.*, 1933, **66**, 1867.
15. H. Yasuo, M. Suzuki and N. Yoneda, *Chem. Pharm. Bull.*, 1979, **27**, 1931; *Chem. Abs.*, 1980, **92**, 76264j.
16. A.M. Likhosherstov, A.M. Kritsyn, A.S. Lebedeva and A.P. Skoldinov, *Zh. Org. Khim.*, 1973, **9**, 2245; *Chem. Abs.*, 1974, **80**, 47819r.
17. J. Bajgrowicz, A. El Achquar, M.-L. Roumestant, G. Pigière and P. Viallefont, *Heterocycles*, 1986, **24**, 2165.
18. D. Seebach, E. Dziadulewicz, L. Behrendt, S. Cantoreggi and R. Fitzi, *Liebigs Ann. Chem.*, 1989, 1215.
19. D. Barbry and B. Hasiak, *Bull. Soc. Chim. Fr.*, 1975, 2315.
20. H. Paulsen and K. Todt, *Chem. Ber.*, 1967, **100**, 512.

21. I.F. Bel'skii and L. Ya. Barkovskaya, *Zh. Org. Khim.*, 1967, **3**, 385; *Chem. Abs.*, 1967, **66**, 115581v.
22. Th. Severin and H. Lerche, *Chem. Ber.*, 1976, **109**, 1171.
23. J.A.M. Hamersma and W.N. Speckamp, *Tetrahedron Lett.*, 1982, **23**, 3811.
24. J.A.M. Hamersma and W.N. Speckamp, *Tetrahedron*, 1982, **38**, 3255.
25. J.A.M. Hamersma and W.N. Speckamp, *Tetrahedron*, 1985, **41**, 2861.
26. F.F. Blicke and E.-P. Tsao, *J. Am. Chem. Soc.*, 1953, **75**, 3999.
27. T.D. Perrine and E.L. May, *J. Org. Chem.*, 1954, **19**, 773.
28. J. Diamond, W.F. Bruce and F.T. Tyson, *J. Org. Chem.*, 1957, **22**, 399.
29. J. Diamond, M. Dymicky and W.F. Bruce, *Br. Pat.*, 843 924; *Chem. Abs.*, 1961, **55**, 2705a.
30. J. Diamond and W.F. Bruce, *US Pat.*, 2 775 588; *Chem. Abs.*, 1957, **51**, 7445a.
31. G.R. Clemo, R. Raper and H.J. Vipond, *J. Chem. Soc.*, 1949, 2095.
32. (a) R. Lukeš and J. Málek, *Chem. Listy*, 1951, **45**, 72; *Chem. Abs.*, 1951, **45**, 9523f; (b) R. Lukeš, V. Dudek, O. Sedláková and J. Korán, *Collect. Czech. Chem. Commun.*, 1961, **26**, 1105; *Chem. Abs.*, 1961, **55**, 18756g.
33. F.F. Blicke and N.J. Doorenbos, *J. Am. Chem. Soc.*, 1954, **76**, 2317.
34. O.M. Friedman, H. Sommer and E. Boger, *J. Am. Chem. Soc.*, 1960, **82**, 5202.
35. S. Shiotani and T. Kometani, *Chem. Pharm. Bull.*, 1973, **21**, 1053; *Chem. Abs.*, 1973, **79**, 66149t.
36. M.E. Garst and D. Lukton, *J. Org. Chem.*, 1981, **46**, 4433.
37. D.L. Lee, C.J. Morrow and H. Rapoport, *J. Org. Chem.*, 1974, **39**, 893.
38. V.G. Smirnova and R.G. Glushkov, *Khim. Farm. Zh.*, 1973, **7**, 3; *Chem. Abs.*, 1974, **80**, 47816n.
39. M. Saburi, K. Miyamura, M. Morita, Y. Mizoguchi, S. Yoshikawa, S. Tsuboyama, T. Sakurai and K. Tsuboyama, *Bull. Chem. Soc. Jpn.*, 1987, **60**, 141; *Chem. Abs.*, 1988, **108**, 105155f.
40. F. Fernández and C. Pérez, *Heterocycles*, 1987, **26**, 2411.
41. T. Nagasaka, H. Tamano and F. Hamaguchi, *Heterocycles*, 1986, **24**, 1231.
42. W.N. Speckamp, *Recl. Trav. Chim. Pays-Bas*, 1981, **100**, 345.
43. T. Shono, H. Hamaguchi and Y. Matsumura, *J. Am. Chem. Soc.*, 1975, **97**, 4264.
44. T. Nagasaka, Y. Koseki, H. Hayashi, Y. Yasuda and F. Hamaguchi, *Yakugaku Zasshi*, 1989, **109**, 823; *Chem. Abs.*, 1990, **112**, 216668.
45. G.I. Georg, X. Guan and J. Kant, *Bioorg. Med. Chem. Lett.*, 1991, **1**, 125; *Chem. Abs.*, 1991, **115**, 92854a.
46. K. Maruoka, T. Miyazaki, M. Ando, Y. Matsumura, S. Sakane, K. Hattori and H. Yamamoto, *J. Am. Chem. Soc.*, 1983, **105**, 2831.
47. H. Yamamoto and K. Maruoka, *J. Am. Chem. Soc.*, 1981, **103**, 4186.
48. K. Hattori, Y. Matsumura, T. Miyazaki, K. Maruoka and H. Yamamoto, *J. Am. Chem. Soc.*, 1981, **103**, 7368.
49. K. Maruoka, S. Nakai and H. Yamamoto, *Org. Synth.*, 1988, **66**, 185.
50. K. Hattori, K. Maruoka and H. Yamamoto, *Tetrahedron Lett.*, 1982, **23**, 3395.
51. P. Krogsgaard-Larsen and T. Roldskov-Christiansen, *Eur. J. Med. Chem. Chim. Ther.*, 1979, **14**, 157; *Chem. Abs.*, 1979, **91**, 123659h.
52. J. Diamond and W.F. Bruce, *US Pat.*, 2 775 589; *Chem. Abs.*, 1957, **51**, 7445b.
53. W. Lwowski and T.J. Maricich, *J. Am. Chem. Soc.*, 1965, **87**, 3630.
54. K. Hafner and C. König, *Angew. Chem.*, 1963, **75**, 89.
55. R.J. Cotter and W.F. Beach, *J. Org. Chem.*, 1964, **29**, 751.
56. M. Anderson and A.W. Johnson, *J. Chem. Soc.*, 1965, 2411.
57. R.F. Childs and A.W. Johnson, *J. Chem. Soc. C*, 1966, 1950.

58. R.J. Sundberg, B.P. Das and R.H. Smith, Jr, *J. Am. Chem. Soc.*, 1969, **91**, 658.
59. S.R. Tanny and F.W. Fowler, *J. Am. Chem. Soc.*, 1973, **95**, 7320.
60. J.H. Boyer and F.C. Canter, *J. Am. Chem. Soc.*, 1955, **77**, 3287.
61. K.A. Maier and O. Hromatka, *Monatsh. Chem.*, 1971, **102**, 513.
62. H.-W. Bersch, R. Rissmann and D. Schon, *Arch. Pharm. (Weinheim)*, 1982, **315**, 749; *Chem. Abs.*, 1982, **97**, 162796u.
63. A. Schousboe, O.M. Larsson, L. Hertz and P. Krogsgaard-Larsen, *Drug Dev. Res.*, 1981, **1**, 115; *Chem. Abs.*, 1981, **95**, 35327a.
64. A.C. Cope and W.D. Burrows, *J. Org. Chem.*, 1966, **31**, 3099.
65. P.A. Grieco and W.F. Fobare, *Tetrahedron Lett.*, 1986, **27**, 5067.
66. D. Damour, J. Pornet and L. Miginiac, *Tetrahedron Lett.*, 1987, **28**, 4689.
67. M. Mori, N. Kanda, Y. Ban and K. Aoe, *J. Chem. Soc., Chem. Commun.*, 1988, 12.
68. D.P. Curran and C.-T. Chang, *Tetrahedron Lett.*, 1990, **31**, 933.
69. S. Miyazawa, K. Ikeda, K. Achiwa and M. Sekiya, *Chem. Lett.*, 1984, 785.
70. F.D. Lewis, G.D. Reddy, S. Schneider and M. Gahr, *J. Am. Chem. Soc.*, 1989, **111**, 6465.
71. Y.T. Jeon, C.-P. Lee and P.S. Mariano, *J. Am. Chem. Soc.*, 1991, **113**, 8847.
72. R.W. Shaw, M. Anderson and T. Gallacher, *Synlett*, 1990, 584.
73. R.W. Shaw, D. Lathbury, M. Anderson and T. Gallacher, *J. Chem. Soc., Perkin Trans. 1*, 1991, 659.
74. T. Gallacher, I.W. Davies, S.W. Jones, D. Lathbury, M.F. Mahon, K.C. Molloy, R. W. Shaw and P. Vernon, *J. Chem. Soc., Perkin Trans. 1*, 1992, 433.
75. M.R. Gagné, C.L. Stern and T.J. Marks, *J. Am. Chem. Soc.*, 1992, **114**, 275.
76. A.D. Yanina and M.V. Rubtsov, *Zh. Obshch. Khim.*, 1962, **32**, 3693; *Chem. Abs.* 1963, **58**, 12512b.
77. C.A. Grob, R.M. Hoegerle and M. Ohta, *Helv. Chim. Acta*, 1962, **45**, 1823.
78. H.T. Nagasawa and J.A. Elberling, *Tetrahedron Lett.*, 1966, 5393.
79. H.T. Nagasawa, J.A. Elberling, P.S. Fraser and N.S. Mizuno, *J. Med. Chem.*, 1971, **14**, 501.
80. N. Shirai, F. Sumiya, Y. Sato and M. Hori, *J. Org. Chem.*, 1989, **54**, 836.
81. A. Ebnöther and E. Jucker, *Helv. Chim. Acta*, 1964, **47**, 745.
82. S. Sakanoue, S. Harusawa, N. Yamazaki, R. Yoneda and T. Kurihara, *Chem. Pharm. Bull.*, 1990, **38**, 2981; *Chem. Abs.*, 1991, **114**, 122029j.
83. N.J. Leonard, K. Jann, J.V. Paukstelis and C.K. Steinhardt, *J. Org. Chem.*, 1963, **28**, 1499.
84. J. Fujiwara, H. Sano, K. Maruoka and H. Yamamoto, *Tetrahedron Lett.*, 1984, **25**, 2367.
85. L. Ferrero, S. Geribaldi, M. Rouillard and M. Azzaro, *Can. J. Chem.*, 1975, **53**, 3227.
86. Y. Ishida, S. Sasatani, K. Maruoka and H. Yamamoto, *Tetrahedron Lett.*, 1983, **24**, 3255.
87. M. Zaidlewicz and I.G. Uzarewicz, *Pol. J. Chem.*, 1988, **62**, 143; *Chem. Abs.*, 1990, **113**, 58530v.
88. M. Zaidlewicz and I.G. Uzarewicz, *Heteroat. Chem.*, 1993, **4**, 73; *Chem. Abs.*, 1993, **119**, 72440s.
89. T. Tsujikawa, Y. Nakagawa, K. Tsukamura and K. Masuda, *Heterocycles*, 1977, **6**, 261.
90. V. Virmani, M.B. Nigam, P.C. Jain and N. Anand, *Indian J. Chem.*, 1979, **17B**, 472; *Chem. Abs.*, 1981, **94**, 15632x.

91. J. Singh, V. Sardana, P.C. Jain and N. Anand, *Indian J. Chem.*, 1983, **22B**, 1083; *Chem. Abs.*, 1984, **101**, 55065j.
92. V.G. Granik and R.G. Glushkov, *Zh. Org. Khim.*, 1971, 7, 1146; *Chem. Abs.*, 1971, **75**, 110176t.
93. V.G. Granik, I.V. Persianova, N.P. Kostyuchenko, R.G. Glushkov and Y.N. Sheinker, *Zh. Org. Khim.*, 1972, **8**, 181; *Chem. Abs.*, 1972, **76**, 139797c.
94. V.G. Granik, N.B. Marchenko, T.F. Vlasova and R.G. Glushkov, *Khim. Geterotsikl. Soedin.*, 1976, 1509; *Chem. Abs.*, 1977, **86**, 139783b.
95. V.G. Granik, O.Y. Belyaeva, R.G. Glushkov, T.F. Vlasova, A.B. Grigor'ev and M.K. Polievktov, *Khim. Geterotsikl. Soedin.*, 1977, 1518; *Chem. Abs.*, 1978, **88**, 74274f.
96. V.G. Granik, A.M. Zhidkova and R.A. Dubinskii, *Khim. Geterotsikl. Soedin.*, 1982, 518; *Chem. Abs.*, 1982, **97**, 55765m.
97. J.-P. Célérier, E. Deloisy, G. Lhommet and P. Maitte, *J. Org. Chem.*, 1979, **44**, 3089.
98. P. Delbecq, J.-P. Célérier and G. Lhommet, *Tetrahedron Lett.*, 1990, **31**, 4873.
99. P. Delbecq, D. Bacos, J.-P. Célérier and G. Lhommet, *Can. J. Chem.*, 1991, **69**, 1201.
100. J.-P. Célérier, E. Deloisy-Marchalant, G. Lhommet and P. Maitte, *Org. Synth.*, 1989, **67**, 170.
101. G. Lhommet, M.G. Richaud and P. Maitte, *J. Heterocycl. Chem.*, 1982, **19**, 431.
102. D.C. Cook and A. Lawson, *J. Chem. Soc., Perkin Trans. 1*, 1973, 465.
103. T. Mukaiyama and S. Ono, *Tetrahedron Lett.*, 1968, 3569.
104. J.-P. Célérier, M.G. Richaud and G. Lhommet, *Synthesis*, 1983, 195.
105. J-C. Pommelet, H. Dhimane, J. Chuche, J.-P. Célérier, M. Haddad and G. Lhommet, *J. Org. Chem.*, 1988, **53**, 5680.
106. J. Barluenga, M. Tomás, V. Kouznetsov and E. Rubio, *Synlett*, 1992, 563.
107. Y. Matsumura, J. Fujiwara, K. Maruoka and H. Yamamoto, *J. Am. Chem. Soc.*, 1983, **105**, 6312.
108. P.H. Lambert, M. Vaultier and R. Carrié, *J. Org. Chem.*, 1985, **50**, 5352.
109. K. Suzuki, T. Ohkuma and G. Tsuchihashi, *J. Org. Chem.*, 1987, **52**, 2929.
110. K. Suzuki, T. Ohkuma and G. Tsuchihashi, *J. Org. Chem.*, 1988, **53**, 4160.
111. A.L.J. Beckwith, P.H. Eichinger, B.A. Mooney and R.H. Prager, *Aust. J. Chem.*, 1983, **36**, 719.
112. F. Hogue and H. Frye, *Inorg. Nucl. Chem. Lett.*, 1974, **10**, 505; *Chem. Abs.*, 1974, **81**, 42211x.
113. H. Bestian, J. Heyna, A. Bauer, G. Ehlers, B. Hirsekorn, T. Jacobs, W. Noll, W. Weibezahn and F. Römer, *Liebigs Ann. Chem.*, 1950, **566**, 210.
114. F.F. Blicke and E.B. Hotelling, *J. Am. Chem. Soc.*, 1954, **76**, 2422.
115. R.P. Mull, P. Schmidt, M.R. Dapero, J. Higgins and M.J. Weisbach, *J. Am. Chem. Soc.*, 1958, **80**, 3769.
116. R.J. Wineman, M.H. Gollis, J.C. James and A.M. Pomponi, *J. Org. Chem.*, 1962, **27**, 4222.
117. A.A.R. Sayigh and H. Ulrich, *J. Chem. Soc.*, 1963, 3144.
118. G. Ehrhart and G. Seidel, *Chem. Ber.*, 1964, **97**, 1994.
119. K.H. Saunders, *J. Chem. Soc.*, 1955, 3275.
120. D. Seebach, D. Enders and B. Renger, *Chem. Ber.*, 1977, **110**, 1852.
121. A.I. Meyers, P.D. Edwards, W.F. Rieker and T.R. Bailey, *J. Am. Chem. Soc.*, 1984, **106**, 3270.

122. A.I. Meyers, P.D. Edwards, T.R. Bailey and G.E. Jagdmann, Jr, *J. Org. Chem.*, 1985, **50**, 1019.
123. P. Beak and W.K. Lee, *J. Org. Chem.*, 1993, **58**, 1109.
124. P. Beak and W.K. Lee, *Tetrahedron Lett.*, 1989, **30**, 1197.
125. M. Murakami, M. Hayashi and Y. Ito, *J. Org. Chem.*, 1992, **57**, 793.
126. K. Nyberg and R. Servin, *Acta Chem. Scand.*, *Ser. B*, 1976, **30**, 640.
127. M. Mitzlaff, K. Warning and H. Jensen, *Liebigs Ann. Chem.*, 1978, 1713.
128. M. Malmberg and K. Nyberg, *J. Chem. Soc., Chem. Commun.*, 1979, 167.
129. M. Malmberg and K. Nyberg, *Acta Chem. Scand., Ser. B*, 1979, **33**, 69.
130. M. Malmberg and K. Nyberg, *Acta Chem. Scand., Ser. B*, 1981, **35**, 411.
131. L. Eberson, M. Malmberg and K. Nyberg, *Acta Chem. Scand., Ser. B*, 1983, **37**, 555.
132. J.R.M. Lundkvist, L.G. Wistrand and U. Hacksell, *Tetrahedron Lett.*, 1990, **31**, 719.
133. J.R.M. Lundkvist, B. Ringdahl and U. Hacksell, *J. Med. Chem.*, 1989, **32**, 863.
134. J.R.M. Lundkvist, H.M. Vargas, P. Caldirola, B. Ringdahl and U. Hacksell, *J. Med. Chem.*, 1990, **33**, 3182.
135. M.J. Fisher and L.E. Overman, *J. Org. Chem.*, 1990, **55**, 1447.
136. R.T. Dean, H.C. Padgett and H. Rapoport, *J. Am. Chem. Soc.*, 1976, **98**, 7448.
137. H.H. Otto and P. Schwenkkraus, *Tetrahedron Lett.*, 1982, **23**, 5389.
138. J.A. Lowe, III, S.E. Drozda, R.M. Snider, K.P. Longo and J.P. Rizzi, *Bioorg. Med. Chem. Lett.*, 1993, **3**, 921; *Chem. Abs.*, 1994, **120**, 54442k.
139. F. Sumiya, N. Shirai and Y. Sato, *Chem. Pharm. Bull.*, 1991, **39**, 36.
140. P.Y. Johnson and J. Lisak, *Tetrahedron Lett.*, 1975, 3801.
141. S. Wawzonek and J.M. Shradel, *J. Org. Chem.*, 1981, **46**, 2410.
142. H.-J. Nitzschke and H. Budka, *Chem. Ber.*, 1955, **88**, 264.
143. A. Archelas, R. Furstoss, D. Srairi and G. Maury, *Bull. Soc. Chim. Fr.*, 1986, 234.
144. N. Floyd, F. Munyemana, S.M. Roberts and A.J. Willetts, *J. Chem. Soc., Perkin Trans. 1*, 1993, 881.
145. L.V. Ershov and V.G. Granik, *Khim. Geterotsikl. Soedin.*, 1985, 929; *Chem. Abs.*, 1986, **104**, 168389y.
146. M. Haddad, J.-P. Célérier, G. Haviari and G. Lhommet, *Heterocycles.*, 1990, **31**, 1251.
147. J.-P. Célérier, G. Lhommet and P. Maitte, *Tetrahedron Lett.*, 1981, **22**, 963.
148. J.-P. Célérier, E. Deloisy-Marchalant and G. Lhommet, *J. Heterocycl. Chem.*, 1984, **21**, 1633.
149. P. Brunerie, J.-P. Célérier, H. Petit and G. Lhommet, *J. Heterocycl. Chem.*, 1986, **23**, 1183.
150. S. Rajappa, B.G. Advani and R. Sreenivasan, *Indian J. Chem.*, 1977, **15B**, 886.
151. S. Rajappa, B.G. Advani and R. Sreenivasan, *Tetrahedron,* 1977, **33**, 1057.
152. S. Hünig, W. Grässman, V. Meuer, E. Lücke and W. Brenninger, *Chem. Ber.*, 1967, **100**, 3039.
153. A. Müller and A. Sauerwald, *Monatsh. Chem.*, 1927, **48**, 727.
154. R.A. Laforge, C.E. Cosgrove and A. D'Adamo, *J. Org. Chem.*, 1956, **21**, 988.
155. O. Cervinka, *Chem. Listy*, 1958, **52**, 1145; *Chem. Abs.*, 1958, **52**, 18406d.
156. G. Baddeley, J. Chadwick and H.T. Taylor, *J. Chem. Soc.*, 1956, 451.

157. F.F. Blicke and E.-P Tsao, *J. Am. Chem. Soc.*, 1954, **76**, 2203.
158. N.J. Leonard, J.W. Curry and J.J. Sagura, *J. Am. Chem. Soc.*, 1953, **75**, 6249.
159. G.S. Kolesnikov and N.N. Mikhailovskaya, *Zh. Obshch. Khim.*, 1957, **27**, 458; *Chem. Abs.*, 1957, **51**, 15532q.
160. N.S. Kozlov, V.A. Tarasevich and S.I. Kozintsev, *Dokl. Akad. Nauk BSSR*, 1984, **28**, 37; *Chem. Abs.*, 1984, **100**, 174237f.
161. G. Opitz and A. Griesinger, *Liebigs Ann. Chem.*, 1963, **665**, 101.
162. R. Wittman, *Angew. Chem.*, 1961, **73**, 219.
163. W. Siefken, *Liebigs Ann. Chem.*, 1949, **562**, 75.
164. R. Curtis and H. Tilles, *Fr. Pat.* 1 327 964; *Chem. Abs.*, 1963, **59**, 6372a.
165. S.I. Burmistrov and V.V. Egorova, *Ukr. Khim. Zh.*, 1958, **24**, 222; *Chem. Abs.*, 1958, **52**, 18466b.
166. C.-H. Cheng and J.-Y. Chi, *Yao Hsueh Hsueh Pao*, 1963, **10**, 655; *Chem. Abs.*, 1964, **60**, 6825c.
167. H. Najer, R. Giudicelli and J. Sette, *Bull. Soc. Chim. Fr.*, 1962, 556.
168. G. D'Alo and M. Perghem, *Boll. Chim. Farm.*, 1960, **99**, 733; *Chem. Abs.*, 1961, **55**, 14356d.
169. F.F. Blicke, *US Pat.* 2 735 847; *Chem. Abs.*, 1956, **50**, 15602h.
170. R.G. Rice, E.J. Kohn and L.W. Daasch, *J. Org. Chem.*, 1958, **23**, 1352.
171. V.M. Solov'ev, A.P. Arendarvik and A.P. Skoldinov, *Zh. Obshch. Khim.*, 1959, **29**, 613; *Chem. Abs.*, 1960, **54**, 389c.
172. W. Reppe, *et al.*, *Liebigs Ann. Chem.*, 1955, **596**, 12.
173. J. Van Alphen, *Recl. Trav. Chim. Pays-Bas*, 1939, **58**, 1105; *Chem. Abs.*, 1940, **34**, 2331.
174. Z. Welvart, *Bull. Soc. Chim. Fr.*, 1955, 218.
175. B.C. Gautam, R.S. Kapil and N. Anand, *Indian J. Chem.*, 1966, **4**, 239.
176. M.A.T. Rogers, *Nature (London)*, 1956, **177**, 128.
177. H. Bock and H. Tom Dieck, *Chem. Ber.*, 1966, **99**, 213.
178. (a) J.M. McManus and C.F. Gerber, *J. Med. Chem.*, 1966, **9**, 256; (b) J.H. Biel, *US Pat.* 2 932 646; *Chem. Abs.*, 1960, **54**, 17436a.
179. F.F. Blicke and G.R. Toy, *J. Am. Chem. Soc.*, 1954, **76**, 4615.
180. E. Keschmann, E. Zbiral and J. Schweng, *Liebigs Ann. Chem.*, 1977, 1508.
181. W.S. Hamama, M. Hammouda, E.M. Kandeel and E.M. Afsah, *Chin. Pharm. J.*, 1992, **44**, 25; *Chem. Abs.*, 1992, **117**, 26178.
182. O. Cervinka and L. Hub, *Collect. Czech. Chem. Commun.*, 1965, **30**, 3111.
183. A.M. Likhosherstov, K.S. Raevskii, A.S. Lebedeva, A.M. Kritsyn and A.P. Skoldinov, *Khim. Farm. Zh.*, 1967, **1**, 30; *Chem. Abs.*, 1967, **67**, 90642w.
184. L. Ya. Barkovskaya, V.M. Shostakovskii and I.F. Bel'skii, *Izv. Akad. Nauk SSSR, Ser. Khim.*, 1967, 2773; *Chem. Abs.*, 1968, **69**, 86806b.
185. P. Krogsgaard-Larsen, K. Thyssen and K. Schaumberg, *Acta Chem. Scand., Ser. B*, 1978, **32**, 327.
186. B.R. deCosta, C. Dominguez, X.-S. He, W. Williams, L. Radesca and W. Bowen, *J. Med. Chem.*, 1992, **35**, 4334.
187. R. Partch, *Tetrahedron Lett.*, 1966, 1361.
188. M.S. Raasch, *US Pat.*, 2 612 500; *Chem. Abs.*, 1953, **47**, 3736c.
189. S. Morosawa, *Bull. Chem. Soc. Jpn.*, 1958, **31**, 418; *Chem. Abs.*, 1959, **53**, 8160g.
190. A.F. Casy and H. Birnbaum, *J. Chem. Soc.*, 1964, 5130.
191. Z.G. Finney and T.N. Riley, *J. Med. Chem.*, 1980, **23**, 895.
192. N.J. Leonard and S. Gelfand, *J. Am. Chem. Soc.*, 1955, **77**, 3269.
193. N.J. Leonard and E. Barthel, Jr, *J. Am. Chem. Soc.*, 1949, **71**, 3098.

194. L.S. Hegedus, M.A. Schwindt, S. Delombaert and R. Imwinkelried, *J. Am. Chem. Soc.*, 1992, **112**, 2264.
195. H. Takahata, K. Takahashi, E.-C. Wang and T. Yamazaki, *J. Chem. Soc., Perkin Trans. 1*, 1989, 1211.
196. M. Yamaguchi and I. Hirao, *Tetrahedron Lett.*, 1983, **24**, 1719.
197. R.G. Glushkov, V.G. Smirnova, K.A. Zaitseva, N.A. Novitskaya, M.D. Mashkovskii and G.N. Pershin, *Khim. Farm. Zh.*, 1974, **8**, 14; *Chem. Abs.*, 1974, **81**, 13368y.
198. L.E. Hecker and J.E. Saavedra, *Carcinogenesis*, 1980, **1**, 101; *Chem. Abs.*, 1981, **94**, 169124v.
199. U. Hacksell, L.-E. Arvidsson, U. Svensson, J.L.G. Nilsson, D. Sanchez, H. Wikström, P. Lindberg, S. Hjorth and A. Carlsson, *J. Med. Chem.*, 1981, **24**, 1475.
200. R.A. Johnson, M.E. Herr, H.C. Murray and L.M. Reineke, *J. Am. Chem. Soc.*, 1971, **93**, 4880.
201. R.A. Johnson, M.E. Herr, H.C. Murray and G.S. Fonken, *J. Org. Chem*, 1968, **33**, 3187.
202. J. De Ruiter, S. Andurkar, T.N. Riley and D.E. Walters, *J. Heterocycl. Chem.*, 1992, **29**, 779.
203. J.B. Hester, *J. Org. Chem.*, 1967, **32**, 4098.
204. J.R. Steiner and J. Clardy, *J. Am. Chem. Soc.*, 1993, **115**, 6452.
205. G. Alvernhe, A. Laurent and K. Touhami, *J. Fluorine Chem.*, 1985, **29**, 363.
206. N. Finch, L. Blanchard and L.H. Werner, *J. Org. Chem.*, 1977, **42**, 3933.
207. C.G. Overberger, J. Reichenthal and J.-P. Anselme, *J. Org. Chem.*, 1970, **35**, 138.
208. K. Ogura, Y. Shimamura and M. Fujita, *J. Org. Chem.*, 1991, **56**, 2920.
209. T. Shono, Y. Matsumura, O. Onomura, M. Ogaki and T. Kanazawa, *J. Org. Chem.*, 1987, **52**, 536.
210. J. Diamond and W.F. Bruce, *US Pat.*, 2 740 777; *Chem. Abs.*, 1956, **50**, 15598h.
211. J. Diamond, W.F. Bruce and F.T. Tyson, *J. Org. Chem.*, 1965, **30**, 1840.
212. H.O. House and H.C. Mueller, *J. Org. Chem.*, 1962, **27**, 4436.
213. P.Y. Johnson and D.J. Kerkman, *J. Org. Chem.*, 1976, **41**, 1768.
214. P.Y. Johnson, I. Jacobs and D.J. Kerkman, *J. Org. Chem.*, 1975, **40**, 2710.
215. W.-H. Lin and R.J. Lagow, *J. Fluorine Chem.*, 1990, **50**, 15.
216. E. Hayashi, T. Abe, H. Baba and S. Nagase, *J. Fluorine Chem.*, 1984, **26**, 417.
217. R. Lukeš and K. Smolek, *Collect. Czech. Chem. Commun.*, 1939, **11**, 506; *Chem. Abs.*, 1940, **34**, 7869.
218. C.A. Zezza, M.B. Smith, B.A. Ross, A. Archin and P.L.E. Cronin, *J. Org. Chem.*, 1984, **49**, 4397.
219. V.G. Granik, M.K. Polievktov and R.G. Glushkov, *Zh. Org. Khim.*, 1971, **7**, 1431; *Chem. Abs.*, 1971, **75**, 129299g.
220. G. Bradley, J.F. Cavalla, T. Edington, R.G. Shepherd, A.C. White, B.J. Bushell, J. R. Johnson and G.O. Weston, *Eur. J. Med. Chem. Chim. Ther.*, 1980, **15**, 375.
221. J. Diamond and W.F. Bruce, *US Pat.*, 2 740 779; *Chem. Abs.*, 1956, **50**, 15599h.
222. R.E. Lyle and G.G. Lyle, *J. Am. Chem. Soc.*, 1954, **76**, 3536.
223. J. Lee and A. Ziering, *US Pat.*, 2 802 822; *Chem. Abs.*, 1958, **52**, 2940f.
224. J. Diamond, W.F. Bruce and F.T. Tyson, *J. Med. Chem.*, 1964, **7**, 57.
225. M. Prost, M. Urbain and R. Charlier, *Helv. Chim Acta*, 1969, **52**, 1134.

226. M. Kimura and S. Morosawa, *Bull. Chem. Soc. Jpn.*, 1979, **52**, 1437; *Chem. Abs.*, 1979, **91**, 157571p.
227. V.G. Granik, A.N. Akalaev and R.G. Glushkov, *Zh. Org. Khim.*, 1971, **7**, 2429; *Chem. Abs.*, 1972, **76**, 59420j.
228. R. Aumann and P. Hinterding, *Chem. Ber.*, 1990, **123**, 611.
229. M. Yamaguchi and I. Hirao, *J. Org. Chem.*, 1985, **50**, 1975.
230. E. Testa, L. Fontanella and V. Aresi, *Liebigs Ann. Chem.*, 1964, **676**, 151.
231. M.I. Iorio and M. Miraglia, *Chim. Ther.*, 1973, **8**, 85; *Chem. Abs.*, 1973, **79**, 78581u.
232. Z. Polivka, J. Metys and M. Protiva, *Collect. Czech. Chem. Commun.*, 1986, **51**, 2034; *Chem. Abs.*, 1987, **107**, 77656d.
233. Y. Terada and J. Okada, *Yakugaku Zasshi*, 1977, **97**, 207; *Chem. Abs.*, 1977, **87**, 39246.
234. M.E. Fox, A.B. Holmes, I.T. Forbes and M. Thompson, *Tetrahedron Lett.*, 1992, 7421.
235. M.E. Fox, A.B. Holmes, I.T. Forbes and M. Thompson, *J. Chem. Soc., Perkin Trans. 1*, 1994, 3379.

CHAPTER 3

Azepanones

I. AZEPAN-2-ONES (CAPROLACTAMS)

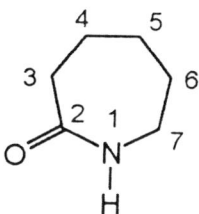

The literature contains an enormous number of papers pertaining to this subject area. This state of affairs is largely due to the fact that caprolactam (above) polymerises to nylon-6, one of the more popular polyamides used in fibre production.[1] The availability of caprolactam has stimulated much organic chemistry research in the last 40 years. It is not our intention to deal with the industrial aspects of caprolactam production which is generally based on the Beckmann rearrangement of cyclohexanone oxime or related variations.[1]

However, it is appropriate to mention here the connection with L-lysine, which arises by hydrolytic ring opening of 3-aminocaprolactam. This has been utilised for the production of both chiral and racemic lysine on a commercial scale.[2,3]

Considering the wealth of papers published in the area, it is surprising that the organic chemistry of caprolactam and its derivatives has not been reviewed previously. However, casual references do appear in a recent text devoted to lactam chemistry.[210]

PREPARATION

1. Beckmann Rearrangements

From the earliest days,[4-6] the Beckmann rearrangement has been the most popular method for construction of the caprolactam skeleton. Sulphuric acid has very frequently been the reagent of choice for this transformation.[7-12]

(1) → (2) and/or (3)

Many other reagents have been used, e.g. benzenesulphonyl chloride in alkaline solution,[13,31] toluene-p-sulphonyl chloride in alkaline solution,[14,15,20] PCl_5,[16,17,21] $POCl_3$,[18,19,22] toluene-p-sulphonyl chloride in pyridine,[23,24] polyphosphoric acid[25-30] and hydroxylamine-O-sulphonic acid.[32] Further modifications have included the use of trimethylsilyl polyphosphate, which was considered superior to polyphosphate ester (PPE) on the grounds of ease of formation,[33] although yields of lactams were similar. More recently, the combined use of tetrabutylammonium perrhenate(VII) and trifluoromethanesulphonic acid has been suggested for inducing the Beckmann rearrangement of oximes,[34] as has montmorillonite clay/KSF in refluxing toluene.[35] Further work may confirm the generality of these new devices.

There is a report that attempted hydrogenation of the oxime **4** over Raney nickel yielded lactam **5**, although the yield was poor.[36] The lactam **6** may be obtained by photolysing the appropriate oxime in methanol using radiation which is 90% 253.7 nm.[37]

(4)   (5)   (6)

The well known dependence of the Beckmann rearrangement upon the stereoselectivity of the oximes used leads to the conclusion that, unless the oximes are purified, mixtures of lactams (e.g. **1** → **2** + **3**) may be expected.

Indeed, in early days, it was very difficult to identify and purify products (e.g. from **1** → **2** + **3**) and this probably explains the discrepancies found in the literature melting points (see tables). The advent of chromatography has largely solved these problems. In some cases, special efforts have been made to pre-purify oximes so that the Beckmann rearrangement gave regiospecific results. Thus, regiospecific methylation of the dilithio salt of cyclohexanone oxime gave the *Z*-isomer **7**, which underwent rearrangement (85% $H_2SO_4/0\,°C$) giving predominantly the 3-methylcaprolactam isomer **8**, while the *E*-isomer **9** (oximation of 2-methylcyclohexanone) gave 7-methylcaprolactam (**10**).[38] The separated 2-methylcyclohexanone oximation products were later shown to behave similarly even using sulphuric acid at 110 °C.[39]

(7) (8) (9) (10)

It has been recognised that in some cases the Beckmann fragmentation reaction intervenes depending on the reagent used. Thus, 2,2-dimethylcyclohexanone oxime (**11**) reacted with PCl_5 in diethyl ether (0 °C) giving 6-methyl-5-heptenenitrile (**12**) in 94% yield, whereas 7,7-dimethylcaprolactam

(11) $CH_3CH=CH(CH_2)_3CN$ (12)

(13) (14)

(13) was obtained in poor yield using thionyl chloride.[17] Other examples are seen when the 2-substituent stabilised a carbocation at C-2 of the oxime[21,23,40] leading to fragmentation products.

Apart from bulky substituents at C-2, Beckmann rearrangement of cyclohexanone oximes tolerates the presence of several substituent groups in various positions round the ring. 2-Aminocyclohexanone oxime hydrochloride (14) underwent rearrangement to 3-aminocaprolactam (15) in 96% H_2SO_4; apparently a pre-isomerisation of (E)- to (Z)-oxime took place.[41] The same rearrangement has been reported using a medium of SO_3 in liquid SO_2.[42] Other

(15)　　(16)　　(17)

(18)　　(19)

groups compatible with the rearrangement are $-CH_2SiMe_3$ and $-SiMe_3$,[43] $-NF_2$,[44] $-CH_2OCOPh$, $-CH_2O^tBu$ and $-CO_2^tBu$,[45] $Ph_2P(O)CH_2-$[46] and $-CN$.[47,48] Several examples involving chiral substrates have been reported: it is possible, with care, to avoid racemisation. Thus, 16 gave 17 and 18 gave 19[42] and (−)-menthone oxime (20) gave 21 with $POCl_3$,[14,19] or benzenesulphonyl chloride/NaOH.[49] A Beckmann rearrangement of an optically active (+)-4-methylcyclohexanone oxime in a chiral host [(−)-1,6-bis(o-chorophenyl)-1,6-diphenylhexa-2,4-diyne-1,6-diol] yielded (S)-(−)-5-methylcaprolactam (22) in 80% ee.[50]

(20)　　(21)　　(22)

Azepanones

(23)   (24)

Caprolactams have been obtained by an ingenious protocol involving Beckmann rearrangements of oximes of thiophenocyclohexanones (23, 24) followed by desulphurisation of the thiophenocaprolactams thus obtained. Since the rearrangement gives, in general, a separable mixture of regioisomers (e.g. 25 and 26 from 23), and thiophenes can carry a variety of substituents (R in

(25)   (26)

23, 24), this is a versatile approach. The following caprolactams are exemplary: 7-ethyl;[51] 3-ethyl, 7- and 3-propyl;[52] 4-ethyl;[53] various 3- and 7-aminoalkyl;[54] 3- and 7-cycloalkyl[55] and 7-ω-carboxyalkyl.[56]

2. Schmidt Reaction

2-Methylcyclohexanone has long been known[57] to react with hydrazoic and sulphuric acids giving 7-methylcaprolactam (10). A more detailed study of the Schmidt reaction applied to 2-alkyl-, 2-cyano-, 2-chloro- and 2-ethoxycarbonyl-cyclohexanones revealed that exclusive formation of 7-substituted caprolactams took place.[58] In this work hydrazoic acid in CHCl$_3$ was used. 7-Cyanocaprolactam was similarly obtained with sulphuric acid/CHCl$_3$/NaN$_3$ mixture.[59] Introduction of polyphosphoric acid (PPA) as both reagent and solvent with sodium azide[60] was shown to be a superior way of conducting Schmidt reactions and most, but not all,[61] subsequent applications for caprolactams have followed this lead:[62] in the case of methylene-2,2-bicyclohexanone (27), the polyphosphoric acid medium was much superior;[63] the products were a pair of diastereomers of 7,7'-methylenebiscaprolactam (28). The 4,4'-isopropylidene compound 29 similarly gave two diastereomers of 5,5'-isopropylidenebiscaprolactam (30).[64] When 2-arylmethylcyclohexanones were treated with sodium azide in PPA, 7-arylmethylcaprolactams were obtained in modest yield,[65] compound 31 again proving to be a diastereomeric mixture. 2-Arylcyclohexanones give either 3- or 7-arylcaprolactams depending upon the conditions used for the Schmidt reaction: low temperatures in PPA favour the formation of the 3-aryl isomers.[65,66]

Fairly recent work[67,68] has revealed that when chiral substituted β-keto esters (e.g. 32) were reacted with sodium azide and methanesulphonic acid in chloroform at reflux, the products (33, R = PhCH$_2$, 2-naphthyl-CH$_2$, p-BrC$_6$H$_4$CH$_2$, and m-ClC$_6$H$_4$CH$_2$) were formed with good retention of configuration and in good yield.

In several cases, Schmidt reactions have yielded different products from those obtained by the Beckmann rearrangement.[63,64] An interesting case involved

2-indolyl-2-cyclohexanone (**34**).[69] Here Beckmann rearrangement of the oxime yielded the 7-isomer **35** whereas the Schmidt reaction gave the 3-isomer **36**.

(**35**)   (**36**)

3. Cyclising Reactions

Since lactams are reversibly related to the appropriate amino acids, it is not surprising that caprolactams have frequently been obtained by cyclisation of 6-aminohexanoic acid or its derivatives. Indeed, this method has served to establish beyond doubt the structure of some substituted caprolactams when Beckmann or Schmidt reactions gave ambiguous mixtures.[70,71] Heating 6-amino-6,6-dimethylhexanoic acid at 280 °C for 30 min gave 7,7-dimethylcaprolactam (**13**);[72] however, heating amino acids over Al_2O_3 is more highly recommended.[73] Amino esters can more easily be thermally cyclised: several caprolactams have been thus obtained; lactams **37** and **38** (R = CH_3, C_2H_5),[74] **39**,[75] **40**,[76] **41**[77] and **42**[91] are examples. The amino acid **43** cyclised to

(**37**)   (**38**)   (**39**)

(**40**)   (**41**)   (**42**)

44 under the influence of diethyl phosphite.[79] More unusual was the thermolysis of **45**, which yielded the lactam **46**.[80]

A popular procedure has been reduction of nitrile esters and subsequent cyclisation either with or without isolation of the intermediate amino esters. These reactions appear to result in good yields of caprolactams.[70,71,81] Reductive cyclisations of various azido-sugar derivatives have been reported;[82–85] this approach makes available polyhydroxycaprolactams of known stereochemistry (see tables for examples). Catalytic reductive cyclisation of a protected chiral amino ester has been shown to give an excellent yield of the chiral aminocaprolactam L-Boc-**15**.[87] Acyclic azido esters may also be reductively cyclised without loss of chirality: **47** and **48** were thus obtained[88] and easily separated chromatographically. Similarly, reduction of ω-oximino esters gives caprolactams,[89] but in the special case of the oximino diester **49**, reduction with reduced iron powder gave the desired hydroxamic acid **50** in only 1.5% yield.[90] 6-Ketohexanoic esters were reduced with hydrogen and ammonia over nickel–aluminium catalyst to provide the caprolactams **38** (R = H, CH$_3$, C$_2$H$_5$).[74] Sodium methoxide was shown to cause cyclisation of methyl lysinate hydrochloride to aminocaprolactam (**15**) in either chiral or racemic form.[91] There is one example of a palladium-catalysed carbonylation of an iodoalkenylamine (**51**) leading to an interesting methylene caprolactam structure

Azepanones

(49) **(50)** **(51)** **(52)**

(52).[92] Likewise, there is a single example of an iminium ion cyclisation (53 → 54) leading to a substituted caprolactam;[93] there seems no reason why other examples of this type should not be uncovered.

(53) **(54)**

4. Miscellaneous Preparations

Spirooxaziridines have proved to be a useful source of caprolactams. Originally these were deduced to lie on the pathway to N-ethylcaprolactam, for example when nitroethane was irradiated in cyclohexane.[94] Later work showed that irradiation of oximes gave caprolactams presumably via oxaziridines (e.g. 55)[95] and subsequently the oxaziridines were isolated (oxidation of imines) and photolysed.[95,96] This led to the first example of a chiral oxaziridine → caprolactam synthesis[96] giving (S)-$(-)$-5-$tert$-butylcaprolactam (56), the structure of which was confirmed by X-ray analysis. The methodology was extended by other workers to several further examples of 5-substituted caprolactams[97] and 3-, 4-, 6- and 7-substituted caprolactams.[98] It was recently disclosed that the ring opening of spirooxaziridines can be catalysed by manganese(III) tetraphenylporphyrin in a regio- and stereospecific manner.[99]

Claisen rearrangement of alkyl imidates (**57**, R = H, Ph) at 210–212 °C gave both 3-allyl and *N*-allyl products (**58** and **59**, respectively). The proportion of the latter increased with increasing temperature.[100,101] A related, but milder, reaction involves reaction of the imidate **60** with lithium salts of allyl alkoxides **61** (R^2, R^3 = H, CH_3, Ph, etc.). The diastereomeric products generally favoured one isomer (e.g. **62**) over the other.[102]

(**55**) (**56**) (**57**) (**58**)

(**59**) (**60**) (**61**) (**62**)

Allylic peroxidation of the endocyclic imine **64** gave **65** with R = either H or OH, depending on the conditions used.[103] Ozonolysis of a dimethoxybenz-azepinone (**66**) gave the muconate **67** in fair yield[104] in a fast (20 min) reaction regulated by addition of boron trifluoride etherate.

(**63**) (**64**) (**65**)

Several ring expansion protocols leading to caprolactams have been published. For example, 1-chloro-1-nitrosocyclohexane (**68**, R = H) reacted

Azepanones

(66) (67) (68)

with triphenylphosphine[105] to give caprolactam in reasonable yield and triphenylaluminium caused the conversion of the tetramethyl analogue (68, R = CH$_3$) into 3,3,7,7-tetramethylcaprolactam 6.[106]

(69) (70) (71)

(72) (73)

1,1-Azidosulphides (e.g. 69) reacted with a four molar proportion of trifluoroacetic acid in chloroform to give the spiro derivative 70.[107] On the other hand, tin(IV) chloride led to the formation of imino thioether structures such as 71, which can be converted into caprolactams or used as activated lactams in reactions with nucleophiles. Ring expansion of cyclohexanone was brought about by the combined action of dimethoxyamine [HN(OCH$_3$)$_2$] and trimethylsilyl triflate which gave the N-methoxycaprolactam 72,[108] presumably via intermediate 73. More unusual and specialised was the conversion (in 13% yield) of the 3-pyrrolidinylindole 74 into the substituted caprolactam 75 by refluxing with diethyl malonate and powdered sodium hydroxide in xylene for 2 days.[109]

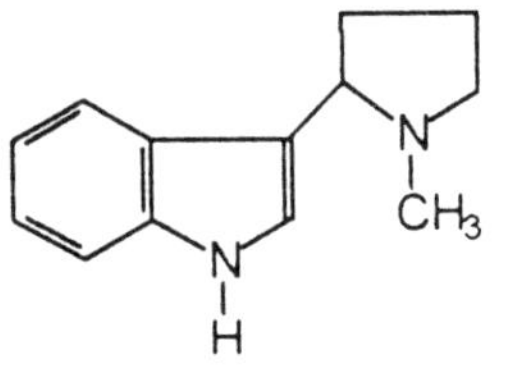

(74) 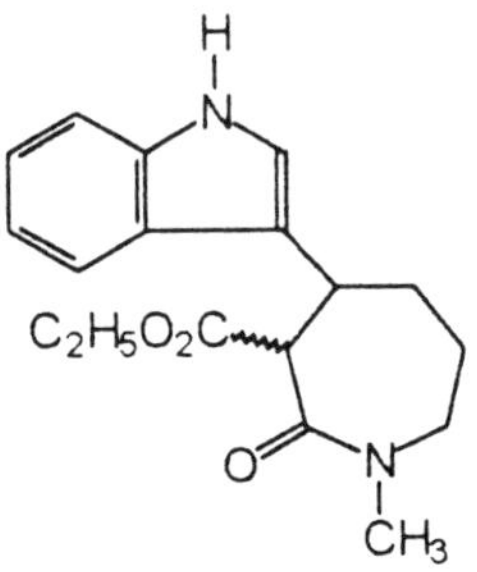 (75)

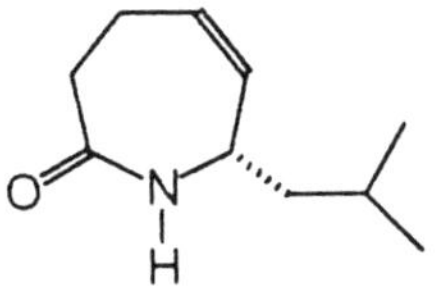

(76) 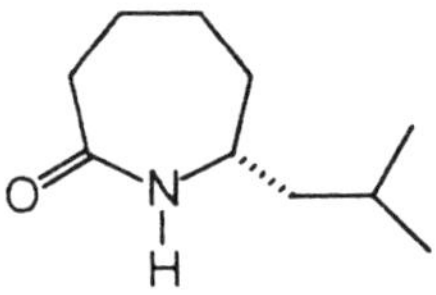 (77)

Not surprisingly, reduction of double bonds in unsaturated lactams leads to caprolactams (hexahydroazepin-2-ones). For example, hydrogen over palladium converted **76** into **77** without loss of chirality.[110] Various tetrahydroazepin-2-ones have been hydrogenated over palladised carbon[111] and dihydroazepinone **78** was likewise reduced to the hexahydro state.[112] Sodium borohydride reduced the α,β-unsaturated ketone **79** to the caprolactam **80**[113] and the imine **81** to the

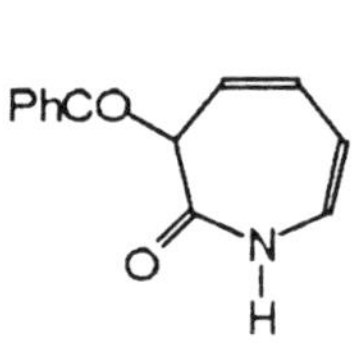

(78) 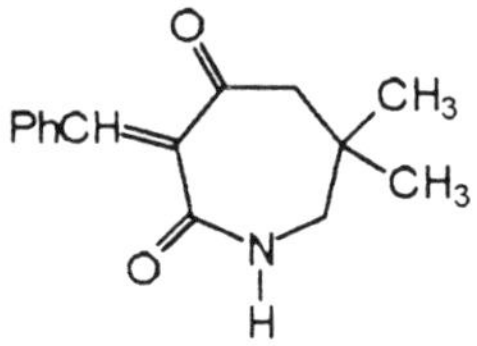 (79) 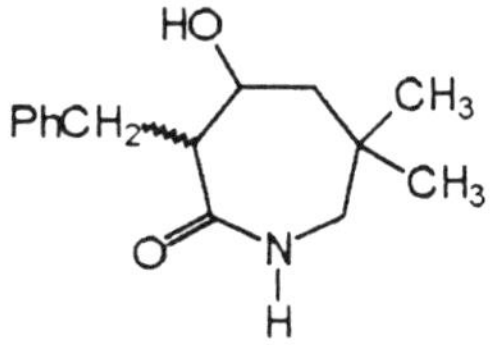 (80)

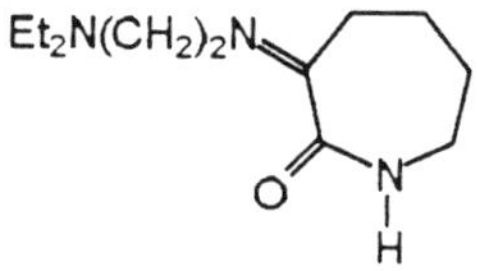

(81) 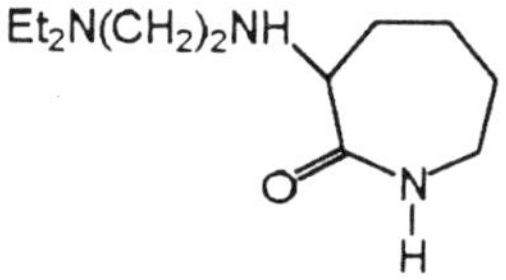 (82)

aminocaprolactam **82**.[114] Addition of methanol across the double bond in **84** was achieved using mercury(II) acetate in methanol and gave **85**.[115]

Azepanones

(84)

(85)

5. Caprolactam Imines

These structures are of interest, partly because as cyclic amidines they may show some biological activities. Reaction of chlorosulphonyl isocyanate with N-methylcaprolactam gives the chlorosulphonyl derivative **86** (R = SO_2Cl), from which the simple imine **86** (R = H) was obtained using aqueous sodium hydroxide.[116]

(86)   (87)   (88)

(89)   (90)

Oximinocaprolactam (**87**, R = H) arises when caprolactam is treated with hydroxylamine hydrochloride and sodium hydrogencarbonate in methanol at reflux[117] or by using phosgene, then methanol (to give the lactim ether **88**) followed by hydroxylamine.[118] Other derivatives (**87**, R = CH_3SO_2-, $-CO_2CH_3$) were made from **87** (R = H) using the relevant reactive acid halides.[118] The tosyl analogue **86** (R = tosyl) arose from treatment of N-methylcaprolactam with tosyl isocyanate.[119] Several methyl-substituted caprolactams were oximated after conversion into the corresponding thiolactams.[120] More complex cyclic amidines have been made from lactim ether **88** by treatment with various nitrogen nucleophiles: examples include **89**,[121] **90** (R^1 = various aryl, R^2 = H, CH_3),[122] **90** (R^1 = various aryl-NH, R^2 = H, CH_3)[123] and **91** (various aryl).[124] Aryl sulphimides **92** were made from **86** (R = SO_2Cl) in Friedel–Crafts-type reactions with arenes and $AlCl_3$.[125] Lactam acetal **93**[126] has

(91)

(92)

(93)

(94)

proved a very popular substrate for making caprolactam imines: it reacts with many different kinds of nitrogen nucleophiles. The following selection gives an idea of the scope. Arylamines yielded products such as **86** (R = various aryl),[127–129] **94** (R^1 = aryl, R^2 = CN, p-$NO_2C_6H_4$),[130] **94** (R^1 = H, R^2 = CO_2CH_3),[131] **86** (R = $CONH_2$, $CO_2C_2H_5$),[132] **92**[133] and **86** (R = o-$CO_2C_2H_5C_6H_4$).[134]

6. Caprolactam Thiones (Thiocaprolactams)

An early paper[135] described how cyclohexanone oxime with benzenesulphonyl chloride in pyridine saturated with hydrogen sulphide gave the parent of this series **95** (R = H). The N-methyl analogue (**95**, R = CH_3) was made from the corresponding N-methylcaprolactam in 20% yield[136] by heating with P_4S_{10} in xylene, although a 68.5% yield had previously been reported.[120] An earlier

(95)

(96)

Azepanones

(97)

(98)

method using CS_2 does not now appear attractive.[137] The use of hydrogen sulphide was recommended when activated iminium species such as **96** (R = Cl or OSO_2Ph) were the substrates.[138,139] Owing to the developing interest in thiolactams, several improved reagents have more recently been introduced for the conversion of caprolactams into their thio analogues. Lawesson's reagent[140] (**97**) and an 'improved' version (**98**) have been published. Other devices include treatment of the caprolactam with bis(tricyclohexyltin) sulphide/boron trichloride, which generates B_2S_3 *in situ*,[142] and treatment of the lactam with P_4S_{10} and organolithiums (^nBuLi, CH_3Li or PhLi) under mild conditions.[143] Alternatively, the fluoroborate salt **96** (R = OEt) could be reacted with NaSH[144] or the lactam with bis(1,5-cyclooctanediylboryl) sulphide [(9BBN)$_2$S], which gave an 88% yield of **95** (R = CH_3).[145] Nowadays, it appears that the choice of reagent for this conversion is a matter of personal preference.

7. Natural Products

Until recent times, the caprolactam nucleus was not a feature of many natural product structures. Mycobactins[146] were exceptional. The appearance of the reduced azepine ring in natural products is due to the connection of 3-aminocaprolactam with lysine; it is indeed 'cyclolysine,' so lysine and δ-hydroxylysine are implicated in the biosynthetic pathways involved.

The bengamides (**99**), isolated from a Fijian sponge show a variety of biological actions, including anthelmintic and bacteriostatic.[147] Some bengamides

(99)

contain a 6-hydroxy-3-aminocaprolactam moiety, e.g. bengamide B [**99**, R^1 = H, R^2 = $CH_3(CH_2)_{12}CO_2-$, R^b = CH_3), whereas others are merely 3-aminocaprolactams such as bengamide E (**99**, R^1 = R^2 = R^b = H).[148] Several total syntheses of bengamide E have been reported,[149-151] including an enantioselective example.[152] Bengamides A[153] and B[152] have also been synthesised. Sponges have yielded other examples.[359,360]

Caprolactins A and B [**100**, R = $-(CH_2)_5CH(CH_3)_2$ and $-(CH_2)_4CH(CH_3)(C_2H_5)$] were isolated from a bacterium in a deep ocean sediment and have been synthesised.[154] The pathogenic fungus *Periconia circinata* has yielded a non-toxic metabolite called circinatin **101**[155] and the cobactins have continued to receive attention;[156] benzyl-cobactin T (**102**, R = CH_2Ph) was synthesised from the corresponding Boc-protected 3-aminocaprolactam and catalytically debenzylated to cobactin T (**102**, R = H).

(**100**)　　　　　(**101**)

(**102**)

Mycobactin S2 has been synthesised.[157] Since the bulk of synthetic problems in this area are concerned with the side chain rather than the heterocyclic nucleus, detailed discussion of the relevant chemistry is outside the scope of this volume.

PROPERTIES AND REACTIONS

Caprolactams are, on the whole, thermally stable solids or distillable oils. They are generally neutral, typical lactams which may be hydrolysed to the corresponding amino acids (see below). In early work the only physical measurements recorded were refractive indices, but gradually infrared and, later, nuclear magnetic resonance spectra have been reported. Accordingly, most authors during the last 30 years have provided useful reference data to which we allude in the tables. The preferred conformations to be found in the ring system have attracted passing interest[158,159,207] and it seems that the conformational preferences of caprolactams are similar to those of cyclohexanes.[159]

Caprolactams undergo a variety of chemical transformations, many of which are predictable but a few are not.

1. Substitution on Nitrogen

As expected, N-substitution with alkyl halides and bases is facile: sodium hydride is a very satisfactory reagent for the deprotonation of caprolactams;[160] many other examples are given in the tables. N-Acylation does not necessarily require any base to be present; for example, N-acetylation[161] and N-benzoylation[162] have been thus carried out. However, it seems more expedient to use a base such as triethylamine or N,N-dimethylaniline.[163] A two-stage haloalkylation has been carried out by heating caprolactam with trimethyl silyl chloride and paraformaldehyde; in this way **103** (R = CH_2Cl) was obtained.[164]

(**103**)

Michael additions are frequently encountered; thus acrylonitrile reacted with caprolactam in the presence of Triton B, giving **103** (R = CH_2CH_2CN),[165] and phenylacetylene reacts in the presence of sodium to yield N-styrylcaprolactam **103** (R = PhCH=CH_2).[166] Styrene reacted with caprolactam in hexamethylphosphoramide to give **103** (R = CH_2CH_2Ph).[167] Carbamates (e.g. **103**, R = PhNHCO—) arise by using isocyanates.[168] Trimethylsilylation [**103**, R = $(CH_3)_3Si$] was achieved with trimethylsilyl chloride in the presence of triethylamine or better with trimethylsilyldiethylamine.[169] N-Nitrosocaprolactam **103** (R = NO) has been made by reaction with nitrosating gases (from $NaNO_2$ and H_2SO_4)[170] or by direct reaction with sodium nitrite and concentrated hydrochloric acid at $-10\,°C$.[171] The chemical behaviour of N-nitrosocaprolactam has been studied,[170-172] particularly with respect to ring opening.

2. Metal Complexes

Copper(II) complexes are known in which both two (**104**, M = Cu, X = Br, Cl)[173] and three (**105**, M = Cu, X = Cl)[174] molecules of caprolactam are incorporated. Manganese(II) forms complexes of type **105** (M = Mn, X = Cl),[173] while zinc(II) is also found as in **104** (M = Zn, X = Br) and **105**

($M = Zn$, $X = I$).[175] Platinum(II) behaves as in **104** ($M = Pt$, $X = Cl$)[176] but rhodium(III) appeared as a three-caprolactam assembly (**106**).[176]

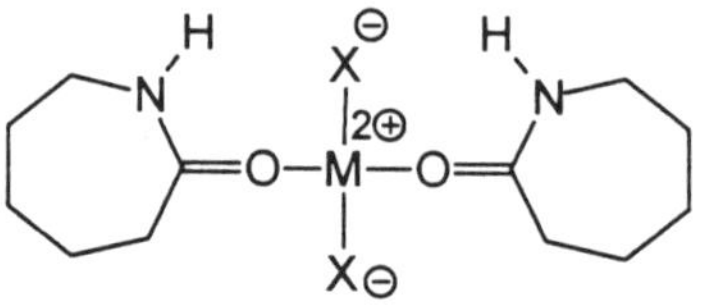

(**104**) 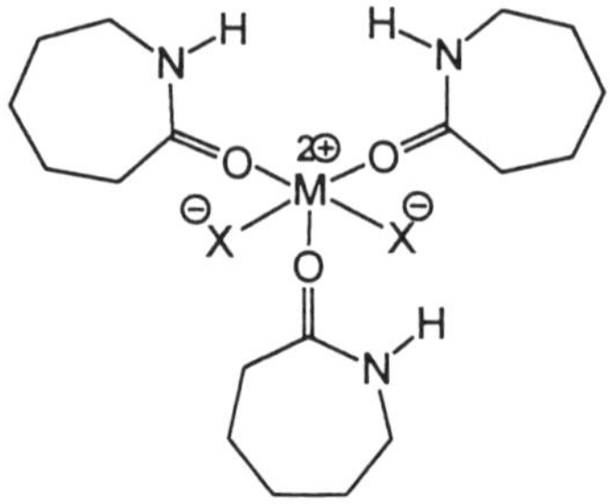 (**105**)

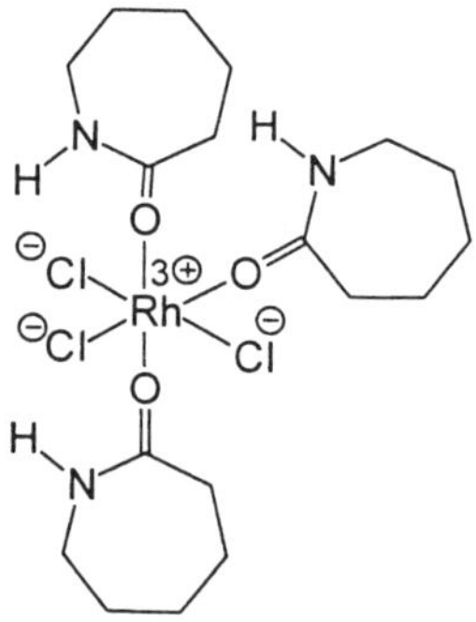

(**106**) 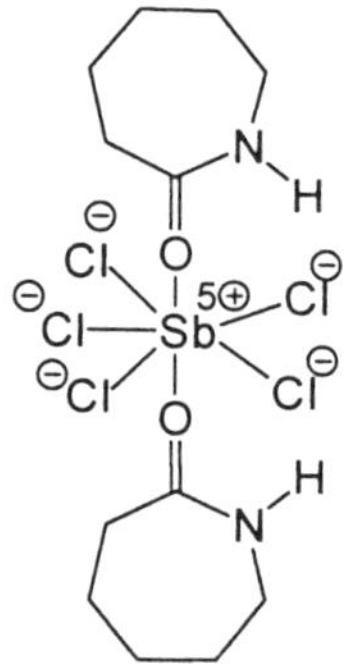 (**107**)

$VOCl_2$ complexes two caprolactams (**104**, $M = VO$, $X = Cl$)[177] and antimony pentachloride gave the complex **107**.[178] Boron compounds associate with caprolactam in different ways depending on the attached ligands. BF_3 forms a [BF_3 + 2-caprolactam] complex[178-180] but the reaction with triethylboron leads to a more fundamental change (**108** and **109** isolated).[181] Triisopropylboron gave the four-membered ring compound **110**.[181] The

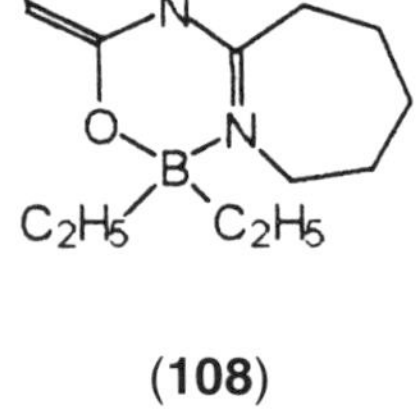

(**108**) 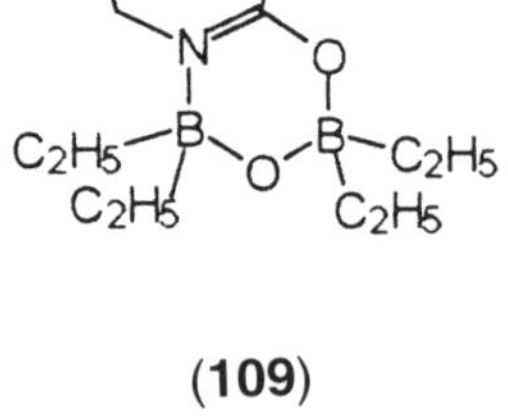 (**109**) 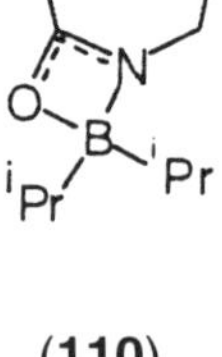 (**110**)

chromium carbonyl carbene compound **111** reacted with *N*-methylcaprolactam to give the product **112**.[182]

(111) (112)

3. Oxidation and Reduction

Oxygenation of caprolactam under the influence of UV irradiation gave a hydroperoxide (**113**);[183] zinc chloride was shown[184] to accelerate this reaction. Cobalt salts induced the conversion of **113** into the dione **114** (R = H) (adipimide). The peracid-, cobalt- or manganese-catalysed oxidation of caprolactam led directly to an adipimide (**114**, R = H).[185] The initially formed radical in this photo process was trapped with benzophenone yielding the tertiary alcohol **115**.[186]

(113) (114) (115)

(116) (117) (118)

Anodic oxidation of caprolactam in the presence of alcohols gives the 7-alkoxylactams **116** (R^1 = H, R^2 = CH_3, C_2H_5);[187] conversion of the latter into the bisalkoxy compound **116** (R^1 = CH_2OH, R^2 = CH_3) followed by $TiCl_4$-induced reaction with allyltrimethylsilane gave the annulated product **117**.[188] However, anodic oxidation of *N*-alkylcaprolactams takes another course; thus *N*-methylcaprolactam gave a mixture of **103** (R = H, CH_2OH, CHO).[189] Similarly, RuO_4 oxidations gave mixtures involving *exo*- and *endo*-carbon atoms: *N*-Methylcaprolactam (**103**, R = CH$_3$) yielded caprolactam (24%), *N*-formyl caprolactam (**103**, R = CHO) (35%) and *N*-methyladipimide (**114**, R = CH$_3$) (38%).[190] The (*S*)-3-*N*-phthaloylcaprolactam **118** (X = Y = H) could be oxidised to the (*S*)-3-*N*-phthalimidoadipimide **118** (X + Y = O) using benzene seleninic anhydride[191] or KMnO$_4$ in CH$_3$CO$_2$H/H$_2$O/H$_2$SO$_4$.[192] The *R*-isomer behaved similarly.[191] Biological oxidation of *N*-benzylcaprolactam (**103**, R = PhCH$_2$) with *Beauveria sulfurescens* gave a mixture of 4- and 5-hydroxy products (**119** and **120**).[193]

(119) (120) (121)

There are really only two useful modes of reduction for caprolactams. First, LiAlH$_4$ has been widely used to convert the lactams into azepanes (Chapter 2).[194–197] On the other hand, sodium borohydride treatment of *N*-substituted caprolactams (**103**, R = CO$_2$C$_2$H$_5$, CO$_2$CH$_2$Ph or CO$_2^t$Bu) gave the very useful ethoxyazepanes **121** (R = CO$_2$C$_2$H$_5$, CO$_2$CH$_2$Ph or CO$_2^t$Bu) (Chapter 2).[198,199]

4. Halogenation

Invariably caprolactams undergo halogenation at C-3. Monochlorination is difficult to achieve, the 3,3-dichlorolactam **122** (R = H, X = Y = Cl) usually being the principal product.[57,200–203] The reagents used have generally been PCl$_5$ alone or in admixture with POCl$_3$; it is advisable to carry out these reactions at low temperatures.[200,205] The dibromo-compound **122** (R = H, X = Y = Br) was similarly (Br + PCl$_5$) obtained[203,204] and used to obtain enamine **123** (R = H) from which dione **124** (R = H) and other derivatives were made.[204] A related series of reactions from the dichloro analogue **122** (R = CH$_3$, X = Y = Cl) gave the 2,3-dione **124** (R = CH$_3$).[205] By contrast, monochlorination of *N*-benzoylcaprolactam is straightforward using sulphuryl chloride;[163] the technical process has been improved by transamidation of the product (**122**, R = PhCO, X = H, Y = Cl) with caprolactam which gave a better and more convenient overall conversion into 3-chlorocaprolactam (**122**, R = X = H, Y = Cl).[206] The latter has been efficiently obtained by catalytic hydrogenolysis of the dihalide **122** (R = H, X = Y = Cl).[201,203] Replacement of halogen atoms with the amino group has been much studied[2,201–203,208] since the

(122)   (123)   (124)

(125)   (126)

3-aminocaprolactam so obtained can be converted into lysine by hydrolysis. 3-Chlorocaprolactam may either be aminated directly with ammonia[202,203] or first displaced with azide ion (to 122, R = X = H, Y = N$_3$) and subsequently reduced catalytically.[201,207,211] Production of L-lysine is an important industrial process: in one case (Toray), racemic 3-aminocaprolactam is hydrolysed enzymatically to L-(S)-lysine (125) using L-caprolactam hydrolase (*Candida humicola*) and simultaneously the unwanted (R)-aminocaprolactam is racemised by D-aminocaprolactam racemase (*Alcaligenes faecalis*). Thus all material is converted into L-lysine.[209] 3-Iodocaprolactam was obtained from the 3-bromo analogue by reaction with sodium iodide.[202]

Not all halogenations proceed uneventfully; it was demonstrated that sulphuryl chloride in CCl$_4$ first converted caprolactam into an intermolecular condensation product (126) which gave 122 (R = H, X = Y = Cl) on treatment with methanol and HCl.[212] POCl$_3$ alone converted caprolactam into 127, which also gave 126 by treatment with sulphuryl chloride.[212]

(127)   (128)   (129)   (130)

Direct displacement of bromine from 122 (R = H, X = Y = Br) with sodium diethylmalonate gave the substitution product 122 [R = X = H,

Y = CH(CO$_2$Et)$_2$],[213] which was converted into several other substances (e.g. **122**, R = X = H, Y = CH$_2$CO$_2$H).

Elimination of halides, on the other hand, is known. The dichlorocaprolactam **122** (R = H, X = Y = Cl), made from caprolactam with COCl$_2$ and Cl$_2$, underwent conversion to the cyclic acrylamide **128** (X = Cl) when heated with AlCl$_3$ in CCl$_4$.[214] Dehydrobromination of 3-bromocaprolactam (**122**, R = X = H, Y = Br) by lutidine gave two products (**128**, X = H, and **129**), both of which reacted with sodium diethylmalonate yielding the 4-substituted ester **130** [R = CH(CO$_2$Et)$_2$].[215,216] The latter reaction is of Michael type, it being considered that the two alkenyl substrates involved were in equilibrium via an enol during the reaction. In similar fashion, 3-bromocaprolactam and sodium methoxide gave **130** (R = OCH$_3$),[217] although earlier workers[162] had mistakenly thought that this treatment comprised a direct displacement of bromide by methoxide leading to **122** (R = X = H, Y = OCH$_3$).

Chlorination of caprolactam with *tert*-butyl hypochlorite gave the *N*-chlorocaprolactam **103** (R = Cl), which underwent photolytic dehydrohalogenation and solvent capture in methanol which led to the methoxy lactam **116** (R^1 = H, R^2 = CH$_3$).[219] If caprolactam is allowed to react with phosgene (COCl$_2$) alone, the imidoyl chloride **131** (R = Cl) is obtained and reacted with Meldrum's acid

(**131**)   (**132**)

leading to 2-alkenylazepanes (Chapter 2).[220] On the other hand, *N*-alkylcaprolactams gave chloroiminium salts (**132**), which also reacted with Meldrum's acid yielding useful 2-alkenyl-*N*-alkyl azepanes[221,222] (Chapter 2).

5. Substitution at C-2

In addition to the cases above, which arise by preliminary halogenation of caprolactams, several other methods are known for achieving substitution at C-2. It has long been known[223] that Grignard reagents add to *N*-methylcaprolactam in bis-fashion (**133**, R = CH$_3$, C$_2$H$_5$, C$_3$H$_7$, C$_4$H$_9$). More recently, it was shown that alkynyl boranes add to *N*-alkyllactams and that after LiAlH$_4$ reduction, alkynylamines were obtained.[244] In this way compound **134** was made.

The behaviour of caprolactam with dimethyl sulphate is interesting: mild treatment brought about conversion to the lactim ether (**131**, R = OCH$_3$),[225] which reacts as previously with Meldrum's acid.[226] However, if more vigorous

Azepanones

(133) R-N(R)(CH₃) structure; **(134)** PhC≡C-N(C₂H₅); **(135)** CH₃O⁺-N(CH₃) · CH₃SO₄⁻

conditions are employed, alkyl O to N migrations take place and *N*-alkyl-caprolactams (**103**, R = CH_3, C_2H_5, etc.) are obtained.[225] On the other hand, *N*-methylcaprolactam forms a salt (**135**) with dimethyl sulphate; application of sodium methoxide then leads to the acetal (**133**, R = OCH_3).[126a] The latter reacts easily with various activated methylene compounds (e.g. CH_3NO_2, malonates); the products are 2-alkenylazepanes (Chapter 2).[228] As stated above (Section 5), lactam acetals are useful starting materials for preparation of lactam imines (**86**).

6. Substitution at C-3

It was to be expected that deprotonation of *N*-substituted caprolactams would be possible and that the resulting anions would react with electrophiles. LDA is a suitable base to achieve this objective; introduction of the diethyl carbonate or carbon dioxide led to isolation of the substituted products (**122**, R = CH_3, X = H, Y = $CO_2C_2H_5$ or CO_2H, respectively).[229,230] Examples of C-3 alkylation have been reported.[244] The *N*-acyl caprolactam **103** (R = CH_3CO) could be

(122) **(103)** **(136)** **(137)**

deprotonated using NaH in THF and was then hydroxylated with veratraldehyde giving **122** $\left(R = CH_3CO, X = H, Y = CH(OH)\text{-}C_6H_3(OCH_3)_2 \right)$.[231]

Caprolactam itself has been C-3 acylated using the combination of BuLi/TMEDA with the heterocycle **136** followed by treatment with HCl,[232] the product being **122** (R = X = H, Y = $COCH_2COCH_3$). *N*-Trimethylsilylcaprolactam [**103**, R = $(CH_3)_3Si$] was reacted with $LiNEt_2$/HMPT and benzophenone to yield **122** (R = X = H, Y = $-C(OH)Ph_2$);[233] other ketones could probably be introduced.[233] *N*-Methylcaprolactam (**103**, R = CH_3)

underwent Vilsmeier reaction to give **122** [R = CH$_3$, X + Y = CHN(CH$_3$)$_2$] and the formyl derivative **122** (R = CH$_3$, X = H, Y = CHO).[234] Nitration of caprolactam[235,236] with fuming nitric acid gave 3-nitrocaprolactam (**122**, R = X = H, Y = NO$_2$), catalytic reduction of which gave 3-aminocaprolactam (**122**, R = X = H, Y = NH$_2$) with poor enantioselectivity using either ruthenium or palladium catalysts.[236,237] CS$_2$ reacted with N-benzoylcaprolactam and ^{t}C$_5$H$_{11}$ONa followed by methyl iodide to give the C-3 substitution product **137**;[238] sodium hydride has also been used in this reaction in place of ^{t}C$_5$H$_{11}$ONa.[239] Conversion of the N-trifluoroacylcaprolactam **103** (R = CF$_3$CO) into the O-trimethylsilyl derivative **138** permitted alkylation in the presence of trimethylsilyl triflate and isolation of the C-3-substituted product (**122**, R = X = H, Y = –CH$_2$N⟩).[240] Lactam acetals (**122**, R = CH$_3$, X = Y = Et) have reacted with electrophiles such as benzoyl chloride giving, for example, 3-alkenyl structures (**139**) as a mixture of isomers from which 3-benzoylcaprolactam (**122**, R = CH$_3$, X = H, Y = COPh) could be

(138)   (139)   (140)

obtained by heating with HCl.[241] The success of such a reaction is considered to be due to the fact that the original acetal comes into equilibrium with the enol ether **140**, which then reacts at C-3 with an electrophilic reagent.[241]

7. Annulations or Cyclisations Involving the Caprolactam Nucleus

There are several modes of cyclisation elaborating from the caprolactam ring system. 1,7-Photocyclisation of N-phenacylcaprolactam (**103**, R = PhCOCH$_2$–) gave diastereomers **141** and **142** in 26% and 17% yield, respectively.[242] The action of sodium on ester **143** in the presence of ethyl acrylate caused N-alkylation (Michael reaction) followed by Dieckmann

(141)   (142)   (143)   (144)

Azepanones

cyclisation to **144**.[243] A very interesting series of 2,3-cyclisations via iminium ions has been reported;[244] both alkynes and alkenes have participated in these reactions. Exemplary is the conversion of **145** [X = CH$_3$, (CH$_3$)$_3$Si, (CH$_3$)$_3$SiCH$_2$, Ph] to mixtures of diastereomers **146** and the alkene analogue **147**, which, when treated with SnCl$_4$ at $-23\,°$C, was converted in good yield exclusively into one diastereomer (**148**, X = Ph, CH$_3$). Several other examples were presented.[244] A 1,3-cyclisation has been contrived by the rhodium acetate-induced conversion of the diazoalkene **149** into an intermediate which on silica was changed into the tricyclic product **150**.[245]

Spiro cyclisation takes place when the imine **151**, generated *in situ*, reacts with a dipolarophile such as *N*-phenylmaleimide to yield **152** and **153**.[246] It is

(153)   (154)

considered that the 1,3-dipolar species involved can be regarded as in **154**, arising from **151** by prototropic shift.

The Russian literature contains examples of annulations via the α-cyanolactam (**122**, R = CH$_3$, X = H, Y = CN) leading to pyrimido[4,5-*b*]azepine derivatives.[351,358] The lactam acetal **133** (R = OC$_2$H$_5$) has also been used to elaborate the pyrimido[4,5-*b*]azepine ring system[132] and the quinolo[2,3-*b*]azepine system.[134]

8. Ring Expansion and Contraction

As far as we are aware there is only one recorded example of a ring expansion involving the caprolactam nucleus. Lithium triphenylsilylacetylide reacted with *N*-methylcaprolactam (**103**, R = CH$_3$) to yield an α,β-unsaturated aldehyde containing the octahydroazocine ring (**155**).[247] Unfortunately, the structure of the above product was incorrectly formulated in the original report.[248]

(155)   (156)   (157)

(158)   (159)

Ring contraction by the Favorskii rearrangement has been known for some time.[249,250] Thus, a series of lactams **156** ($n = 1$–6; X = Cl, Br) gave the amino acids or esters (depending on work-up conditions) (**157**, $n = 1$–6: R = H or CH_3) by refluxing with NaOH in dioxane. Although details for conversion of the halocaprolactams **156** ($n = 1$, X = Cl, Br) are sketchy (the main interest was in larger rings), this does seem to be a method applicable to seven-membered ring halolactams. This view is confirmed by the report[251] that $KO^tBu/^tBuOH$ converted the halolactam **158** into the substituted pipecolic acid **159**. It is surprising that more use has not been made of this conversion; overall it represents a possible method for conversion of substituted cyclohexanones (via Beckmann or Schmidt reactions) into substituted pipecolic acids.

9. Hydrolysis of Caprolactams

It is clear that lactams and the corresponding amino acids are interrelated by hydrolysis and dehydrations. Caprolactam on hydrolysis yields 6-aminocaproic acid (6-aminohexanoic acid).[252] By substitution of the caprolactam nucleus followed by hydrolysis, one has a useful method for synthesising some unusual amino acids. N-Alkylcaprolactams such as N-hexadecylcaprolactam were hydrolysed by 48% HBr to give hexadecylaminocaproic acid hydrobromide in good yield.[254] 6-Aminostearic acid was made similarly.[254] Amino diacids can be obtained by hydrolysis of 7-carboxyalkylcaprolactams (**160**); in this way, amino acid derivatives (**161**, R = H or CO_2CH_2Ph) were made.[255] Since the 7-substituent on the caprolactam ring was introduced by a combination of alkylation of cyclohexanone enamine + Schmidt reaction chemistry, this approach is obviously open to further expansion.

(160) $HO_2C(CH_2)_2\overset{NHR}{\underset{|}{C}H}(CH_2)_4CO_2H$ (162)

(161)

Hydrolysis of 3-aminocaprolactam to lysine was mentioned at the beginning of this chapter. N-Methyl-L-lysine has also been obtained by N-hydroxyalkylation (**162**, R = H → **162**, R = CH_2OH) followed by reduction ($Et_3SiH/CF_3CO_2H/CHCl_3$) to **162** (R = CH_3) and hydrolysis.[256] The conversion of 3-aminocaprolactam into L-lysine by microbial enzymes was reviewed in 1974.[257]

10. Thiocaprolactams

Thiocaprolactam is less basic than caprolactam[258] and has a greater dipole moment and lower infrared stretching frequency which confirms that C-2 is more positive in the thiolactam.[259] Hydrogen peroxide oxidation of thiocaprolactam in neutral solution led to the formation of the corresponding S-oxide.[260] Trimethylsilylthiolactam [**95**, R = $(CH_3)_3Si$] reacts with alkyl halides yielding compounds **163** (R = CH_3, C_2H_5, C_3H_7, etc.), which on further treatment give N-alkylthiolactams (**95**, R = CH_3, C_2H_5, C_3H_7, etc.).[261] Alkylation and acylation of thiocaprolactams is a facile process:

(**95**) (**163**) (**164**) (**165**)

the N-benzyl analogue **95** (R = CH_2Ph) was C-3 formylated with $^tBuOCH(NMe_2)_2$ and acylated with $^iPr_2NMgBr/(MeO)_2C=O$ in good yield. The N-unsubstituted thiolactam **95** (R = H) may be alkylated using 2 equivalents of ^nBuLi yielding products **164** ($R^1 = R^2 = H$; $R^1 = H$, $R^2 = CH_3$; $R^1 = R^2 = CH_3$).[264]

(**166**) (**167**) (**168**) (**169**)

N-Methylthiocaprolactam (**95**, R = CH_3) reacted with methyl iodide yielding the iminium salt **165**, which gave intermediates **166** (R = Ph, C_5H_{11}, CH_3 or TMS) on treatment with the appropriate alkynes; $LiAlH_4$ converted these into the 2-alkynylazepanes **167**.[265] Rhodium acetate served to convert the diazothiolactam **168** into the thiol **169** (R = SH) and thence to the enone **169** (R = H) with Raney nickel.[266] S-Allylthioimidates (**163**, R = $CH_2CX=CH_2$, X = $HNCO_2Et$) undergo rearrangements to N-allylthiolactams **95** (R = $CH_2CX=CH_2$, X = $HNCO_2Et$) either on treatment with $^iPr_2NEt$[267] or by palladium(II) catalysis.[268,269]

CAPROLACTAM TABLES

The literature contains a very large number of structures involving the caprolactam ring; we have had to be selective, omitting those examples in which the seven-membered ring comprised a minor part of the total structure. The tables present details of more than 450 substances whose structures appear authentic.

In most cases we present the structures as the authors have published them. The usual conventions of stereochemical representation (wedge and dashed bonds) are retained where appropriate. If any doubt exists as to stereochemistry, bonds of average thickness are employed. Absolute (*R* and *S*) stereochemical representations are shown where claimed by authors and these may be assumed to be relative configurations unless supported by rotation measurements indicating relationship to chiral material of known absolute configuration. In only a few cases has the optical purity been established; these are shown in the tables where available.

Table 1. *N*-Substituted caprolactams

Structure (R)	M.p. (°C)/b.p. (°C/mmHg)	Derivatives and m.p. (°C)	Spectroscopic evidence/data	Ref.
H	127–133/7 65–68			7
CH_3CO	134–136/26–27 134–135/26 130–131/13 126/0.14 85/0.05		n_D^{25} 1.4885, d_4^{25} 1.094 IR n_D^{25} 1.4868, d^{25} 1.1335 IR, ^{1}H NMR	225 15 161 270 231
C_2H_5CO	85/0.1			270
n-C_3H_7CO	100/0.1		n_D^{25} 1.4818, d^{25} 1.0845	270
n-$C_9H_{19}CO$	175/0.1		n_D^{25} 1.4740	270
CH_3–C$_6$H$_4$–SO_2–	123–124.5		IR	160
$(CH_3)_3Si$	111–111.5/16		n_D^{20} 1.4700, d_4^{20} 0.9573	169

Table 1 (continued)

Structure (R)	M.p. (°C)/b.p. (°C/mmHg)	Derivatives and m.p. (°C)	Spectroscopic evidence/data	Ref.
CO$_2$Et	90/0.2	Amide 101	IR, ^{1}H NMR	113
	118/5		IR, ^{1}H NMR	199
CHO	148/16		IR, ^{1}H NMR	190
PhCO	67–69.5			163, 271
	68–69			162
CH$_3$O—C$_6$H$_4$—CO—	76			270
ON	12.8–14			170
	11.0			171
HO	80–81			272
	82–83		IR, MS	90
CH$_3$	120/19		n_D^{25} 1.4818	273
	128/32		n_D^{25} 1.4812, d_4^{25} 1.0154	225
	100–102/18			113
C$_2$H$_5$	97/5.5		n_D^{25} 1.4777, d_4^{25} 0.9850	225, 274
	110–113/8			113
	140–142/55		IR, ^{1}H NMR, MS	94
n-C$_3$H$_7$	122/14			356
i-C$_3$H$_7$	87–89/0.7			94
n-C$_4$H$_9$	137–140/17		n_D^{20} 1.4782, IR	160
	145/14			276, 356
	80–82/0.6			113
	110–112/6	Dimethyl acetal 105–115/11		275
(CH$_3$)$_2$CHCH$_2$	148–150/25			113
	141/14			356
CH$_3$CH$_2$CH$_2$CH(CH$_3$)–	82–83/0.01			94

Table 1 (continued)

Structure (R)	M.p. (°C)/b.p. (°C/mmHg)	Derivatives and m.p. (°C)	Spectroscopic evidence/data	Ref.
$(CH_3)_2CHCH_2CH_2$	148–150/18			113
$n\text{-}C_6H_{13}$	190–197/30 167–174/20		n_D^{20} 1.4754, IR	160
cyclo-C_6H_{11}	50–51			94
$n\text{-}C_8H_{17}$	152–154/2.5 164–167/7 150–160/5	 Diethyl acetal 155–170/12		113 275 253
$CH_3(CH_2)_5\overset{\underset{\mid}{CH_3}}{CH}-$	130–132/0.05			94
$n\text{-}C_{16}H_{33}$	215–223/2–3			253
$PhCH_2$	56–57 110/0.01 145–149/1 55–57		 IR, MS, ^{1}H NMR	277 221 278 94
$PhCH_2CH_2$	152–154/0.01			94, 167
$CH_2=CHCH_2$	76–78/0.4		IR, MS, ^{1}H NMR n_D^{20} 1.4960	101 279
Ph	75 75–76.5		^{1}H NMR	113 280, 281
Cl–C$_6$H$_4$–	67			113
$ClCH_2$	140–143/13 132–134/8 118–120/3.5–4		n_D^{20} 1.5085 n_D^{20} 1.5098	164 322
$Cl(CH_2)_3$	99/0.01		IR, ^{1}H NMR	222
$PhCHCH_2-$ $\underset{\mid}{}$ OH	88–89			166

Table 1 (continued)

Structure (R)	M.p. (°C)/b.p. (°C/mmHg)	Derivatives and m.p. (°C)	Spectroscopic evidence/data	Ref.
$C_2H_5SCH_2-$	138–141/5–6		n_D^{25} 1.5189, d_4^{25} 1.0689	225
PhCH=CH–	cis 47–48 trans 117–118			166
CH₃_/Ph (C=C with H)	Oil		IR, ^{1}H NMR, MS	101
$CH_2=CH(CH_2)_9-$	160–165/0.8 165–168/1		IR	160
PhNHCO–	67–69			168
$CH_2=CHCO-$	95/0.25			282
$NCCH_2$	Chromatography, oil		IR, ^{1}H NMR, MS	283
$H_2NCH_2CH_2$		HCl 130–131 N-COCH₃ 140/0.001	IR, MS IR, ^{1}H NMR, MS	283
$NCCH_2CH_2$	150–156/2 153–158/1.5–1.8 32–34		n_D^{25} 1.4903, d_4^{25} 1.074	225
$(CH_3)_2NCH_2$	Chromatography, oil		IR, ^{1}H NMR, MS	283
PhCOCH– (with OH)	110–111	O-COCH₃ 89–90	IR, ^{1}H, ^{13}C NMR	242
$CH_3COCCO-$ (with N_2)	Oil		IR, ^{1}H NMR	245

Table 2. 3-Substituted caprolactams

Structure (R)	M.p. (°C)/b.p. (°C/mmHg)	Derivatives and m.p. (°C)	Spectroscopic evidence/data	Ref.
CH_3	97–98			39
	92–93			113
	95.5–96.5		^{1}H NMR, IR	38
	96–97			70
CH_3 (S)	89–92		$[\alpha]_D$ +31.6° $(c = 0.5, CHCl_3)$ ^{1}H, ^{13}C NMR, 84% ee	98
CH_3 (R)	89–92		$[\alpha]_D$ − 26.2° $(c = 0.5, CHCl_3)$ ^{1}H NMR, 74% ee	98
	86–90		$[\alpha]_D^{25}$ − 26.77° $(c = 4.4, CHCl_3)$	77
C_2H_5	99–100			52
	98–99			113
n-C_3H_7	80–81			52
	79–80			113
t-C_4H_9	154.3–154.8		MS, ^{13}C NMR	159
cyclohexenyl	146–148		IR	284
cyclohexyl	107.5–108.5			55
Ph	183–185		IR, ^{1}H NMR	113, 66
CH_3O-C$_6$H$_4$-	198–200		^{1}H NMR	66
CH_3O_2C	79–80	N-Phenyl 104–105	^{1}H NMR, MS	285
Cl	97.2–98			58
	91–93			201
	92–93			162

Table 2 (continued)

Structure (R)	M.p. (°C)/b.p. (°C/mmHg)	Derivatives and m.p. (°C)	Spectroscopic evidence/data	Ref.
Br	111–112 113–115 109–113			162 203 202
I	127–128			202
NO_2	163–164		IR, ^{1}H NMR	236
N_3	107 105–108		IR, ^{1}H NMR	211 201
$CH_3O\diagdown$ (S)	54–55		IR, ^{1}H NMR	98
NC	159–160/0.5 105.5 92–96 98–99 94–98		 IR, UV ^{1}H, ^{13}C NMR, MS, IR	286 48 285 358
PhCO	203 186–188		IR, ^{1}H NMR	232 112
CH_3COCH_2CO	133–134		IR	232
H_2N	167/12	HCl 294–296		78, 202
	68–71	Picrate 223 N^3COCH_3 162		201
	61–65 68–70 75–79.5	HCl 291–295		73 287 91, 288
PhCONH	208–210			289
PhCH=N	125–126		^{1}H NMR, IR	246
⌬N—	Oil	Fumarate 215	^{1}H NMR, MS	290
$H_2N\diagdown$ (S)	71–72	HCl 300–305(d) $[\alpha]_D^{25}$ − 26.4	$[\alpha]_D^{25}$ − 34.0° (c = 4, 1 M HCl) ee 100% $[\alpha]_D^{25}$ − 24.5 ± 1.2° (c = 4, 1 M HCl)	91 201
		$BzOCO^3NH$ 145–148	^{1}H NMR	291

Azepanones

Table 2 (continued)

Structure (R)	M.p. (°C)/b.p. (°C/mmHg)	Derivatives and m.p. (°C)	Spectroscopic evidence/data	Ref.
Phthalimido- (S)	250–255 (d), no melt		^{1}H NMR, $[\alpha]_D^{20}+64.5°$ ($c=1$, CHCl$_3$), ^{1}H NMR	191, 256
Ph$_3$CNH– (S)	187		$[\alpha]_D^{20}+28°$ ($c=1$, CHCl$_3$), ^{1}H NMR	256
tBuOCONH– (S)	152–153	N^{1}CH$_2$COMe oil $[\alpha]_D^{25}+25.3$	$[\alpha]_D^{25}+39°$ ($c=1$, CHCl$_3$), MS, ^{1}H NMR	292
HCONH– (S)				
3,4,5-(CH$_3$O)$_3$C$_6$H$_2$CO-NH– (S)	176	N^{1}CH$_2$CO$_2$Me 60–62; N^{1}CH$_2$CONH$_2$ 208–210; N^{1}CH$_2$CH$_2$OH 89–92	$[\alpha]_D^{25}+83°$ ($c=1$, CHCl$_3$), ^{1}H NMR; $[\alpha]_D^{25}+1.2°$ ($c=1$, MeOH); $[\alpha]_D^{25}+12.4°$ ($c=1$, MeOH); $[\alpha]_D^{25}+6.5°$ ($c=1$, MeOH)	293
3-pyridyl-CO-NH– (S)	95–97		$[\alpha]_D^{25}+35.7°$ ($c=1$, MeOH), ^{1}H NMR	293
PhCO-NH– (S)	177		$[\alpha]_D^{25}+88°$ ($c=1$, MeOH), ^{1}H NMR	293
3-CF$_3$-C$_6$H$_4$-CO-NH– (S)	218–219		$[\alpha]_D^{25}+26.9°$ ($c=1$, MeOH), ^{1}H NMR	293

Table 2 (continued)

Structure (R)	M.p. (°C)/b.p. (°C/mmHg)	Derivatives and m.p. (°C)	Spectroscopic evidence/data	Ref.
4-CH$_3$O-C$_6$H$_4$-C(O)-NH- (S)	170–172		$[\alpha]_D^{25}$ +42.2° (c = 1, MeOH), ^{1}H NMR	293
CH$_3$CONH- (S)	147–149		$[\alpha]_D^{25}$ +13.7° (c = 1, MeOH), ^{1}H NMR	293, 148
N-Cbz-prolyl-NH- (S)	156–158		$[\alpha]_D^{25}$ −69.7° (c = 1, MeOH), ^{1}H NMR	293
(2-oxopyrrolidin-1-yl)-CH$_2$-C(O)-NH- (S)	245–247		$[\alpha]_D^{25}$ +7.1° (c = 1, MeOH), ^{1}H NMR	293
Ph(CH$_2$)$_3$-C(O)-NH- (S)	106–108	N^{1}CH$_2$CO$_2$Me oil; N^{1}CH$_2$CONH$_2$ 153–155; N^{1}CH$_2$CH$_2$OH oil	$[\alpha]_D^{25}$ +10.2° (c = 1, MeOH), ^{1}H NMR; $[\alpha]_D^{25}$ −9.7° (c = 1, MeOH); $[\alpha]_D^{25}$ +8.2° (c = 1, MeOH); $[\alpha]_D^{25}$ −1° (c = 2.2, MeOH)	293
NH$_2$·HCl (R)	Not quoted		$[\alpha]_D^{25}$ +26.4° (c = 4, 1 M HCl), ee > 99%	91
phthalimido (R)	248–253			191

Table 2 (continued)

Structure (R)	M.p. (°C)/b.p. (°C/mmHg)	Derivatives and m.p. (°C)	Spectroscopic evidence/data	Ref.
tBuOCONH...... (R)	150–151	N-CH$_2$CO$_2$Me oil	$[\alpha]_D^{25}$ − 39.2° (c = 1.2, CHCl$_3$), MS $[\alpha]_D^{25}$ − 22.8° (c = 0.143, CHCl$_3$) MS, ^{1}H NMR	292
3-CF$_3$-C$_6$H$_4$-C(O)-NH...... (R)	218–219		$[\alpha]_D^{25}$ − 26.3° (c = 1, MeOH), ^{1}H NMR	293
Et$_2$N(CH$_2$)$_2$NH–	149–156/1–2 57–60	Tartrate 128–129	IR	114
3,4,5-(CH$_3$O)$_3$-C$_6$H$_2$-C(S)-NH–	206–207		IR	294
(HO$_2$CCH$_2$)$_2$N–			^{13}C NMR	295
CH$_3$CH(OH)-CH$_2$CONH–	183–185		IR, ^{1}H NMR	146
(3-amino-azepan-2-one)-NH–	220/0.25		IR, ^{1}H NMR	296
(HO$_2$C)$_2$CH–	160	Monoethyl ester 76–77 Diethyl ester 94–96		213
HO$_2$CCH$_2$	176–177	Ethyl ester 76.5–77.5 Amide 202		213
HOCH$_2$CH$_2$	89–91			297

Table 2 (continued)

Structure (R)	M.p. (°C)/b.p. (°C/mmHg)	Derivatives and m.p. (°C)	Spectroscopic evidence/data	Ref.
$CH_2=CHCH_2-$	87		IR, ^{1}H NMR, MS	101
$CH_2=CHCH-$ with Ph substituent, 2 diastereomers	116–117		UV, ^{1}H NMR	101
$EtO_2C-C(=CH_2)-$	87–88			213
(Z)-Ph-CH=CH-CH$_2$CH$_3$	Oil		^{1}H NMR, IR, MS	244
$CH_3COCH_2CH_2$	91–93		IR, MS, ^{1}H, ^{13}C NMR	103
$CH_3C\equiv C(CH_2)_2$	78–79		IR, MS, ^{1}H NMR	244
2-oxopyrrolidin-1-yl-CH$_2$–	173–175		IR, ^{1}H, ^{13}C NMR	240
$Ph_2C(OH)-$	260		IR, ^{1}H NMR	233
2-indolyl (1H-indol-2-yl)	176		IR	69
PhC(OH)=N–C(CH$_3$)=CH–	204–206		MS, IR, ^{1}H NMR	232
CH_3SSC-	144–145		IR, MS	239
$HSSC-$	103–105		IR, ^{1}H NMR	239
$Ph_2P(=O)CH_2$	Crystalline no m.p. given		IR, ^{1}H NMR	46
$(CH_3)_3SiCH_2$	VPC pure		IR, ^{1}H NMR, MS	43

Table 2 (continued)

Structure (R)	M.p. (°C)/b.p. (°C/mmHg)	Derivatives and m.p. (°C)	Spectroscopic evidence/data	Ref.
[structure: 2-methyl-2-phenyl-6-methyl-4H-1,3-oxazin-4-one]	178–179		IR, ^{1}H NMR	232
[structure: 5-methyluracil / thymine]	>300		^{1}H NMR	298
$CH_2=CHCH_2CH_2$	66–67		IR, ^{1}H NMR	245
[structure: 2-methoxy-5-nitroanilino, NH–]	194–195		IR, ^{1}H, ^{13}C NMR	299
$CH_3CH_2CH(CH_3)(CH_2)_4NH-$	white powder, no m.p. given		IR, ^{1}H, ^{13}C NMR, MS	154
$(CH_3)_2CH(CH_2)_5NH-$	white powder, no m.p. given		IR, ^{1}H, ^{13}C NMR, MS	154
[structure: 5-fluoro-2,3-dimethylquinazolin-4(3H)-one]	193–194		^{1}H NMR, IR	300
[structure: 7-membered lactam with CH$_2$– substituent]	174–182		IR	63

Table 2 (continued)

Structure (R)	M.p. (°C)/b.p. (°C/mmHg)	Derivatives and m.p. (°C)	Spectroscopic evidence/data	Ref.
azepan-2-one-3-yl-CH₂–	247		IR	63
H₂NCH₂CH₂	82–84	Maleate 99–100 Oxalate 158–161		54

Table 3. 4-Substituted caprolactams

Structure (R)	M.p. (°C)/b.p. (°C/mmHg)	Derivatives and m.p. (°C)	Spectroscopic evidence/data	Ref.
CH$_3$	78		IR, ^{1}H NMR	111
	100–101			6
	97.7–98			71
	104–106			49, 218
CH$_3$ (R)	98–103		$[\alpha]_D$ − 34.0° (c = 0.7, H$_2$O) ^{1}H, ^{13}C NMR, IR	98
		N-(R)-CH(CH$_3$)Ph oil	^{1}H, ^{13}C NMR, IR, MS	49
CH$_3$(CH$_2$)$_3$	118–119/0.8		IR, ^{1}H NMR, MS	65
CH$_3$O	62		IR, ^{1}H NMR	217
(HO$_2$C)$_2$CH	162 (d)	Diethyl ester 49–50		216 215
HO$_2$CCH$_2$-	193–194	Ethyl ester 78.5–80		216
		Amide 206–208		215
(CH$_3$)$_3$Si	Chromatography		IR, ^{1}H NMR, MS	43
3-indolyl	149–150.5		UV, IR	109
C$_2$H$_5$O	54.5–56			53

Table 4. 5-Substituted caprolactams

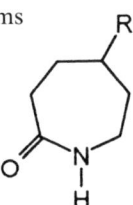

Structure (R)	M.p. (°C)/b.p. (°C/mmHg)	Derivatives and m.p. (°C)	Spectroscopic evidence/data	Ref.
CH$_3$	41–42			9, 301
	46–47		IR, ^{1}H NMR	45
CH$_3$ (R)	Oil		$[\alpha]_D$ + 18.0° (c = 1.4, MeOH) IR, ^{1}H, ^{13}C NMR, MS	9, 97
	44–46		$[\alpha]_D^{25}$ + 29.44° (c = 2.4, CHCl$_3$)	77
CH$_3$ (S)	45–46		$[\alpha]_D^{25}$ − 28.9° (c = 2.4, CHCl$_3$)	45
	44–46		$[\alpha]_D^{25}$ − 20.8° (c = 2.1, MeOH) IR, UV	302
C$_2$H$_5$	56–57			303
	124–130/0.5			304
C$_2$H$_5$ (R)	61–62		$[\alpha]_D$ + 14.3° (c = 0.46, MeOH) IR, ^{1}H, ^{13}C NMR, MS	65, 97
CH$_3$CH$_2$CH$_2$	82			305
	81–82			304
	145–149/0.75			
^{t}Bu (R)	151–152		$[\alpha]_D$ + 14.7° (c = 0.5, MeOH) IR, ^{1}H, ^{13}C NMR, MS	97
^{i}Pr	82.5–83.5			304
^{t}Bu	155–156			304

Table 4 (continued)

Structure (R)	M.p. (°C)/b.p. (°C/mmHg)	Derivatives and m.p. (°C)	Spectroscopic evidence/data	Ref.
(cyclohexyl)	156–157			304
$^nC_7H_{15}$	71.5–72			304
Ph	129–134			281
	199–200		IR	28
Ph,,,, (S)	194–196		$[\alpha]_D - 45.3°$ ($c = 0.73$, MeOH) ^1H, ^{13}C NMR	97
CF_3	101–102.5			306
$HOCH_2$	127–129	O-Tosyl 109–110	IR, ^1H NMR	45
		O-^tBu 77–78 OBzOCO– 116		307
$HOCH_2$ (R)	73		$[\alpha]_D^{25} + 16.0°$, IR, ^1H NMR	45
		O-^tBu 73	$[\alpha]_D^{25} + 15.9°$ ($c = 2.7$, MeOH)	
$HOCH_2$,,, (S)	73		$[\alpha]_D^{25} - 16.2°$	45
		O-^tBu 73	$[\alpha]_D^{25} - 16.1°$ ($c = 2.1$, MeOH)	
		O-Tosyl 109–111	$[\alpha]_D^{25} - 13.25$ ($c = 1.2$, MeOH)	
ICH_2	127–128		IR, ^1H NMR	45
ICH_2,,, (S)	121–122		$[\alpha]_D^{23} - 16.7°$ ($c = 2.3$, MeOH) IR, ^1H NMR	45
HO	138–141	O-$COCH_3$ 62		309, 310
PhCOO	133.5–135		IR, ^1H NMR	30
HO (R)	130–132		$[\alpha]_D + 18.9°$ ($c = 0.46$, MeOH) IR, ^1H, ^{13}C NMR, MS	30, 97

Table 4 (continued)

Structure (R)	M.p. (°C)/b.p. (°C/mmHg)	Derivatives and m.p. (°C)	Spectroscopic evidence/data	Ref.
HO$_2$C	129–130	HCl 157 Methyl ester 76 ^{t}But ester 92–93 (HCl 109)	IR, ^{1}H NMR	45
HO$_2$C (R)	168–169	Methyl ester 93	$[\alpha]_D^{25}$ + 31.9, IR, ^{1}H NMR	45
HO$_2$C (S)	168–169	Methyl ester 93–94	$[\alpha]_D^{25}$ − 31.9° (c = 3.0, MeOH)	45
CH$_3$O, OCH$_3$	181–183		IR	26
CO$_2$H, CH, NHCOPh 2 diastereomers	268–272		IR, ^{1}H NMR	31
(7-methyl azepanone)	288.5–289.5		IR	64

Table 5. 6-Substituted caprolactams

Structure (R)	M.p. (°C)/b.p. (°C/mmHg)	Derivatives and m.p. (°C)	Spectroscopic evidence/data	Ref.
CH$_3$	60			6
	64–65		IR, ^{1}H NMR	111
	69–70			49
	140–142/10			71
	70–72			
CH$_3$ (R)	99–100		$[\alpha]_D$ − 21.2° ($c = 0.69$, EtOH) IR, MS, ^{1}H, ^{13}C NMR	98
	68–69		$[\alpha]_D$ − 22.2°	308
Ph	150		IR, ^{1}H NMR	207
BzOCO	128–131		IR, ^{1}H NMR	65
EtO$_2$C	160–165/0.6		IR, ^{1}H NMR	65
Me$_3$Si	91–92		IR, ^{1}H NMR, MS	43

Table 6. 7-Substituted caprolactams

Structure (R)	M.p. (°C)/b.p. (°C/mmHg)	Derivatives and m.p. (°C)	Spectroscopic evidence/data	Ref.
CH$_3$	90–91		IR, ^{1}H NMR	8, 39, 60
	91–92		MS, IR, ^{1}H NMR	12, 190
	90.5–91.5			58
	89.5–90.5			9, 305
C$_2$H$_5$	91.5–92			51, 58
CH$_3$CH$_2$CH$_2$	100.5–101.5			10
	98.6–99.2			58
	97–98			60
	101–102			74
	180–185/14			
CH$_3$CH$_2$CH$_2$CH$_2$	73–73.5			65
	70			311
(CH$_3$)$_2$CHCH$_2$	80–81		^{1}H, ^{13}C NMR, IR, MS	110a
(CH$_3$)$_2$CH$_2$CH$_2$ (R)	Oil: Chromatography		$[\alpha]_D^{20}$ − 9.3° (c = 0.53, MeOH) IR, ^{1}H NMR	110, 110a
^{t}Bu	127.3		^{13}C NMR, MS	159
cyclohexyl	134–135		IR, MS	21
	139		^{1}H NMR	65
	146–147			55
cyclohexenyl	111		IR, MS	21
	116–118		IR	284
	118–119		^{1}H NMR, IR	312
cycloheptyl	128–130			55

Table 6 (continued)

Structure (R)	M.p. (°C)/b.p. (°C/mmHg)	Derivatives and m.p. (°C)	Spectroscopic evidence/data	Ref.
$nC_{12}H_{25}$	59–60		IR	254
CH_3O	69–70		^{1}H NMR, IR	219
	106–108/0.18		^{1}H NMR	187
CH_3O (R)			IR, ^{1}H NMR	98
C_2H_5O	106/0.18		^{1}H NMR	187
$C_2H_5O_2C$	97.2–98			58
	96.5–98			60
	48–49.5		IR, ^{1}H NMR, MS	90
	150–160/1–2		IR	227
H_2NOC	239–241			60
	217			79
NC	126.6–127.4			58
	126–127			60
	127–128			59
$C_2H_5O_2CCH_2$	113–114		^{1}H NMR	313
$HO_2CCH_2CH_2$	152–153	Methyl ester 79–80 Amide 182–183		255
	154			314
$HO_2CCH_2CH_2CH_2$	103	Ethyl ester 96.5		314
$C_2H_5O_2CCH=CHCH_2$	102			314
$CH_3O_2C(CH_2)_4$	55–56			56
$CH_3O_2C(CH_2)_5$	64–65			56
$CH_3O_2C(CH_2)_6$	55–57			56
$CH_3O_2C(CH_2)_7$	65–66			56
$(CH_3)_3SiCH_2$	75–76		IR, MS, ^{1}H NMR	43
CH_3CO	104–105		IR, MS, ^{1}H NMR	62

Table 6 (continued)

Structure (R)	M.p. (°C)/b.p. (°C/mmHg)	Derivatives and m.p. (°C)	Spectroscopic evidence/data	Ref.
2-methylcyclohexanone	161–163			315
1-phenylcyclohexyl	118–120			20
BrCH$_2$CH$_2$	121–122		IR, MS, ^{1}H NMR	313
HOCH$_2$CH$_2$	138		MS, ^{1}H NMR, IR	313
4-methyl-3-hydroxy-2-nitrobenzamide derivative	220–222			316
Ph$_2$C(OH)–	260		IR, ^{1}H NMR	186
1-chlorocyclohexyl	185–187 (d)		IR, MS	21
PhCH$_2$–	157–160			65
CH$_3$O–C$_6$H$_4$–CH$_2$–	153–155			65
2-OCH$_3$–C$_6$H$_4$–CH$_2$–	120–122			65

Table 6 (continued)

Structure (R)	M.p. (°C)/b.p. (°C/mmHg)	Derivatives and m.p. (°C)	Spectroscopic evidence/data	Ref.
CH₃–C₆H₄–CH₂–	174–175			65
H₂N	156–158/10 23			51
H₂N(CH₂)₂	181–186/8 98–99	Oxalate 180–181(d) Maleate 129.5–131		54
CH₃CH(NH₂)–	70–73			56
H₂N(CH₂)₄		HCl salt 198–200	IR, MS (base)	310
Ph	135–136 138–139		^{1}H NMR, IR	207, 66, 317
3-CH₃O-C₆H₄–	99–100			81
4-Cl-C₆H₄–	172–174		IR	23
2-CH₃-C₆H₄–	109–115		IR	23
4-CH₃-C₆H₄–	153–154 150–151		MS, IR UV, ^{1}H NMR	23 66
2-NO₂-C₆H₄–	182–183		^{1}H NMR	66

Table 6 (continued)

Structure (R)	M.p. (°C)/b.p. (°C/mmHg)	Derivatives and m.p. (°C)	Spectroscopic evidence/data	Ref.
O₂N–C₆H₄– (para)	141–142		¹H NMR	66
O₂N–C₆H₃(NO₂)–	~210 (d)		¹H NMR, UV	66
CH₃O–C₆H₄– (para)	152–153		¹H NMR	66
5-fluorouracil-N-yl	183		¹H NMR	318
2-pyridyl	96–96.5			40
2-quinolyl	190–192			40
2-indolyl	196		IR	69
3-indolyl	235		IR	319
7-oxo-hexahydroazepin-2-yl	235–236 Diastereoisomers 195–197		IR	63
O₂N–C₆H₄–CH₂–	153–154		IR, ¹H NMR	65

Azepanones 167

Table 7. 1,3-Disubstituted caprolactams

Structure	M.p. (°C)/b.p. (°C/mmHg)	Derivatives and m.p. (°C)	Spectroscopic evidence/data	Ref.
3-CH₃, N-CH₃	96–97/13		IR, ¹H NMR, MS	113, 95
3-C₂H₅, N-CH₃	64/0.5 110–111/28			320 81 113
3-ⁿPr, N-CH₃	116–118/23			113
3-ⁿBu, N-CH₃	124/23			113
3-H₂N, N-CH₃		N-Phthalimido (d) N-Trityl 89	$[\alpha]_D^{20} + 69.0°$ $(c = 1, CHCl_3)$ ¹H NMR $[\alpha]_D^{20} + 17.5°$ $(c = 1, CHCl_3)$ ¹H NMR	256
3-(3-Cl-1-OH-cyclohexenyl), N-CH₃	126–127			81
3-(3-ⁱPrO-1-OH-cyclohexenyl), N-CH₃	119–121			81

Table 7 (continued)

Structure	M.p. (°C)/b.p. (°C/mmHg)	Derivatives and m.p. (°C)	Spectroscopic evidence/data	Ref.
(3-oxocyclohex-1-enyl-azepanone, N-CH₃)	109–110			81
(2-ethyl-3-oxocyclohex-1-enyl-azepanone, N-CH₃)	114–116			81
(dibromo-hydroxyphenyl-azepanone, N-CH₃)	205–207			81
(bromo-hydroxyphenyl-azepanone, N-CH₃)	160–163			81
(HO₂C-azepanone, N-CH₃)	118–119		IR, ¹H NMR	229
(3-hydroxyphenyl-azepanone, N-CH₃)	192–193	OMe 74–75		81

Azepanones

Table 7 (continued)

Structure	M.p. (°C)/b.p. (°C/mmHg)	Derivatives and m.p. (°C)	Spectroscopic evidence/data	Ref.
3-HO, N-CH$_3$ azepanone	94–94.5/2–3		IR	205
3-Cl, N-COCH$_3$ azepanone	94.5–96/0.5		n_D^{25} 1.5104	203
3-Br, N-COCH$_3$ azepanone	109.5–110.5/0.8		n_D^{25} 1.5321	203
3-Cl, N-COPh azepanone	118 120–122 121 122–123		IR, ^{1}H NMR	162 163 206 203
3-NC, N-COCH$_3$ azepanone	150–151/0.7		n_D^{20} 1.5101	286
3-(but-2-ynyl), N-CO$_2^t$Bu azepanone	64–65		^{1}H NMR, IR, MS	244
3-Br, N-OH azepanone	Chromatography		MS, ^{1}H NMR	90

Table 7 (continued)

Structure	M.p. (°C)/b.p. (°C/mmHg)	Derivatives and m.p. (°C)	Spectroscopic evidence/data	Ref.
H₂N-azepanone, N-OCH₂Ph	Oil	N-BOC 102.5–103.5 N-COCH₂-CH(OH)CH₃ 130.5–131.5	$[\alpha]_D^{23} - 11.3°$ (c = 2.2, MeOH) ¹H NMR $[\alpha]_D^{23} - 16.35°$ (c = 5.5, MeOH)	157
ᵗBuO₂CNH-azepanone, N-CH₂CO₂H	Foam		$[\alpha]_D^{24} - 3.06°$ (c = 0.98, MeOH) IR, ¹H NMR	321
CH₃-azepanone, N-ⁿHex	Chromatography		IR, ¹H NMR	95
CH₃O-azepanone, N-CH₂Ph	Chromatography		IR, ¹H NMR	95
PhCH₂-azepanone, N-CH₂Ph	Chromatography		IR, ¹H NMR	95

Azepanones 171

Table 8. 1,n-Disubstituted caprolactams

Structure	M.p. (°C)/b.p. (°C/mmHg)	Derivatives and m.p. (°C)	Spectroscopic evidence/data	Ref.
4-HO, N-CH$_2$Ph	98–99		IR, ^{1}H, ^{13}C NMR, MS	193
5-OH, N-CH$_2$Ph	83–84		IR, ^{1}H, ^{13}C NMR, MS	193
(S)-^{t}Bu, N-CH$_3$CHPh (S)	90–95		$[\alpha]_D^{20}$ −143° (c = 0.32, MeOH) $[\alpha]_{578}^{20}$ −153° (c = 0.32, MeOH) ^{1}H NMR, IR, X-ray	96
4-CH$_3$, N-CH$_2$CH$_2$CN	136–139/1		n_D^{20} 1.4786	165
(S)-CO$_2$C$_2$H$_5$, N-CH$_3$CHPh (R)	Chromatography		$[\alpha]_D$ +98.0° (c = 1.0, MeOH) ^{1}H, ^{13}C NMR	97

Table 8 (continued)

Structure	M.p. (°C)/b.p. (°C/mmHg)	Derivatives and m.p. (°C)	Spectroscopic evidence/data	Ref.
(R) CO₂C₂H₅ / CH₃CHPh (R) / (R)	Chromatography		$[\alpha]_D$ +81.3° (c = 1.3, MeOH) ^{1}H, ^{13}C NMR, MS, IR	97
(S) C₂H₅ / CH₃CHPh (R)	Chromatography		$[\alpha]_D$ +126.9° (c = 1.3, MeOH) IR, ^{1}H, ^{13}C NMR	97
(R) C₂H₅ / CH₃CHPh (R)	Chromatography		$[\alpha]_D$ +111.5° (c = 1.8, MeOH) IR, ^{1}H, ^{13}C NMR	97
(R) Ph / CH₃CHPh (R)	Chromatography		$[\alpha]_D$ +133.5° (c = 0.93, MeOH) IR, ^{1}H, ^{13}C NMR	97
(S) Ph / CH₃CHPh (R)	Chromatography		$[\alpha]_D$ −63.1° (c = 1.12, MeOH) IR, ^{1}H, ^{13}C NMR	97

Azepanones

Table 8 (continued)

Structure	M.p. (°C)/b.p. (°C/mmHg)	Derivatives and m.p. (°C)	Spectroscopic evidence/data	Ref.
(S) CH₃ azepanone with N-CH₃CHPh (R)		Chromatography	$[\alpha]_D$ +140.9° (c = 0.97, MeOH) IR, ^{1}H, ^{13}C NMR	97
(R) CH₃ azepanone with N-CH₃CHPh (R)		Chromatography	$[\alpha]_D$ +125.9° (c = 1.3, MeOH) IR, ^{1}H, ^{13}C NMR	97
(S) OCH₂Ph azepanone with N-CH₃CHPh (R)		Chromatography	$[\alpha]_D$ +86.8° (c = 1.1, MeOH) ^{1}H, ^{13}C NMR	97
(R) OCH₂Ph azepanone with N-CH₃CHPh (R)		Chromatography	$[\alpha]_D$ +73.4° (c = 1.0, MeOH) IR, ^{1}H, ^{13}C NMR	97
Azepanone with OCH₃ and N-CH₂OCH₃				188

Table 8 (continued)

Structure	M.p. (°C)/b.p. (°C/mmHg)	Derivatives and m.p. (°C)	Spectroscopic evidence/data	Ref.
azepinone with CH₃ and (CH₂)₃NH₂ on N	125–126/3		¹H NMR	39
azepinone with CH₃ and (CH₂)₂CN on N	150–151/3		¹H NMR	39
azepinone with CH₃ and CHO on N	135/13		IR, ¹H NMR	190
azepinone with OCH₃ and N-CH₂-(5-fluorouracil)	156–157		IR, ¹H NMR, UV	318
azepinone with CO₂C₂H₅ and N-OH	187		IR, MS	90
azepinone with CH₃ and N-CH₃	Chromatography		¹H NMR	95

Table 8 (continued)

Structure	M.p. (°C)/b.p. (°C/mmHg)	Derivatives and m.p. (°C)	Spectroscopic evidence/data	Ref.
2-methyl-1-benzyl azepanone (CH₃, CH₂Ph substituents)	Chromatography		^{1}H NMR	95
2-methoxy-1-benzyl azepanone (OCH₃, CH₂Ph substituents)	Chromatography		^{1}H NMR	95

Table 9. Ring-disubstituted caprolactams

Structure	M.p. (°C)/b.p. (°C/mmHg)	Derivatives and m.p. (°C)	Spectroscopic evidence/data	Ref.
3,6-diMe caprolactam (cis)	122–123			9
3-N₃, 7-Ph (S,R)	149	N-CH₂CO₂ᵗBu 115–116	¹H NMR, IR	207
3-N₃, 7-Ph (S,S)	124.5		¹H NMR, IR	207
3-Br, 6-Ph	229		¹H NMR, IR	207
3-NH₂, 6-OH	226–228	HCl + H₂O 222 Picrate 223	IR	323 324
3-iPr, 6-CH₃ (S,R)	100–100.5		$[\alpha]_D^{17} +6.37°$ ($c = 0.92$, EtOH) IR, MS	29
3-N₃, 7-Ph (S,S)	117–118	N-CH₂CO₂ᵗBu 131–133	¹H NMR, IR	207

Table 9 (continued)

Structure	M.p. (°C)/b.p. (°C/mmHg)	Derivatives and m.p. (°C)	Spectroscopic evidence/data	Ref.
N₃-, Ph azepanone	Syrup		¹H NMR, IR	207
Br-, Ph azepanone	80–100		¹H NMR, IR	207
CH₃-, CH₃ azepanone	No data			335
CF₃-, CF₃ azepanone	128–130			306
CH₃-, R, S, CH₃ azepanone	134–136		$[\alpha]_D - 15.1°$ (c = 0.57, MeOH) IR, ¹H, ¹³C NMR, MS	98, 65
CH₃-, C₂H₅ azepanone	100–101			11
C₂H₅-, CH₃ azepanone	105.5–106			11

Table 9 (continued)

Structure	M.p. (°C)/b.p. (°C/mmHg)	Derivatives and m.p. (°C)	Spectroscopic evidence/data	Ref.
CH₃, CH₃ azepanone (trans)	73–73.5		IR, MS	111
CH₃, CH₃ azepanone (cis)	130 124–126 122–123			281 111 9
CH₃, CH₃ azepanone	125–126			9
CH₃, iPr azepanone (−)	118.7–119.9 120–122 119–121 119–121.5 118–120		$[\alpha]_D^{27} - 56.22°$ ($c = 4.61$, EtOH), IR, MS, ¹H NMR $[\alpha]_D - 51.0°$ $[\alpha]_D^{23} - 56.5°$ ($c = 1$, EtOH), ¹H NMR, MS, IR $[\alpha]_D^{20} - 51.9°$ ($c = 2.25$, MeOH) $[\alpha]_D^{20} - 56.7°$ ($c = 5.5$, MeOH) IR	29 49 19 14 158 61, 65, 281 4, 5
iPr, CH₃ azepanone (−)	101		$[\alpha]_D^{20} - 58.8°$ ($c = 5.5$, MeOH), IR	158

Azepanones

Table 9 (continued)

Structure	M.p. (°C)/b.p. (°C/mmHg)	Derivatives and m.p. (°C)	Spectroscopic evidence/data	Ref.
(+) isopropenyl, CH₃ azepanone	88		$[\alpha]_D^{20}$ +85.0° (c = 5.9, MeOH), IR	158
dibromo azepanone	147.5–148.5			325
hydroxy, methyl azepanone	115–118/0.05 75–78 Diastereo- isomers 98–101		IR, ^{1}H NMR	326
dimethyl azepanone	133.5–134.5			9
methyl, tBu azepanone	105–106			9
methyl, nPr azepanone	184–189/8 6–8			74

Table 9 (continued)

Structure	M.p. (°C)/b.p. (°C/mmHg)	Derivatives and m.p. (°C)	Spectroscopic evidence/data	Ref.
7-membered lactam with C$_2$H$_5$ and ^{n}Pr substituents	175–180/8			74
7-membered lactam with Ph and CO$_2$H substituents	237–240	Ethyl ester 94–96	^{1}H NMR, IR	195
7-membered lactam with Si(CH$_3$)$_3$ and CH$_3$ substituents (trans)	71–72		IR, ^{1}H NMR, MS	43

Table 10. *gem*-Disubstituted caprolactams

Structure	M.p. (°C)/b.p. (°C/mmHg)	Derivatives and m.p. (°C)	Spectroscopic evidence/data	Ref.
(NHPh, NC, O, N-H azepanone)	185–186		IR	327
(4-Cl-C₆H₄-NH, NC, O, N-H azepanone)	130–132		IR	327
(4-CH₃O-C₆H₄-NH, NC, O, N-H azepanone)	142–144		IR	327
(CH₃, CH₃-C≡C-CH₂CH₂-, azepanone)	88.5–90.5		IR, ¹H, ¹³C NMR, MS	244
(CH₃, n-butyl, azepanone)			IR, ¹H, ¹³C NMR, MS	444
(OH, CH₃-C(=O)-CH₂-, bicyclic azepanone)	100/0.001		MS, ¹H NMR, IR	103

Table 10 (continued)

Structure	M.p. (°C)/b.p. (°C/mmHg)	Derivatives and m.p. (°C)	Spectroscopic evidence/data	Ref.
3-Ph, 3-OH azepan-2-one	139–140 141–142	N-CO_2Et 72–74	^{13}C NMR, IR	114 204
3,3-Br$_2$ azepan-2-one	162–163.5 160–162		^{13}C NMR, IR	203 204
3,3-Cl$_2$ azepan-2-one	124.5–126.5 124–126 123.5–125.5 126		IR IR, ^1H NMR, MS	201 202 212 214
3,3-Ph$_2$ azepan-2-one	214			75
3-(CH_3O_2CNH), 3-Ph azepan-2-one	149–150		^1H NMR, IR, MS	285
3-OH, 3-CN azepan-2-one	132–134		IR	327
3-Ph, 3-CN azepan-2-one	136–137		^1H, ^{13}C NMR, MS	285

Table 10 (continued)

Structure	M.p. (°C)/b.p. (°C/mmHg)	Derivatives and m.p. (°C)	Spectroscopic evidence/data	Ref.
3-Ph, 3-CONH₂ azepan-2-one	192–193		¹H, ¹³C NMR, IR, MS	285
3-(3,4-dimethoxybenzyl) azepan-2-one	166–168		IR, ¹H NMR	231
3-(bis(methylthio)methylene) azepan-2-one	145–147 152–153		¹H NMR IR, ¹H NMR	238 239
4,4-dimethyl azepan-2-one	140–141 (subl.) 103–104		¹H NMR ¹H NMR, IR	27 65
5,5-dimethyl azepan-2-one	90.5–92.5 88		¹H, ¹³C NMR	328 65
5-CH₃, 5-CHCl₂ azepan-2-one	152–154		IR, ¹H NMR	16

Table 10 (continued)

Structure	M.p. (°C)/b.p. (°C/mmHg)	Derivatives and m.p. (°C)	Spectroscopic evidence/data	Ref.
(CH₃, CCl₃-substituted azepinone)	160–161		IR, ¹H NMR	16
(di-CO₂C₂H₅-substituted azepinone)	84–85		IR, ¹H NMR	36
(Ph, CO₂C₂H₅-substituted azepinone)	107–108	HCl salt 128–131		329
(CN, aryl-OCH₃ substituted azepinone)	146–147	N-CH₃ 106–108	IR	47
(CH₃, Ph-substituted azepinone)	134–135		$[\alpha]_D$ + 24.6° (c = 0.79, MeOH) IR, ¹H, ¹³C NMR, MS	97
(di-NF₂-substituted azepinone)	100–102			44

Table 10 (continued)

Structure	M.p. (°C)/b.p. (°C/mmHg)	Derivatives and m.p. (°C)	Spectroscopic evidence/data	Ref.
3,3-dimethyl azepanone	81–82 100–101		^{1}H NMR ^{1}H NMR, IR	27 65
3-Ph-3-C$_2$H$_5$ azepanone	172/0.3 96–97			76
3-C$_2$H$_5$-3-(m-OCH$_3$-C$_6$H$_4$) azepanone	87–88			81
3-C$_2$H$_5$-3-^{n}Bu azepanone	115–117/0.1		IR	25
3,3-di-C$_2$H$_5$ azepanone	101–104/0.05			255
3-CH$_3$-3-CO$_2$CH$_3$ azepanone	128		^{1}H NMR, MS	80
3-CH$_2$Ph-3-CO$_2$C$_2$H$_5$ azepanone			$[\alpha]_D$ + 4.6° (c = 1, CHCl$_3$) (ee > 95%)	67

Table 10 (continued)

Structure	M.p. (°C)/b.p. (°C/mmHg)	Derivatives and m.p. (°C)	Spectroscopic evidence/data	Ref.
7,7-dimethyl azepanone	90.5–91 135/13 102–103			17 72
spirocyclopentane azepanone	108–110			107 331
7-Ph-7-C$_2$H$_5$ azepanone	150–160/0.01		IR	25
7-C$_2$H$_5$-7-CH$_2$CH(CH$_3$)$_2$ azepanone	120–132/0.01		IR	25
7-CO$_2$C$_2$H$_5$-7-isopropyl azepanone	109–110		IR, MS	332
7,7-bis(NF$_2$) azepanone	70–71		^{19}F NMR	333
7-(=CHPh) azepanone	124		IR, ^{1}H NMR	24
enol OCH$_3$ azepinone	Oil		IR	330

Table 11. Trisubstituted caprolactams

Structure	M.p. (°C)/b.p. (°C/mmHg)	Derivatives and m.p. (°C)	Spectroscopic evidence/data	Ref.
3,3-dichloro-1-acetyl caprolactam	100.5–102/0.5		n_D^{25} 1.5177	203
3,3-dibromo-1-acetyl caprolactam	126–127/0.5		n_D^{25} 1.5640	203
3-amino-6-OM-1-methyl caprolactam (M = myristoyl)			IR, ^{1}H, ^{13}C NMR, MS	152
3-(bis(methylthio)methylene)-1-(methylthio)thiocarbonyl caprolactam	54–56		^{1}H NMR	238
3-ethyl-3-(3-hydroxyphenyl)-1-methyl caprolactam	172–180	OMe 68–69 140/0.1	IR, ^{1}H NMR	81
3-deutero-3-(3-methoxyphenyl)-1-methyl caprolactam	82–83		MS, ^{1}H NMR, IR	81

Table 11 (continued)

Structure	M.p. (°C)/b.p. (°C/mmHg)	Derivatives and m.p. (°C)	Spectroscopic evidence/data	Ref.
(structure with C₂H₅, OH, Cl, N-CH₃, C=O)	138–139			81
(structure with NC, PhNH, N-CH₃, C=O)	160–162		IR, UV	205
(structure with Cl, Cl, N-CH₃, C=O)	128–129/1–2		n_D^{20} 1.5172, IR	205
(structure with NC, HO, N-CH₃, C=O)	95–95.5		IR	205
(structure with N(CH₃)₂, OCH₃ aryl, N-CH₃, C=O)	Oil	Picrate 248–249	IR, ^{1}H NMR	47
(structure with CO₂H, OCH₃ aryl, N-CH₃, C=O)	113–114		IR	47

Table 11 (continued)

Structure	M.p. (°C)/b.p. (°C/mmHg)	Derivatives and m.p. (°C)	Spectroscopic evidence/data	Ref.
H₂N–(S,R)–Ph azepanone, N–CH₂CO₂ᵗBu	112		^{1}H NMR, IR	207
H₂N–(S,S)–Ph azepanone, N–CH₂CO₂ᵗBu	Syrup	Maleate 138–140	^{1}H NMR, IR	207
H₂N–(S)–, (S)–Ph azepanone, N–CH₂CO₂ᵗBu	Syrup	Maleate 160–161	^{1}H NMR, IR	207
H₂N–(S)–, (R)–Ph azepanone, N–CH₂CO₂ᵗBu	Syrup	Maleate 149–150.5	^{1}H NMR, IR	207
CH₃, CH₃, CH₃ azepanone (NH)	73.5–75 73–75		IR	9 334
CH₃, CH₃, CH₃ azepanone (NH)	137–138			9

Table 11 (continued)

Structure	M.p. (°C)/b.p. (°C/mmHg)	Derivatives and m.p. (°C)	Spectroscopic evidence/data	Ref.
	109–110		^{1}H NMR, MS, IR	115
	72–78		IR, MS	111
	133–134		$[\alpha]_D^{18} - 29.6°$ ($c = 1.2$, CHCl$_3$) (ee > 80%), IR, ^{1}H, ^{13}C NMR, MS	22
	113–114		$[\alpha]_D^{18} + 1.1°$ ($c = 1.2$, CHCl$_3$) IR, ^{1}H, ^{13}C NMR, MS	22
	172–174		IR, ^{1}H NMR	16
	123.5–124.5	O-acetate 153–4		338

Table 11 (continued)

Structure	M.p. (°C)/b.p. (°C/mmHg)	Derivatives and m.p. (°C)	Spectroscopic evidence/data	Ref.
4-(4-methoxyphenyl)-6,6-dimethyl azepanone	134–136		IR, MS	111
4-phenyl-6,6-dimethyl azepanone	189–193		IR, MS	111
4-hydroxy-6,6-dimethyl azepanone	184–185		IR, ^{1}H NMR, MS	115
4,4-dimethyl-6-ethyl azepanone	100–106 (subl.)		IR, MS	111
4,4-dimethyl-6-phenyl azepanone	169–173		IR, MS	111

Table 11 (continued)

Structure	M.p. (°C)/b.p. (°C/mmHg)	Derivatives and m.p. (°C)	Spectroscopic evidence/data	Ref.
3-azido-6,6-dimethyl-hexahydro-2H-azepin-2-one	100–101		IR	331
3-amino-6,6-dimethyl-hexahydro-2H-azepin-2-one	116–118		IR	331
3-chloro-6,6-dimethyl-hexahydro-2H-azepin-2-one	101.5–103		^{1}H NMR	331
3-chloro-5-methyl-5-phenyl-hexahydro-2H-azepin-2-one	179–180		MS, IR, ^{1}H NMR	251
3-(benzylamino)-5-hydroxy-5-methyl-hexahydro-2H-azepin-2-one	141–153 135/0.001		IR, ^{1}H NMR	337
3-bromo-5-hydroxy-5-methyl-hexahydro-2H-azepin-2-one	181.5–184 (d) (subl.) 155		IR, ^{1}H NMR, MS	337

Table 11 (continued)

Structure	M.p. (°C)/b.p. (°C/mmHg)	Derivatives and m.p. (°C)	Spectroscopic evidence/data	Ref.
4-Br, 3-OH, 3-OCH₃ azepanone	124–126		IR, UV	336
4-Br, 3,3-di-OH azepanone	120–121		IR, UV	336
4,5,6-tri-OH azepanone	202–203		$[\alpha]_D^{20} - 41.0°$ ($c = 0.8$, H₂O) ^{13}C NMR	85
4-(2-CH₃-C₆H₄-NH), 6,6-di-CH₃ azepanone	174–175		IR	65
4-PhNH, 6,6-di-CH₃ azepanone	214–215		IR	65
4-methylene, N-CH₂Ph azepanone			IR, ¹H NMR, MS	92

Table 11 (continued)

Structure	M.p. (°C)/b.p. (°C/mmHg)	Derivatives and m.p. (°C)	Spectroscopic evidence/data	Ref.
3-methyl-7-methyl-N-methyl azepan-2-one	Chromatography		IR, ^{1}H NMR	95
3-isopropyl-7-methyl-N-benzyl azepan-2-one	Chromatography		IR, ^{1}H NMR	95
3-methyl-7-methyl-N-benzyl azepan-2-one	Chromatography		IR, ^{1}H NMR	95
3-methyl-7-methyl-N-^{n}hexyl azepan-2-one	Chromatography		IR, ^{1}H NMR	95

Azepanones

Table 12. Tetrasubstituted caprolactams

Structure	M.p. (°C)/b.p. (°C/mmHg)	Derivatives and m.p. (°C)	Spectroscopic evidence/data	Ref.
3,3,7,7-tetramethyl caprolactam	108 113.5–114.5		^{1}H NMR, IR, MS	106 37
4,4,6,6-tetramethyl caprolactam	145–151		^{1}H NMR, IR, MS	111
4,4-dimethyl-5,6-dimethoxy caprolactam	79–80		^{1}H NMR, IR, MS	115
4,4-dimethyl-5,6-diethoxy caprolactam	36–38		^{1}H NMR, IR, MS	115
4,4-dimethyl-5,6-diisopropoxy caprolactam	Oil		^{1}H NMR	115
4,4-dimethyl-6,6-dimethoxy caprolactam	70		IR, ^{1}H NMR	65

Table 12 (continued)

Structure	M.p. (°C)/b.p. (°C/mmHg)	Derivatives and m.p. (°C)	Spectroscopic evidence/data	Ref.
(3-OH, 4-OH, 5-CO₂CH₃, 7-Ph azepan-2-one)	212–214		¹H NMR, IR, MS	339
(3,3-Cl₂, 6,6-(CH₃)₂ azepan-2-one)	129–129.5		IR	27
(3-(3,4-dimethoxyphenyl)-3-CH₃-5-OCHO-1-CH₃ azepan-2-one)	109–110		IR, ¹H NMR	93
(3,3-Cl₂, 5-Ph, 5-CH₃ azepan-2-one)	144–145		¹H NMR, IR, MS	251
(5,5-(OCH₃)₂, 6-CH₃, 6-CH₂OH azepan-2-one)	160–161	O-acetate 124–126	IR, ¹H NMR	338
(4-CH₃, 6,6-(CH₃)₂, 1-CO₂C₂H₅ azepan-2-one)	125/0.05		¹H NMR, MS	111

Table 12 (continued)

Structure	M.p. (°C)/b.p. (°C/mmHg)	Derivatives and m.p. (°C)	Spectroscopic evidence/data	Ref.
[structure]	178–180	Tetraacetate 195.5–196.5	$[\alpha]_D^{22} + 24.0°$ ($c = 1.06$, H_2O) $[\alpha]_D^{20} + 18.1°$ ($c = 1.8$, $CHCl_3$)	83b
[structure]	235	Tetraacetate 207	$[\alpha]_D^{20}$ 0°, IR, $[\alpha]_{365}^{20} - 17.4°$ ($c = 1.1$, H_2O) $[\alpha]_D^{22} - 9.4°$ ($c = 1.1$, $CHCl_3$)	83b
[structure]	Syrup	Tetraacetate 204	$[\alpha]_D^{20} +36.81 \pm 1°$ ($c = 1.1$, $CHCl_3$), IR	84
[structure]	179–180	Tetraacetate 181 and 190.5 (d)	$[\alpha]_D^{20} + 2.3°$ $[\alpha]_{436}^{20} - 4.0°$ ($c = 3.55$, H_2O) $[\alpha]_D^{25} - 21.8°$ ($c = 1.1$, $CHCl_3$)	83b
[structure]	175–177		$[\alpha]_D - 22°$ ($c = 0.13$, H_2O), IR	86
[structure]	212–214 (d) 214		$[\alpha]_D - 71°$ ($c = 0.4$, H_2O), IR	86 82 83a

Table 12 (continued)

Structure	M.p. (°C)/b.p. (°C/mmHg)	Derivatives and m.p. (°C)	Spectroscopic evidence/data	Ref.
(CH₃O, OCH₃, CH₃O, OCH₃ azepinone)	69–71		$[\alpha]_{365}^{20}$ −22.2° (CHCl$_3$) $[\alpha]_D^{20}$ 0°, IR, ^{1}H NMR	83a
(tetrahydroxy azepinone)	167–170		$[\alpha]_D$ −91.4° (H$_2$O), IR	82
(bis-CO₂CH₃ azepinone)	150–152		^{1}H NMR, IR, MS	104
(tetrahydroxy azepinone)	195 (d)	Tetraacetate 214–5	$[\alpha]_D^{20}$ +49° ($c = 1$, H$_2$O) $[\alpha]_D^{20}$ +75° ($c = 1$, H$_2$O)	340
(dimethyl, CH₃, CH₂Ph azepinone)	Chromatography		IR, ^{1}H NMR	95

Table 13. Thiocaprolactams

Structure	M.p. (°C)/b.p. (°C/mmHg)	Spectroscopic evidence/data	Ref.
thiocaprolactam, N-H	105–106 106 101–104	IR, MS, ^{1}H, ^{13}C NMR	139 140, 141 142, 181
N-CH$_3$	49–51 148–151/12 159–163/13 48–50	IR, MS, ^{1}H, ^{13}C NMR	139 136 138 144 145
S-oxide, N-H	106–108		260
N-Si(CH$_3$)$_3$	97/1	n_D^{20} 1.5414, ^{1}H, ^{13}C NMR	261
4,6-dimethyl, N-H	130		281
7-methyl, N-H	83 87–88		120 65
5-methyl, N-H	76–78 76–77		120, 281

Table 13 (continued)

Structure	M.p. (°C)/b.p. (°C/mmHg)	Spectroscopic evidence/data	Ref.
(thiolactam, N-propenyl-CH₃)	103.5–105/1.0 41–42		341
(thiolactam, N-allyl)	83/0.3	IR, ^{1}H NMR	268, 269, 343b
(thiolactam, N-(CH₂)₂S–^{n}Pr)	134–135/1	n_D^{20} 1.5630	342
(thiolactam, N-(CH₂)₂S–C₂H₅)	36.5–37	n_D^{20} 1.5730	342
(3-allyl thiolactam, N-H)	58–60 60–60.5 127–129/0.1 67.5–68.5 150–155/0.6	IR, ^{1}H NMR IR, ^{1}H NMR	264 343a 101
(thiolactam, N-vinyl)			342, 347

Azepanones

Table 13 (continued)

Structure	M.p. (°C)/b.p. (°C/mmHg)	Spectroscopic evidence/data	Ref.
(thioxoazepane N-CH₂-C(=CH₂)-CO₂C₂H₅)	150–153/0.23	^{1}H NMR	267, 269
(3-substituted thioxoazepane with CH₂-C(=CH₂)-CO₂C₂H₅, NH)	107–109	^{1}H NMR	269
(3-(1-phenylallyl)-N-methyl thioxoazepane)			344
(3-CO₂CH₃, N-CH₃ thioxoazepane)	128/0.05	IR, ^{1}H NMR	230
(N-CH₂Ph, 7-^{n}Pent thioxoazepane)	59–61	$[\alpha]_D^{25}$ +123.0° ($c = 5$, MeOH)	345
(3-CO₂CH₃, N-CH₂Ph, 7-^{n}Pent thioxoazepane) 2 diastereomers	Oil Chromatography	IR, ^{1}H NMR	263

Table 13 (continued)

Structure	M.p. (°C)/b.p. (°C/mmHg)	Spectroscopic evidence/data	Ref.
CHO, S=, N-CH₂Ph, C₅H₁₁ (n) — 2 diastereomers	Oil Chromatography	IR, ^{1}H NMR	263
3-(3-methyl-2-butenyl) azepane-2-thione (CH₃, CH₃ group)	91–94	IR, ^{1}H NMR	264
3-(2-butenyl) azepane-2-thione (H, CH₃)	97–104	IR, ^{1}H NMR	264
(CH₃)₂NCH= azepane-2-thione, N-CH₃	Unstable oil	^{1}H NMR, UV	346
N-(CH₂)₃CH₃ azepane-2-thione	110/0.01		65
N-CH₂CH(CH₃)CH₃ azepane-2-thione	118–120/0.01		65

Table 13 (continued)

Structure	M.p. (°C)/b.p. (°C/mmHg)	Spectroscopic evidence/data	Ref.
azepane-2-thione, N-(CH$_2$)$_2$CH(CH$_3$)$_2$	120/0.01		65
azepane-2-thione, N-^{n}Oct	110–112/0.01		65
4-CH$_3$, 7-CH(CH$_3$)$_2$ azepane-2-thione, N-H	106–107		65
7-^{n}Pr azepane-2-thione, N-H	83–83.5		65
3-CH$_3$, 3-(but-3-enyl) azepane-2-thione, N-H		^{1}H, ^{13}C NMR	444

Table 14. Caprolactam imines

Structure	M.p. (°C)/b.p. (°C/mmHg)	Derivatives and m.p. (°C)	Spectroscopic evidence/data	Ref.
(7-membered ring, =NH, NH)		HCl 159–160.5 HCl 161–163 HCl 165–167 Benzene-sulphonate 133		225 123 348, 349 13
(7-membered ring, =NOH, NH)	165–165.5 169–170	O-Tosyl 128 O-Mesyl 89–90 O-CO₂Me. HCl 158–160	IR	117 350 118
(ring with CH₃, =NOH, NH)	154			120
(ring with CH₃, =NOH, NH)	128			120
(ring with CH₃, CH(CH₃)₂, =NOH, NH)	85			120
(ring with CH₃, =NH)	143–145		¹H NMR, IR	123
(ring with CH₃, =NTosyl)	109–111		¹H NMR	119

Table 14 (continued)

Structure	M.p. (°C)/b.p. (°C/mmHg)	Derivatives and m.p. (°C)	Spectroscopic evidence/data	Ref.
				351
	65–66			352
	215–216		IR, UV, ^{1}H NMR	121
	91			128
	210			128
	197–198/3	Acid 144		134 357

Table 14 (continued)

Structure	M.p. (°C)/b.p. (°C/mmHg)	Derivatives and m.p. (°C)	Spectroscopic evidence/data	Ref.
(2,4-dinitrophenylhydrazone of N-methyl azepanone)	174–175			357
(thiosemicarbazone, N-methyl azepane)	130		IR, ^{1}H NMR	128
(N,N-dimethylamino, cyanoimino azepane)	156/9		IR, UV	351
(3-(hydroxy(phenyl)methyl)-N-methyl-N'-tosyl azepan-2-imine)	170–171		^{1}H NMR	119
(diastereomer)	156–157		^{1}H NMR	119
(N-methyl-N'-(phenylcarbamoyl) azepan-2-imine)	132–134 136–137		MS, IR IR, ^{1}H NMR	122 123

Table 14 (continued)

Structure	M.p. (°C)/b.p. (°C/mmHg)	Derivatives and m.p. (°C)	Spectroscopic evidence/data	Ref.
3-methylbenzoyl azepanone imine (NH)	64–66		MS, IR, ^{1}H NMR	122
benzoyl azepanone imine (N-CH$_3$)	66–68		MS, IR, ^{1}H NMR	122
3-methylbenzoyl azepanone imine (N-CH$_3$)	52–53		MS, IR, ^{1}H NMR	122
4-methoxybenzoyl azepanone imine (N-CH$_3$)	100–102		MS, IR, ^{1}H NMR	122
4-methoxybenzoyl azepanone imine (NH)	135–137		MS, IR, ^{1}H NMR	122
4-cyanobenzyl thio-N-ethyl azepanone isothiourea	109–111		MS, IR, ^{1}H NMR	353

Table 14 (continued)

Structure	M.p. (°C)/b.p. (°C/mmHg)	Derivatives and m.p. (°C)	Spectroscopic evidence/data	Ref.
O_2N-C$_6$H$_4$-N=azepine-N-CH$_3$ (para)		HCl 244–245		127
CH_3O-C$_6$H$_4$-N=azepine-N-CH$_3$ (para)		HCl 162–163		127
Cl-C$_6$H$_4$-N=azepine-N-CH$_3$ (para)		HCl 228–229		127
NO_2-C$_6$H$_4$-N=azepine-N-CH$_3$ (meta)	194–196/1	HCl 203–205		127
F-C$_6$H$_4$-N=azepine-N-CH$_3$ (meta)		HCl 177.0–177.5	pK_a 7.89, ^{1}H NMR, IR	127, 129
NC\C(CO$_2$C$_2$H$_5$)=CH-N=azepine-N-CH$_3$	82–83		^{1}H NMR	131
H_2N-C(=O)-N=azepine-N-CH$_3$	83–86			132

Table 14 (continued)

Structure	M.p. (°C)/b.p. (°C/mmHg)	Derivatives and m.p. (°C)	Spectroscopic evidence/data	Ref.
	151–152			132
	190–192/3 101–102		IR, ^{1}H NMR, pK_a 7.21	129
	198–200/3		pK_a 7.44, IR, ^{1}H NMR	129
	239–240			129
	145–147		IR, ^{1}H NMR, MS	354
	153–154		^{1}H NMR	130

Table 14 (continued)

Structure	M.p. (°C)/b.p. (°C/mmHg)	Derivatives and m.p. (°C)	Spectroscopic evidence/data	Ref.
(4-methoxyphenyl)(dicyanomethylene)-N-methylazepane imine	131–132		^{1}H, ^{13}C NMR	130
PhNHC(O)NH-azepanimine	132–134		IR, ^{1}H NMR	123
CH$_3$CH(Ph)–N=azepanimine		HCl 245–247		124
(2-phenylcyclopropyl)-N=azepanimine		HCl 232–233		355
[S=C(–N=N-methylazepanimine)]$_2$	135–138		IR, ^{1}H NMR	133
3,3-bis(F$_2$N)-2-(FN=)azepane	75–77		^{19}F NMR	333

II. AZEPAN-3-ONES

This group is sparsely represented in the chemical literature.

PREPARATION

The first reference to an azepan-3-one[361] described the Dieckmann (NaOEt/xylene) cyclisation of diester **170** to give a β-keto ester which hydrolysed to **171** (R = H). This approach was later used to obtain the amino ketones **171** (R = CH_3, C_2H_5).[362,363] Diazoacetate ring expansion applied to N-methoxycarbonyl-3-piperidone (**172**) yielded a mixture of azepane-3-keto and -4-keto esters from which the latter was removed as a copper chelate (**173**). The 3-keto ester **174** (R = $CO_2C_2H_5$) was then hydrolysed to the amino ketone and N-reprotected to give **174** (R = H).[364]

It has been reported that iminium ion vs. alkyne cyclisations are possible if mediated by external nucleophile intervention (e.g. I⁻). Thus aminoalkynes **175** (R = H, CH_3, C_2H_5, Ph) reacted with formaldehyde and Bu_4NI to cyclise in moderately good yield to 3-alkylideneazepanes (**176**, R = H, CH_3, C_2H_5,

Ph).³⁶⁵ Ozonolysis of **176** (R = CH₃), after iodine/hydrogen exchange, gave the amino ketone **177** (R = CH₂Ph). Intramolecular Friedel–Crafts cyclisation of *N*-tosylchloroalkenols **178** (R¹ = R² = CH₃ and R¹ = Ph, R² = H) was effected in 90% H₂SO₄ at −15 to 0 °C providing ketones **179** (R¹ = R² = CH₃ and R¹ = Ph, R² = H) (30% and 71% yields, respectively).³⁶⁶ Ring expansion of the phenyl selenide **180** was achieved using ⁿBu₃SnH; in this way the keto ester **181** was obtained in 71% yield.³⁶⁷

Microorganisms have, in a few cases, been shown to oxygenate *N*-substituted azepines; mixtures of products have included azepan-3-ones. The earliest examples demonstrated that *Sporotrichum sulfurescens* converted *N*-benzoylazepane into several products including **177** (R = PhCO), the 4-keto isomer and the corresponding alcohols.³⁶⁸ The overall conversion to ketones was improved by treating the crude bioconversion mixture with Jones reagent.³⁶⁹ *N*-Nitrosoazepan-3-one (**177**, R = NO) has been detected in the urine of rats given *N*-nitrosoazepane.³⁷⁰ 3-Ketomolinate **177** (R = COSC₂H₅) was detected when molinate (a herbicide) came in contact with microorganisms of a *Fusarium* species³⁷¹ and a more highly oxygenated product (**182**) appeared in Kazakhstan rice paddies when ordram (≡ molinate) was metabolised by organisms present.[372]

Some further activity in the azepan-3-one area was seen in 1994. First, electrochemical oxidation of *N*-carbomethoxyazepane followed by *m*-chloroperbenzoic acid oxidation yielded 50% of the ketone **174** (R = H)[441] and second, two separate groups of workers disclosed that generation and rearrangement of certain ammonium ylides could give β-amino ketones.[442,443] Thus α-diazo ketones **182A** (R = PhCH$_2$ or CH$_2$CH=CH$_2$) with Cu(acac)$_2$

(182A)   (182B)

gave the azepan-3-ones **171** (R = PhCH$_2$ or CH$_2$CH=CH$_2$) in acceptable yields by a process thought to involve intramolecular formation of ammonium ylides (**182B**; R = PhCH$_2$ or CH$_2$CH=CH$_2$) followed by [2,3]-sigmatropic rearrangement.

REACTIONS

Because of their rarity, not much work has been done on azepan-3-ones. Reduction of the carbonyl group with NaBH$_4$/C$_2$H$_5$OH appears to proceed normally to give a 3-hydroxyazepane; for example, **177** (R = NO) gave the corresponding alcohol (Chapter 2).[373] It has long been known[362,363] that Clemmensen reduction of bases such as **171** (R = CH$_3$, C$_2$H$_5$) gave rearranged products

(183)   (184)   (185)

(**183**, R = H, CH$_3$ respectively) but Wolff–Kishner reduction was thought to proceed more or less as expected to give azepanes. However, it was later disclosed[374] that such reactions also produced alkenes. For example, 2-methyl-*N*-methylazepan-3-one (**171**, R = CH$_3$) gave, in addition to 2-methyl-*N*-methylazepane, 36.5% of unsaturated material considered to be **184**. All of the *N*-alkylazepan-3-ones reported, being α-amino ketones, are of dubious stability but form salts without apparent difficulty.

Notwithstanding their instability, azepan-3-ones have been successfully employed to annulate further rings to the structure. Thus the keto ester **174** (R = $CO_2C_2H_5$) reacted with hydroxylamine followed by hydrogen chloride to provide the zwitterionic compound **185**.[326,375] Other examples can be found in the patent literature.[377,378]

Table 15. Azepan-3-ones

Structure	M.p. (°C)/b.p. (°C/mmHg)	Derivatives and m.p. (°C)	Spectroscopic evidence/data	Ref.
N-CH₃		HCl 195–196		361
N-CH₂Ph			IR	365
N-CO₂CH₃	140/40		IR, ¹H NMR MS	364 441
N-COPh	113–114		IR	368
PhCH₂, N-CH₃		Supplement	Supplement	443

Table 15 (continued)

Structure	M.p. (°C)/b.p. (°C/mmHg)	Derivatives and m.p. (°C)	Spectroscopic evidence/data	Ref.
allyl-N-methyl azepanone				442
2-methyl-N-methyl azepanone	72–73/11	Picrate 174	n_D^{20} 1.4700	363, 374
2-ethyl-N-methyl azepanone	91–92/13	HCl 162.5–163 Picrate 169–169.5 Picrolonate 164.5–165.5	n_D^{20} 1.4696	362
4-CO$_2$C$_2$H$_5$, N-CO$_2$CH$_3$ azepanone	160/55		IR, ^{1}H NMR	364, 376
4-CO$_2$C$_2$H$_5$, N-CH$_2$Ph azepanone	Colourless oil		^{1}H ^{13}C NMR, IR, MS	367
4-CONHOH, N-CO$_2$tC$_4$H$_9$ azepanone	145 (d)		IR, ^{1}H NMR	375

Table 15 (continued)

Structure	M.p. (°C)/b.p. (°C/mmHg)	Derivatives and m.p. (°C)	Spectroscopic evidence/data	Ref.
[structure: azepanone with $CO_2C_2H_5$ and N-$CO_2{}^tC_4H_9$]	147–148/60			375
[structure: azepanone with Ph and N-Tosyl]	115–117		IR, MS, ^{1}H NMR	366
[structure: azepanone with two CH_3 and N-Tosyl]	100–102		IR, MS, ^{1}H NMR	366

III. AZEPAN-4-ONES

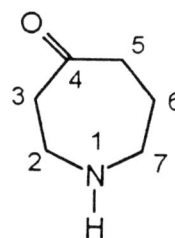

The members of this series have been thoroughly researched.

PREPARATION

The hydrochloride of the parent amino ketone (**186**, R = H) was first reported by Yokoo and Morosawa in 1956:[379] the N-benzyl ester **187** was subjected to the Dieckmann cyclisation (NaOC$_2$H$_5$/xylene), the intermediate cyclic β-keto esters were hydrolysed yielding a modest 21.7% yield of the hydrochloride of **186** (R = CH$_2$Ph). The latter on hydrogenolysis (H$_2$/PdO) gave **186** (R = H), isolated as the hydrochloride. The corresponding N-benzoyl compound **186** (R = COPh) was not obtainable by Dieckmann cyclisation.[379] It was later revealed[380,383] that difficulties with the production of the diester **187** made this

(186) (187) (188) (189)

synthesis less attractive than that involving diazomethane ring expansion of N-benzylpiperid-4-one (see below). Later work on the Dieckmann reaction favoured use of potassium ethoxide[381] or sodium hydride[382] as reagent, but the yields were not much improved. In contrast, the Thorpe–Ziegler cyclisation of dinitriles has been recommended for the preparation of both **186** (R = CH$_2$Ph)[384] and **186** (R = CH$_3$)[385] on the grounds that starting materials were more easily obtained; the yields, however, were not improved. The 3-methyl analogue (**188**) was obtained by a mixed Thorpe–Dieckmann cyclisation of the cyano ester **189** in 33% yield (NaOCH$_3$/xylene).[386]

As inferred above, ring expansion of 4-piperidones has become a popular methodology for the production of azepan-4-ones. Originally,[350] diazomethane (from N-nitrosomethylurethane) reacted with N-benzylpiperid-4-one at −15 °C giving a 43.6% yield of the ketone (as hydrochloride) (**186**, R = CH$_2$Ph). However, it was possible to obtain better yields (not precisely recorded) using preformed ethereal diazomethane.[387] In the case of the N-benzenesulphonyl derivative (**186**, R = PhSO$_2$), in situ generation of diazomethane (from N-nitrosomethylurethane) gave a very poor yield,[388] so one has to conclude that preformed diazomethane is preferable.[390] In most cases the expanded products (**186**) are accompanied by variable amounts of epoxide (**190**);[388,389] it appears that the precise experimental conditions employed are crucial.

Diazo ester ring expansions have become popular: this protocol has the advantage of generating cyclic β-keto esters (**191**) in one step from piperid-4-ones. In the earlier work,[391] keto esters **191** (R = PhCH$_2$, CO$_2$C$_2$H$_5$; R^1 = C$_2$H$_5$) were obtained directly in around 30–34% yields. Later work[392] revealed that, using the same technique (N$_2$CH$_2$CO$_2$C$_2$H$_5$/BF$_3$/−25 °C), hydrolysis of the unisolated β-keto ester and reprotection of the nitrogen atom led to a 50% yield of the ketone **186** (R = CO$_2$C$_2$H$_5$). This work has been confirmed.[364,393] The 5-phenyl analogue **192** was likewise obtained from

N-benzylpiperid-4-one using ethyl N-nitrosobenzyl carbamate and potassium carbonate in methanol at <25 °C, the yield being 21%.[395] Diazoacetic ester ring expansion was particularly successful for the N-*tert*-butoxycarbonyl case (**191**, R = CO$_2^t$Bu, R^1 = C$_2$H$_5$),[396] which, after hydrolysis, provided a 67% overall yield of **186** (R = H, as hydrochloride). Again there are reported cases (e.g. **191**, R = PhCH$_2$, CO$_2$C$_2$H$_5$; R^1 = C$_2$H$_5$) where greatly improved yields arose from process modifications.[397] The N-nitroso diester **193** was shown to react with N-benzylpiperid-4-one providing a 51% yield of **194**.[398]

(**190**)   (**191**)   (**192**)

(**193**)   (**194**)

Other types of ring expansion are known. Thus, a Wagner–Meerwein type of rearrangement ensued when the tertiary dialcohol **195** (R = H) was treated with concentrated H$_2$SO$_4$ at 0 °C: the product **196** (R = H) was obtained in 79% yield.[399] Similarly, the diol **195** (R = CH$_3$) gave **196** (R = CH$_3$) and

(**195**)   (**196**)   (**197**)   (**198**)

197.[400] Mechanistically similar was treatment of the amine **198** with nitrous acid at 0 °C which gave the N-nitroso ketone **199** in 20% yield.[401] A free-radical ring expansion (e.g. **200** → **201**) has been reported.[367] A yield of 66% was achieved

Azepanones

(^{n}Bu$_3$SnH/AIBN/benzene/reflux) in the case of **200** (R = Ph$_3$C); a poorer yield in the case of **200** (R = CH$_2$Ph) is attributable to transannular hydrogen atom abstraction from the benzyl group (CH$_2$Ph). This protocol has the advantage of employing readily accessible piperidones to give azepan-4-ones carrying functionality (CO$_2$CH$_3$) at a normally inaccessible (C-6) position.

(199)   (200)   (201)

The acyloin reaction was first applied to azepan-4-one synthesis in 1954:[402] conversion of diester **202** into hydroxy ketone **203** (R = Ph) proceeded in only

(202)   (203)   (204)

10% yield (Na/xylene). The *N*-ethyl diester corresponding to **202** failed to give an acyloin although all of the higher homologues studied gave larger ring acyloins in satisfactory yields.[402] On the other hand, when the acyloin reaction was conducted in the presence of trimethylsilyl chloride as trapping agent, the protected product (**204**, R = PhCH$_2$) related to acyloin **203** (R = CH$_2$Ph) was isolated in good yield.[403] It is now standard procedure to include trimethylsilyl chloride in acyloin reactions[404,405] designed to provide 5-hydroxyazepan-4-ones. It should be noted that where alkyl groups replace hydrogen atoms α- to ester functionality (e.g. in **202**), the acyloin reaction proceeds well under classical conditions, for example hydroxy ketone **205** was obtained in 80% yield[406] from the corresponding diester using sodium in toluene and without trimethylsilyl chloride. This outcome is attributable to the suppression of Dieckmann reactions which require hydrogen atoms α- to ester groups.

An intramolecular Friedel–Crafts type of electrophilic cyclisation has been uncovered in the case of the chloroalkenyl-protected amino acid chloride **206**, which cyclised in alcohol-free dichloromethane at ambient temperature yielding the dichloro ketone **207** in 56% yield.[366] This procedure also has the merit of providing access to an azepan-4-one substituted at C-6; moreover, **207** was easily converted into the tetrahydroazepinones **208** (R = Cl and OCH$_3$)^{366} (Chapter 5). Flash vacuum thermolysis of the spirooxazolidine **209** gave the

(205) (206) (207) (208)

2-phenylazepan-4-one **210** in 42% yield, accompanied by other identified products.^{408,409}

(209) (210) (211) (212)

Various N-protected bisallylamines (**211**) have been shown to undergo hydroboration/carbon monoxide insertion using chloroborane/dimethyl sulphide.^{410} Generally two or more products were isolated; the most successful in the present context involved the diene **211** (R = PhSO$_2$), which gave ketone **212** in 40% yield along with the corresponding perhydroazocin-5-one. Other N-protecting groups (**211**, R = CO$_2$CH$_3$, CO$_2$CH$_2$Ph) were much less efficient.^{410} This is an interesting approach worthy of further study, perhaps utilising chloroallyl groups as in **206**.

A logical adaptation of standard cyclisations for azepanes (Chapter 2) involved ketone protection (as acetal), which allowed the reductive amination of dialdehyde **213** with various amines (CH$_3$NH$_2$, PhNH$_2$, cyclohexyl-NH$_2$)

(213) (214) (215)

using NaCNBH$_3$.^{411} The products were the acetal-protected azepan-4-ones **186** (R = CH$_3$, Ph and c-C$_6$H$_{11}$); after deprotection (10% HCl) the yields overall of azepan-4-ones from cyclohex-2-enone were fairly poor; however, this has the merit of novelty.

Microbiological oxygenation of both N-benzoyl and N-tosylazepane using *Sporotrichum sulfurescens* led to mixtures of azepan-3-ones (Section II) and azepan-4-ones along with the corresponding alcohols.[368] After CrO_3 oxidation of the mixtures, azepan-4-ones (**186**, R = PhCO, tosyl) were isolated and identified (yields ~50%). N-Benzoyl-4-methylazepane, on the other hand, gave only 11% of the 5-methylazepan-4-one **214**; the major product involved oxidation to the tertiary alcohol **215**.[412] Several other microorganisms have the ability to oxygenate azepanes.[371] The herbicide molinate has been extensively studied. 4-Ketomolinate (**186**, R = $COSC_2H_5$) has been detected in various soils,[371,372,413] rivers[414] and fish,[415,416] while N-nitrosoazepan-4-one (**186**, R = NO) has been found in rats urine[370] after administration of N-nitrosoazepane. However, none of these studies has yet been developed into the status of a synthesis.

REACTIONS

Most reactions reported for azepan-4-ones are predictable. It was demonstrated more than 40 years ago that transannular electron donation from nitrogen to carbonyl in a seven-membered acyloin (**203**, R = Ph) was unlikely since the infrared stretching frequency of the carbonyl group (1701 cm^{-1}) was, unlike those of nine-membered analogues (1666, 1671 cm^{-1}), normal.[402,417] N-Benzyl-azepan-4-one (**186**, R = CH_2Ph) forms an ethylene ketal whose ^{13}C NMR spectrum has been recorded and compared with those of similar structures.[418] Simple azepan-4-ones (**186**, R = H, CH_3, C_2H_5) are fairly unstable bases, usually best handled as salts; the N-acyl derivatives do not appear to involve significant problems.

Not surprisingly, reduction of the carbonyl group in this series can be brought about using $NaBH_4$; in the case of the β-keto ester **191** (R = CO_2CH_3, $R^1 = C_2H_5$), the two stereoisomers (**216**, **217**) were isolated.[419] Similar molecules have been reduced catalytically (H_2/Ni)[420] to the corresponding alcohols. The vinyl-oxycarbonyl group in **186** (R = CH_2=CHOCO–) survived

[(±)**216**]   [(±)**217**]   (**218**)

NaBH$_4$ reduction of the carbonyl group to an hydroxy group.[421] LiAlH$_4$ served to reduce the carbonyl group in **186** (R = PhCH$_2$).[421] The expected addition of organometal reagents to azepan-4-ones has been reported; for example, the 3-methyl analogue **188** underwent addition with phenyllithium.[422] Oxidation of N-substituted azepan-4-ones with selenium dioxide has been claimed to yield mixtures of 3,4- and 4,5-diones.[423] Although the latter were not isolated, subsequent reactions with 2-aminoacetamidine led to N-substituted 2-amino-6,7,8,9-tetrahydro-5H-pyrazino[2,3-d]azepines (**218**); accordingly, the structures of 4,5-diones are thus confirmed. It has to be inferred that the above-mentioned 3,4-diones were minor products.[423]

Bromination of **186** (R = CO$_2$C$_2$H$_5$) with N-bromosuccinimide was originally thought to give 3-bromoazepan-4-one (**219**),[424] while bromine in acetic acid gave a mixture of 3- and 5-bromo (**220**) isomers and the 3,5-dibromo

(**219**)   (**220**)   (**221**)   (**222**)

compound.[424] However, later work[425] suggested that bromination generally gave the 5-bromo isomer as the major product, since reaction with thioacetamide gave thiazoloazepine (**221**), contrary to the original conclusion.[426] On the other hand, reaction of **186** (R = CO$_2$C$_2$H$_5$) with ethyl formate gave a hydroxymethylene compound,[427] subsequently proved by NMR spectroscopy to have structure **222**.[425]

A number of heterocycles have been fused to the azepane ring using azepan-4-one derivatives as starting materials. The 5-bromo derivative **220** has been used to obtain a number of thiazolo- and oxazoloazepines,[428] while the hydroxymethylene derivative **222** served to generate a pyrazoloazepine structure.[427] β-Keto esters (**191**) are useful precursors for pyrimido[4,5-d]azepines[429] by reaction with thiourea, acetamidine, guanidine or cyanoguanidine. Isoxazolo[3,4-d]azepine derivatives[430] and the corresponding isothiazolo[5,4-c]azepines[431] were made as potential GABA-mimetic compounds also commencing with β-keto esters. 1,2,3-Selenadiazole rings can be annulated to the azepane nucleus by reaction of semicarbazones of ketones (**186**, R = CH$_2$Ph or CO$_2$C$_2$H$_5$) with SeO$_2$ or by selenoyl dichloride (SeOCl$_2$) treatment of the corresponding ketone tosylhydrazones.[432] The selenadiazole isomer **223** was generally the preferred product, although the ratio of **223** to **224** was drastically affected by changes in solvent. In this work ^{77}Se NMR was usefully deployed.[432] The

Azepanones

ketone **186** (R = CH$_2$Ph), when subjected to Schmidt reaction conditions, gave 7% of a tetrazolodiazocine (**225**), but the major product was the diazocinone **226** (64%).[61]

(**223**)   (**224**)   (**225**)   (**226**)   (**227**)

The relative position of carbonyl group and heterocyclic nitrogen atom in azepan-4-ones (i.e. β-amino ketones) makes Hofmann elimination reactions likely; these have been demonstrated for the *N*-acylium salts of **186** (R = CH$_3$) which gave enones **227** (R = CH$_3$ and –CH=CH$_2$) when refluxed with Hünig's base (^{i}Pr$_2$EtN) in toluene.[433]

Ring contraction reactions are known in this series. Thus the alcohol **228** gave the pyrrolidine derivatives **229** on being refluxed with propionic anhydride;[422] a transannular electron donation (N → C-4) was postulated to explain this phenomenon. A similar behaviour was seen[421] when ketone **186** (R = CH$_2$Ph)

(**228**)   (**229**)

(**230**)   (**231**)

was reduced to the corresponding alcohol and the latter reacted with triphenylphosphine/carbon tetrachloride giving **230** (see Chapter 2). Attempted Wolff–Kishner reaction on the hydrazine of ketone **199** (H for NO) led to the pyrrole **231**.[401]

Table 16. Monosubstituted azepan-4-ones

Structure	M.p. (°C)/b.p. (°C/mmHg)	Derivatives and m.p. (°C)	Spectroscopic evidence/data	Ref.
Azepan-4-one, N–H	105–108/0.25	HCl 161–163 HCl 171	MS, IR ^{1}H NMR	393 396 379, 380, 427
Azepan-4-one, N–CH$_3$	115–120/45 97–100/18	Picrate 171–172 Picrate 151–152 HCl 166.5–167 HCl 160 (d)	n_D^{23} 1.4893, d_4^{23} 0.963	385, 434 435 390 381
Azepan-4-one, N–CH$_2$Ph	115–118/0.01 129/0.05 120/0.3 98–100/0.1	HCl 186–187 2,4-DNPH·HCl 206 HCl 192–194 HCl 188–191	 n_D^{20} 1.5539 IR	379, 427 380 381 384 428
Azepan-4-one, N–Ph		HCl 144	^{13}C NMR, IR, MS	411 428, 436
Azepan-4-one, N–cyclohexyl	114		IR, ^{1}H, ^{13}C NMR, MS	411

Table 16 (continued)

Structure	M.p. (°C)/b.p. (°C/mmHg)	Derivatives and m.p. (°C)	Spectroscopic evidence/data	Ref.
1-(3-methylbenzyl)azepan-4-one		HCl 169		428
1-(2-methylbenzyl)azepan-4-one		HCl 165		428
1-(3-trifluoromethylbenzyl)azepan-4-one		HCl 176–178		428
1-(cyclohexylmethyl)azepan-4-one		HCl 180		428
1-(2-bromobenzyl)azepan-4-one		HCl 179		428

Table 16 (continued)

Structure	M.p. (°C)/b.p. (°C/mmHg)	Derivatives and m.p. (°C)	Spectroscopic evidence/data	Ref.
1-(3-bromobenzyl)-4-oxoazepane		HCl 192		428
1-(4-bromobenzyl)-4-oxoazepane		HCl 215		428
1-(3-chlorobenzyl)-4-oxoazepane		HCl 212		428
1-(4-chlorobenzyl)-4-oxoazepane		HCl 210		428
1-(4-fluorobenzyl)-4-oxoazepane		HCl 208		428

Table 16 (continued)

Structure	M.p. (°C)/b.p. (°C/mmHg)	Derivatives and m.p. (°C)	Spectroscopic evidence/data	Ref.
N-CH₂CH₂Ph azepan-4-one	130–134/0.4	HCl 195		387, 428
N-CH₂CH=CH₂ azepan-4-one	52–54/0.2			428
N-CH(CH₃)Ph azepan-4-one	105/0.1			428
N-CH(CH₃)C₂H₅ azepan-4-one	58–60/0.1			428
N-NO azepan-4-one	160/0.05		IR, MS, ^{1}H NMR	389

Table 16 (continued)

Structure	M.p. (°C)/b.p. (°C/mmHg)	Derivatives and m.p. (°C)	Spectroscopic evidence/data	Ref.
4-oxo-azepane-N-CO$_2$CH$_3$	140/40		IR, ^{1}H NMR	364
4-oxo-azepane-N-CO$_2$C$_2$H$_5$	117/1	Semicarbazone 184.5–186	^{1}H NMR	393, 427 432
4-oxo-azepane-N-COPh	170/0.25 173/0.3		IR	373 368
4-oxo-azepane-N-SO$_2$Ph	121–122	Oxime 172–173 2,4-DNPH 197		388
4-oxo-azepane-N-SO$_2$-C$_6$H$_4$-CH$_3$	89–90		IR	368
4-oxo-azepane-N-COO-CH=CH$_2$	Oil		IR, ^{1}H NMR, MS	421

Table 16 (continued)

Structure	M.p. (°C)/b.p. (°C/mmHg)	Derivatives and m.p. (°C)	Spectroscopic evidence/data	Ref.
4-oxo-1-(4-nitrobenzoyl)azepane	118.5–119	2,4-DNPH 123–125 Oxime 203.5–204		388
4-oxo-5-(ethoxycarbonyl)azepane	112–114/0.6	HCl 133–134		391, 397

Table 17. Disubstituted azepan-4-ones

Structure	M.p. (°C)/b.p. (°C/mmHg)	Derivatives and m.p. (°C)	Spectroscopic evidence/data	Ref.
3,5-dibromo-azepan-4-one		HBr 186 (d)		424
2,7-diphenyl-azepan-4-one	110–111			401
2,7-dimethyl-azepan-4-one (N-CH$_3$)	100–120/27			382
3-methyl-azepan-4-one (N-CH$_3$)	110/35	Picrate 192–193 n_D^{29} 1.4656 HCl 158–160		434 385 386
2-phenyl-azepan-4-one (N-Ph)	93–94		^{1}H, ^{13}C NMR, MS, IR	408
3-CO$_2$C$_2$H$_5$ azepan-4-one (N-CH$_2$Ph)			IR, ^{1}H NMR, MS	367

Table 17 (continued)

Structure	M.p. (°C)/b.p. (°C/mmHg)	Derivatives and m.p. (°C)	Spectroscopic evidence/data	Ref.
4-oxo-3-(CO₂CH₃)-1-CPh₃ azepane			¹H, ¹³C NMR, MS, IR	367
4-oxo-3-(CO₂C₂H₅)-1-(CO₂CH₃) azepane	210/0.3		IR, ¹H NMR	420
4-oxo-5-Ph-1-CH₂Ph azepane		HCl 225	IR, ¹H NMR, X-ray	395, 407
4-oxo-5-Ph-1-(CO₂C₂H₅) azepane	171/0.15		IR, ¹H NMR	395
3-Br-4-oxo-1-(CO₂C₂H₅) azepane	140–145/1	2,4-DNPH 149–152(d)		424
3-allyl-4-oxo-1-CH₃ azepane	123–124/35	Picrate 138–139		394

Table 17 (continued)

Structure	M.p. (°C)/b.p. (°C/mmHg)	Derivatives and m.p. (°C)	Spectroscopic evidence/data	Ref.
4-(4-hydroxy-4-methylpentyl)-1-benzyl-hexahydroazepin-3-one	Oil		^{1}H, ^{13}C NMR, MS	398
4-(3-oxobutyl)-1-benzyl-hexahydroazepin-3-one	Oil		^{1}H, ^{13}C NMR, MS, IR	398
4-(methoxycarbonylethyl)-1-benzyl-hexahydroazepin-3-one	Oil		^{1}H, ^{13}C NMR, IR	398
3-methyl-1-(2-phenylethyl)hexahydroazepin-4-one		Picrate 124–130		382
2-methyl-1-(2-phenylethyl)hexahydroazepin-4-one				382
3-hydroxymethylene-1-ethoxycarbonyl-hexahydroazepin-4-one (tautomers)	127–130/0.7	2,4-DNPH 172		427

Table 17 (continued)

Structure	M.p. (°C)/b.p. (°C/mmHg)	Derivatives and m.p. (°C)	Spectroscopic evidence/data	Ref.
4-oxo-3-carbamoyl-1-(methoxycarbonyl)azepane	130.5–132			431
3-methyl-4-oxo-1-(phenylsulfonyl)azepane			^{1}H NMR, MS, IR	410
3-methyl-4-oxo-1-(methoxycarbonyl)azepane	Chromatography		^{1}H NMR, MS, IR	410
3-methyl-4-oxo-1-(benzyloxycarbonyl)azepane	Chromatography			410
3-(ethoxycarbonyl)-4-oxo-1-acetylazepane	154/0.28		IR	391, 429, 430
3-(ethoxycarbonyl)-4-oxo-1-(methoxycarbonyl)azepane	160/55		^{1}H NMR, IR	364, 419

Table 17 (continued)

Structure	M.p. (°C)/b.p. (°C/mmHg)	Derivatives and m.p. (°C)	Spectroscopic evidence/data	Ref.
4-methyl-5-oxo-azepane-1-carboxylate ethyl ester	93–95/0.1	2,4-DNPH 145–146		391
4-ethoxycarbonyl-5-oxo-azepane-1-carboxylate ethyl ester	133–135/0.15	2,4-DNPH 119.5		391–393, 397, 430, 437
4-ethoxycarbonyl-1-benzyl-5-oxo-azepane	160–162/0.4	HCl 164.5 2,4-DNPH.HCl 179		391, 397, 438 430
3-methyl-4-oxo-azepane-1-carboxylate ethyl ester	132–134/0.7	2,4-DNPH 177–179		425
3-(n-butylthiomethylene)-4-oxo-azepane-1-carboxylate ethyl ester	85–86	2,4-DNPH 117–119		425
4-(methoxycarbonylmethyl)-1-methyl-5-oxo-azepane				439
4-hydroxy-1-phenyl-5-oxo-azepane	112–113	p-Nitrobenzoate 146–147		402, 439

Azepanones 235

Table 18. Tri- and polysubstituted azepan-4-ones

Structure	M.p. (°C)/b.p. (°C/mmHg)	Derivatives and m.p. (°C)	Spectroscopic evidence	Ref.
(1-benzyl-3-methyl-3-ethoxycarbonyl-azepan-4-one)	171–173/0.2			391
(2,7-diphenyl-1-nitroso-azepan-4-one)	138–142	2,4-DNPH 224–226		401
(5,5-dichloro-1-tosyl-azepan-4-one)	109–111		IR, ^{1}H NMR, MS	366
(5,5-diphenyl-1-methyl-azepan-4-one)	91–92	HCl 244–246 MeI 201–203		399
(2,2,6,6-tetramethyl-5-hydroxy-1-methyl-azepan-4-one)			^{1}H NMR	440
(2,2,6,6-tetramethyl-1-(1,1-dimethyl-2-hydroxyethyl)-azepan-4-one)	78–80 (subl.)		IR, ^{1}H, ^{13}C NMR, MS	405

Table 18 (continued)

Structure	M.p. (°C)/b.p. (°C/mmHg)	Derivatives and m.p. (°C)	Spectroscopic evidence	Ref.
(structure)	67–69 (subl.)		IR, ^{1}H, ^{13}C NMR, MS	405
(structure)	76–79 (subl.)		IR, ^{1}H NMR, MS	405
(structure)	78–80		IR	406
(structure)	37–38 (subl.)		IR, ^{1}H NMR, MS	406
(structure)	57–60 (subl.)		IR, ^{1}H NMR, MS	406

Table 18 (continued)

Structure	M.p. (°C)/b.p. (°C/mmHg)	Derivatives and m.p. (°C)	Spectroscopic evidence	Ref.
[structure: azepanone with OCOCH₃, CH₃ groups, N-C(CH₃)₃]	86–88 (subl.)		IR, ¹H NMR, MS	406
[structure: azepanone with OH, CH₃ groups, N-C(CH₃)₃]	55–56 (subl.)		IR, ¹H NMR, UV, MS	406, 440

References

1. A.H. Jubb, *Basic Organic Chemistry, Part 5*, Wiley, New York, 1975, p. 187.
2. S. Sifniades, in A.N. Collins, G.N. Sheldrake and J. Crosby (Eds), *Chirality in Industry*, Wiley, New York, 1992, pp. 79–86.
3. A. Frankowski and A. Redlinski, *Zesz. Nauk-Politech Lodz, Chem.*, 1980, 37; *Chem. Abs.*, 1981, **94**, 83640y.
4. O. Wallach, *Liebigs Ann. Chem.*, 1893, **278**, 304.
5. O. Wallach, *Liebigs Ann. Chem.*, 1900, **312**, 171.
6. O. Wallach, *Liebigs Ann. Chem.*, 1906, **346**, 249.
7. C.S. Marvel and J.C. Eck, *Org. Synth.*, 1937, **17**, 60.
8. J.G. Hildebrand and M.T. Bogert, *J. Am. Chem. Soc.*, 1936, **58**, 650.
9. H.E. Ungnade and A.D. McLaren, *J. Org. Chem.*, 1945, **10**, 29.
10. A.D. McLaren and G. Pitzl, *J. Am. Chem. Soc.*, 1945, **67**, 1625.
11. R.L. Burke and R.M. Herbst, *J. Org. Chem.*, 1955, **20**, 726.
12. A. Schäffler and W. Ziegenbein, *Chem. Ber.*, 1955, **88**, 767, 1374.
13. P. Oxley and W.F. Short, *J. Chem. Soc.*, 1948, 1514.
14. K. Derdzinski and A. Zabza, *Bull. Acad. Pol. Sci. Ser. Sci. Chim.*, 1977, **25**, 529; *Chem. Abs.*, 1978, **88**, 74466.
15. W.Z. Heldt, *J. Am. Chem. Soc.*, 1958, **80**, 5880.
16. J.R. Merchant and D. Roy, *Curr. Sci.*, 1975, **44**, 263.
17. R.T. Conley and B.E. Nowak, *J. Org. Chem.*, 1962, **27**, 3196.
18. F. Fernandez and C. Perez, *J. Chem. Res. (S)*, 1987, 340.
19. F. Fernandez and C. Perez, *Heterocycles*, 1987, **26**, 2411.
20. K. Miyano and T. Taguchi, *Chem. Pharm. Bull.*, 1970, **18**, 1799.

21. K.K. Kelly and J.S. Matthews, *Tetrahedron*, 1970, **26**, 1555.
22. M. Sainsbury, M.F. Mahon, C.S. Williams, A. Naylor and D.I.C. Scopes, *Tetrahedron*, 1991, **47**, 4195.
23. A.C. Huitric and S.D. Nelson, Jr, *J. Org. Chem.*, 1969, **34**, 1230.
24. T. Sato, H. Wakatsuka and K. Amano, *Tetrahedron*, 1971, **27**, 5381.
25. P. Nedenskov, W. Taub, and D. Ginsburg, *Acta Chem. Scand.*, 1958, **12**, 1405.
26. K. Itoh and Y. Oka, *Chem. Pharm. Bull.*, 1980, **28**, 2862.
27. F.N. Shirota, H.T. Nagasawa and J.A. Elberling, *J. Med. Chem.*, 1977, **20**, 1623.
28. K. Mitsuhashi, S. Shiotani, R. Oh-Uchi, and K. Shiraki, *Chem. Pharm. Bull.*, 1969, **17**, 434.
29. N. Komatsu, S. Simizu and T. Sugita, *Synth. Commun.*, 1992, **22**, 277.
30. Z. Majer, M. Kajtar, M. Tichy and K. Blaha, *Collect. Czech. Chem. Commun.*, 1982, **47**, 950.
31. K. Blaha, A.M. Farag, D. Van der Helm, M.B. Hossain, M. Budesinsky, P. Malon, J. Smolikova and M. Tichy, *Collect. Czech. Chem. Commun.*, 1984, **49**, 712.
32. E.F. Novoselov, S.D. Isaev, A.G. Yurchenko, L. Vodichka and Yu. Trshiska, *Zh. Org. Khim.*, 1981, **17**, 2558; *Chem. Abs.*, 1982, **96**, 122607f.
33. T. Imamoto, H. Yokoyama and M. Yokoyama, *Tetrahedron Lett.*, 1981, **22**, 1803.
34. K. Narasaka, H. Kusama, Y. Yamashita and H. Sato, *Chem. Lett.*, 1993, 489.
35. H.M. Meshram, *Synth. Commun.*, 1990, **20**, 3253.
36. M. Fetizon and S. Nanthavong, *Bull. Soc. Chim. Fr.*, 1969, 194.
37. G. Just and M. Cunningham, *Org. Photochem. Synth.*, 1976, **2**, 96.
38. M.-K. Yeh, *J. Chin. Chem. Soc.*, 1978, **25**, 83; *Chem. Abs.*, 1979, **90**, 22385.
39. K. Suzuki, S.-I. Kiyooka, K. Hidaka, S. Yakushiji, J.-I. Yamamoto and Y. Higashi, *Mem. Fac. Sci. Kochi Univ. Ser. C*, 1987, **8**, 39; *Chem. Abs.*, 1988, **108**, 131737.
40. M. Hamana, H. Noda and J.-I. Uchida, *Yakugaku Zasshi*, 1970, **90**, 991; *Chem. Abs.*, 1970, **73**, 130926.
41. W. Kessler and M. Brenner, *Helv. Chim. Acta*, 1969, **52**, 901.
42. R. Fuhrmann, A.A. Tunick and S. Sifniades, *US Pat.*, 3 922 265; *Chem. Abs.*, 1976, **84**, 121193h.
43. P.F. Hudrlik, M.A. Waugh and A.M. Hudrlik, *J. Organomet. Chem.*, 1984, **271**, 69.
44. T.E. Stevens, *Org. Prep. Proced.*, 1970, **2**, 53.
45. C.G. Overberger, J.H. Kozlowski and E. Radlmann, *J. Polym. Sci., Part A1*, 1972, **10**, 2265.
46. K.A. Petrov, V.A. Chauzov and V.P. Pokatun, *Zh. Org. Khim.*, 1983, **53**, 541; *Chem. Abs.*, 1983, **99**, 53858.
47. T. Yashiro and H. Shirai, *Nagoya Shiritsu Daigaku Yakugakubu Kenkyu Neinpo*, 1975, **23**, 49; *Chem. Abs.*, 1976, **85**, 192528.
48. F. Korte and F.F. Wiese, *Chem. Ber.*, 1964, **97**, 1970.
49. G.G. Lyle and R.M. Barrera, *J. Org. Chem.*, 1964, **29**, 3311.
50. F. Toda and H. Akai, *J. Org. Chem.*, 1990, **55**, 4973.
51. Ya.L. Gol'dfarb, B.P. Fabrichnyi and I.F. Shalavina, *Zh. Obshch. Khim.*, 1961, **31**, 2057; *Chem. Abs.*, 1961, **55**, 27050d.
52. B.P. Fabrichnyi, I.F. Shalavina and Ya.L. Gol'dfarb, *Zh. Org. Khim.*, 1965, **1**, 1507; *Chem. Abs.*, 1966, **64**, 586c.
53. B.P. Fabrichnyi, I.F. Shalavina, S.E. Zurabyan, Ya.L. Gol'dfarb and S.M. Kostrova, *Zh. Org. Khim.*, 1968, **4**, 680; *Chem. Abs.*, 1968, **69**, 18565.
54. B.P. Fabrichnyi, I.F. Shalavina and Ya.L. Gol'dfarb, *Zh. Org. Khim.*, 1969, **5**, 361; *Chem. Abs.*, 1969, **70**, 114910.
55. B.P. Fabrichnyi, I.F. Shalavina, Ya.L. Gol'dfarb and S.M. Kostrova, *Zh. Org. Khim.*, 1974, **10**, 1956; *Chem. Abs.*, 1975, **82**, 57495.

56. Ya.L. Gol'dfarb, B.P. Fabrichnyi, I.F. Shalavina and S.M. Kostrova, *Zh. Org. Khim.*, 1975, **11**, 2400; *Chem. Abs.*, 1976, **84**, 121060.
57. J. von Braun and A. Heymons, *Chem. Ber.*, 1930, **63**, 502.
58. H. Shechter and J.C. Kirk, *J. Am. Chem. Soc.*, 1951, **73**, 3087.
59. R. Fusco and S. Rossi, *Gazz. Chim. Ital.*, 1951, **81**, 511.
60. R.T. Conley, *J. Org. Chem.*, 1958, **23**, 1330.
61. Y. Sakakida, A.S. Kumanireng, H. Kawamoto and A. Yokoo, *Bull. Chem. Soc. Jpn.*, 1971, **44**, 478.
62. K. Mitsuhashi, K. Nomura, I. Watanabe and N. Minami, *Chem. Pharm. Bull.*, 1969, **17**, 1572.
63. J. Kondelikova, J. Kralicek, J. Smolikova and K. Blaha, *Collect. Czech. Chem. Commun.*, 1973, **38**, 523.
64. J. Kralicek, J. Kondelikova and V. Kubanek, *Collect. Czech. Chem. Commun.*, 1974, **39**, 249.
65. T. Duong, R.H. Prager, J.M. Tippett, A.D. Ward and D.I.B. Kerr, *Aust. J. Chem.*, 1976, **29**, 2667.
66. R.H. Prager, J.M. Tippett, and A.D. Ward, *Aust. J. Chem.*, 1978, **31**, 1989.
67. G.I. Georg, X. Guan and J. Kant, *Tetrahedron Lett.*, 1988, **29**, 403.
68. G.I. Georg, X. Guan and J. Kant, *Bioorg. Med. Chem. Lett.*, 1991, **1**, 125.
69. P. Rosenmund, D. Sauer and W. Trommer, *Chem. Ber.*, 1970, **103**, 496.
70. P. Cefelin, A. Frydrychova, J. Labsky, P. Schmidt and J. Sebenda, *Collect. Czech. Chem. Commun.*, 1967, **32**, 2787.
71. P. Cefelin, J. Labsky and J. Sebenda, *Collect. Czech. Chem. Commun.*, 1968, **33**, 1111.
72. J.L. Charlish, W.H. Davies and J.D. Rose, *J. Chem. Soc.*, 1948, 227.
73. A. Bladé-Font, *Tetrahedron Lett.*, 1980, **21**, 2443.
74. L. Ya. Barkovskaya, V.M. Shostakovskii and I.F. Bel'skii, *Izv. Akad. Nauk SSSR, Ser. Khim.*, 1967, 2773; *Chem. Abs.*, 1968, **69**, 86806b.
75. F. Salmon-Legagneur and C. Neveu, *C.R. Acad. Sci.*, 1964, **259**, 1878.
76. E. Testa, L. Fontanella and V. Aresi, *Liebigs Ann. Chem.*, 1964, **676**, 151.
77. C.G. Overberger and G.M. Parker, *J. Polym. Sci., Part A-1*, 1968, **6**, 513.
78. D.W. Adamson, *J. Chem. Soc.*, 1943, 39.
79. C. Bischoff and E. Schröder, *J. Prakt. Chem.*, 1991, **333**, 181; *Chem. Abs.*, 1991, **115**, 92050.
80. A. Mkhairi and J. Hamelin, *Tetrahedron Lett.*, 1986, **27**, 4435.
81. G. Bradley, J.F. Cavalla, T. Edington, R.S. Shepherd, A.C. White, B.J. Bushell, J.R. Johnson and G.O. Weston, *Eur. J. Med. Chem. Chim.Ther.*, 1980, **15**, 375.
82. K. Kefurt, K. Capek, Z. Kefurtova and J. Jary, *Collect. Czech. Chem. Commun.*, 1979, **44**, 2526.
83. (a) K. Kefurt, Z. Kefurtova and J. Jary, *Collect. Czech. Chem. Commun.*, 1984, **49**, 2665; (b) K. Kefurt, Z. Kefurtova and J. Jary, *Collect. Czech. Chem. Commun.*, 1988, **53**, 1795.
84. K. Kefurt, K. Capek, Z. Kefurtova and J. Jary, *Collect. Czech. Chem. Commun.*, 1986, **51**, 391.
85. K. Bock, I. Lundt and C. Pedersen, *Acta Chem. Scand., Ser. B*, 1987, **41**, 435.
86. S. Hanessian, *J. Org. Chem.*, 1969, **34**, 675.
87. H. Ogura and K. Takeda, *Heterocycles*, 1981, **15**, 467.
88. J.A. Robl and M.P. Cimarusti, *Tetrahedron Lett.*, 1994, **35**, 1393.
89. T. Nagasaka, H. Kato, H. Hayashi, M. Shioda, H. Hikasa and F. Hamaguchi, *Heterocycles*, 1990, **30**, 561.
90. D. St.C. Black, R.F.C. Brown and A.M. Wade, *Aust. J. Chem.*, 1972, **25**, 2429.

91. W.J. Boyle, S. Sifniades and J.F. Van Peppen, *J. Org. Chem.*, 1979, **44,** 4841.
92. M. Mori, Y. Washioka, T. Urayama, K. Yoshiura, K. Chiba and Y. Ban, *J. Org. Chem.*, 1983, **48**, 4058.
93. S. Kano, T. Yokomatsu, Y. Yuasa and S. Shibuya, *Chem. Pharm. Bull.*, 1985, **33**, 340.
94. S.T. Reid, J.N. Tucker and E.J. Wilcox, *J. Chem. Soc., Perkin Trans. 1*, 1974, 1359.
95. E. Oliveros, M. Rivière and A. Lattes, *Nouv. J. Chim.*, 1979, **3**, 739.
96. A. Lattes, E. Oliveros, M. Rivière, C. Belzecki, D. Mostowicz, W. Abramskj, C. Piccinni-Leopardi, G. Germain and M. Van Meerssche, *J. Am. Chem. Soc.*, 1982, **104**, 3929.
97. J. Aubé, Y. Wang, M. Hammond, M. Tanol, F. Takusagawa and D. Van der Velde, *J. Am. Chem. Soc.*, 1990, **112**, 4879.
98. J. Aubé, M. Hammond, E. Gherardini and F. Takusagawa, *J. Org. Chem.*, 1991, **56**, 499, 4086.
99. K. Suda, M. Sachima, M. Izutsu and F. Hino, *J. Chem. Soc., Chem. Commun.*, 1994, 949.
100. D. St.C. Black and A.M. Wade, *J. Chem. Soc., Chem. Commun.*, 1970, 871.
101. D. St.C. Black, F.W. Eastwood, R. Okraglik, A.J. Poynton, A.M. Wade and C.H. Welker, *Aust. J. Chem.*, 1972, **25**, 1483.
102. B. Coates, D. Montgomery and P.J. Stevenson, *Tetrahedron Lett.*, 1991, **32**, 4199.
103. D. Schumann and A. Naumann., *Chem. Ber.*, 1982, **115**, 1626.
104. K. Isobe, K. Mohri, K. Tokoro, C. Fukushima, F. Higuchi, J.-I. Taga and Y. Tsuda, *Chem. Pharm. Bull.*, 1988, **36**, 1275.
105. I. Sakai, N. Kawabe and M. Ohno, *Bull. Chem. Soc. Jpn.*, 1979, **52**, 3381.
106. J. Lub, M.L. Beekes and Th.J. de Boer, *Recl. Trav. Chim. Pays-Bas*, 1986, **105**, 22.
107. B.M. Trost, M. Vaultier and M. Santiago, *J. Am. Chem. Soc.*, 1980, **102**, 7929.
108. E. Vedejs and H. Sano, *Tetrahedron Lett.*, 1992, **33**, 3261.
109. J.B. Hester, Jr, *J. Org. Chem.*, 1967, **32**, 4098.
110. (a) P.A. Evans, A.B. Holmes and K. Russell, *Tetrahedron: Asymmetry*, 1990, **1**, 593; (b) P.A. Evans, A.B. Holmes and K. Russell, *J. Chem. Soc., Perkin Trans. 1*, 1994, 3397.
111. G.I. Hutchison, R.H. Prager and A.D. Ward, *Aust. J. Chem.*, 1980, **33**, 2477.
112. M. Ogata, H. Kano and H. Matsumoto, *J. Chem. Soc., Chem. Commun.*, 1968, 397.
113. T. Duong, R.H. Prager, A.D. Ward and D.I.B. Kerr, *Aust. J. Chem.*, 1976, **29**, 2651.
114. V.G. Smirnova and R.G. Glushkov, *Khim.-Farm. Zh.*, 1973, **7**, 3; *Chem. Abs.*, 1974, **80**, 47816n.
115. M.R. Hatswell, R.H. Prager and A.D. Ward, *Aust. J. Chem.*, 1993, **46**, 135.
116. K.R. Rao, Y.V.D. Nageswar, T.N. Srinivasan and P.B. Sattur, *Synth. Commun.*, 1988, **18**, 877.
117. P.V. Thakore, *J. Indian Chem. Soc.*, 1988, **65**, 800.
118. H. Ulrich, B. Tucker and R. Richter, *J. Org. Chem.*, 1978, **43**, 1544.
119. P. Magnus and J. Moursounidis, *J. Org. Chem.*, 1991, **56**, 1529.
120. H. Behringer and H. Meier, *Liebigs Ann. Chem.*, 1957, **607**, 67.
121. H. Wamhoff, *Chem. Ber*, 1969, **102**, 2739.
122. J. Liebscher, M. Pätzel and Y.F. Kelboro, *Synthesis*, 1989, 672.
123. C.R. Rasmussen, J.F. Gardocki, J.N. Plampin, B.L. Twardzik, B.E. Reynolds, A.J. Molinari, N. Schwartz, W.W. Bennetts, B.E. Price and J. Marakowski, *J. Med. Chem.*, 1978, **21**, 1044.
124. J.M. Grisar, G.P. Claxton and R.D. Mackenzie, *J. Med. Chem.*, 1976, **19**, 503.

125. K.R. Rao, Y.V.D. Nageswar, T.N. Srinivasan and P.B. Sattur, *Indian J. Chem.*, 1990, **29B**, 1041.
126. (a) V.G. Granik, M.K. Polievktov and R.G. Glushkov, *Zh. Org. Khim.*, 1971, **7**, 1431; *Chem. Abs.*, 1971, **75**, 129299s; (b) H. Bredereck, F. Effenberger and H.P. Beyerlin, *Chem. Ber.*, 1964, **97**, 3081.
127. A.M. Zhidkova, V.G. Granik, N.S. Kuryatov, V.P. Pakhomov, R.G. Glushkov and B.A. Medvedev, *Khim.-Farm. Zh.*, 1974, **8**, 21; *Chem. Abs.*, 1975, **82**, 118779y.
128. P. Ahuja, J. Singh, M.B. Nigam, V. Sardana, K. Kar and N. Anand, *Indian J. Chem.*, 1982, **21B**, 849.
129. V.G. Granik, A.M. Zhidkova, R.G. Glushkov, I.V. Persianova, E.M. Peresleni, A. P. Engoyan and Yu.N. Sheinker, *Khim. Geterotsikl. Soedin.*, 1974, 1220; *Chem. Abs.*, 1975, **82**, 30783t.
130. M. Pätzel, A. Ushmajev and J. Liebscher, *Synthesis*, 1993, 525.
131. V.G. Granik, N.B. Marchenko, E.O. Sochneva, T.F. Vlasova, A.B. Grigor'ev, M.K. Polievktov and R.G. Glushkov, *Khim. Geterotsikl. Soedin.*, 1976, 1505; *Chem. Abs.*, 1977, **86**, 139814n.
132. S.I. Kaimanakova, E.F. Kuleshova, N.P. Solov'eva and V.G. Granik, *Khim. Geterotsikl. Soedin.*, 1982, 1553; *Chem. Abs.*, 1983, **98**, 89306y.
133. P. Ahuja, J. Singh, M.B. Asthana, V. Sardana and N. Anand, *Indian J. Chem.*, 1989, **28B**, 1034.
134. A.M. Zhidkova, V.G. Granik, R.G. Glushkov, T.F. Vlasova, O.S. Anisimova, T.A. Gus'kova, and G.N. Pershin, *Khim. Geterotsikl. Soedin.*, 1974, 670; *Chem. Abs.*, 1974, **81**, 77820x.
135. R. Huisgen and J. Witte, *Chem. Ber.*, 1958, **91**, 972.
136. W. Walter, J. Voss, J. Curts and H. Pawelzik, *Liebigs Ann. Chem.*, 1962, **660**, 60.
137. J. Procházka, *Collect. Czech. Chem. Commun.*, 1947, **12**, 305.
138. H. Eilingsfeld, M. Seefelder and H.Weidinger, *Chem. Ber.*, 1963, **96**, 2671.
139. J. Witte and R. Huisgen, *Chem. Ber.*, 1958, **91**, 1129.
140. S. Scheibye, B.S. Pedersen and S.-O. Lawesson, *Bull. Soc. Chim. Belg.*, 1978, **87**, 229.
141. B. Yde, N.M. Yousif, H. Pedersen, I. Thomsen and S.-O. Lawesson, *Tetrahedron*, 1984, **40**, 2047.
142. K. Steliou and M. Mrani, *J. Am. Chem. Soc.*, 1982, **104**, 3104.
143. O.P. Goel and U. Krolls, *Synthesis*, 1987, 162.
144. J.J. Bodine and M.K. Kaloustian, *Synth. Commun.*, 1982, **12**, 787.
145. R. Köster and R. Kucznierz, *Liebigs Ann. Chem.*, 1992, 1081.
146. J.G.D. Carpenter and J.W. Moore, *J. Chem. Soc. C*, 1969, 1610.
147. E. Quiñoà, M. Adamczeski, P. Crews and G.J. Bakus, *J. Org. Chem.*, 1986, **51**, 4494.
148. M. Adamczeski, E. Quiñoà and P. Crews, *J. Am. Chem. Soc.*, 1989, **111**, 647.
149. J.A. Marshall and G.P. Luke, *J. Org. Chem.*, 1993, **58**, 6229.
150. N. Chida, T. Tobe and S. Ogawa, *Tetrahedron Lett.*, 1991, **32**, 1063.
151. H. Kishimoto, H. Ohrui and H. Meguro, *J. Org. Chem.*, 1992, **57**, 5042.
152. C.A. Broka and J. Ehrler, *Tetrahedron Lett.*, 1991, **32**, 5907.
153. N. Chida, T. Tobe, S. Okada and S. Ogawa, *J. Chem. Soc., Chem. Commun.*, 1992, 1064.
154. B.S. Davidson and R.W. Schumacher, *Tetrahedron*, 1993, **49**, 6569.
155. V. Macko, M.B. Stimmel, H. Peeters, T.J. Wolpert, L.D. Dunkle, W. Acklin, R. Bänteli, B. Jaun and D. Arigoni, *Experientia*, 1990, **46**, 1206.
156. P.J. Maurer and M.J. Miller, *J. Org. Chem.*, 1981, **46**, 2835.
157. P.J. Maurer and M.J. Miller, *J. Am. Chem. Soc.*, 1983, **105**, 240.

158. A. Zabza, C. Wawrzenczyk and H. Kuczynski, *Bull. Acad. Pol. Sci.*, 1972, **20**, 631; *Chem. Abs.*, 1972, **77**, 140307.
159. A. Matallana, A.W. Kruger and C.A. Kingsbury, *J. Org. Chem.*, 1994, **59**, 3020.
160. C.S. Marvel and W.W. Moyer, Jr, *J. Org. Chem.*, 1957, **22**, 1065.
161. H.A. Offe, *Z. Naturforsch.*, *Teil B*, 1947, **2**, 182; *Chem. Abs.*, 1948, **42**, 4548c.
162. J. Kondelikova, J. Kralicek and D. Krivinkova, *Collect. Czech. Chem. Commun.*, 1971, **36**, 3391.
163. R. Tull, R.C. O'Neill, E.P. McCarthy, J.J. Pappas and J.M. Chemerda, *J. Org. Chem.*, 1964, **29**, 2425.
164. A.G. Shipov, N.A. Orlova and Yu.I. Baukov, *Zh. Obshch. Khim.*, 1984, **54**, 2645; *Chem. Abs.*, 1985, **102**, 78704b.
165. R.P. Mull, P. Schmidt, M.R. Dapero, J. Higgins and M.J. Weisbach, *J. Am. Chem. Soc.*, 1958, **80**, 3769.
166. W. Ziegenbein and W. Franke, *Chem. Ber.*, 1957, **90**, 2291.
167. A. Ts. Malkhasyan, G.G. Sukiasyan, S.G. Matinyan and G.T. Martirosyan, *Arm. Khim. Zh.*, 1976, **29**, 458; *Chem. Abs.*, 1976, **85**, 142966y.
168. P.F. Wiley, *J. Am. Chem. Soc.*, 1949, **71**, 3746.
169. K. Rühlmann and B. Rupprich, *Liebigs Ann. Chem.*, 1965, **686**, 226.
170. R. Huisgen and J. Reinertshofer, *Liebigs Ann. Chem.*, 1952, **575**, 174.
171. K. Heyns and O.-F. Woyrsch, *Chem. Ber.*, 1953, **86**, 76.
172. W. Pritzkow and P. Dietrich, *Liebigs Ann. Chem.*, 1963, **665**, 88.
173. S.S. Kukalenko, Yu.T. Struchkov, S.I. Shestakova, A.G. Tsybulevskii, A.S. Batsanov and E.B. Nazarova, *Koord. Khim.*, 1983, **9**, 312; *Chem. Abs.*, 1983, **98**, 226875.
174. B. Dusek and F. Kutek, *Sb. Vys. Sk. Chem.-Technol. Praze, Anorg. Chem. Technol.*, 1980, **B24**, 33; *Chem. Abs.*, 1981, **95**, 107656.
175. B. Dusek and F. Kutek, *Zh. Neorg. Khim.*, 1980, **25**, 2926; *Chem. Abs.*, 1981, **94**, 24236.
176. P. Dini, J.C.J. Bart, E. Santoro, G. Cum and N. Giordano, *Inorg. Chim. Acta*, 1976, **17**, 97.
177. J.G.H. Du Preez and M. L Gibson, *J.S. Afr. Chem. Inst.*, 1970, **23**, 184; *Chem. Abs.*, 1971, **74**, 49201.
178. M. Masaki, K. Fukui, M. Uchida, K. Yamamoto and I. Uchida, *Bull. Chem. Soc. Jpn.*, 1973, **46**, 3179.
179. M. Rothe, G. Reinisch, W. Jaeger and I. Schopov, *Makromol. Chem.*, 1962, **54**, 183.
180. J. Duynstee, W. Van Raayen, J. Smidt and Th.A. Veerkamp, *Recl. Trav. Chim. Pays-Bas*, 1961, **80**, 1323.
181. R. Köster, R. Kucznierz, W. Schüsler, D. Bläser and R. Boese, *Liebigs Ann. Chem.*, 1993, 189.
182. R. Aumann and P. Hinterding, *Chem. Ber.*, 1990, **123**, 611.
183. A. Rieche and W. Schön, *Chem. Ber.*, 1966, **99**, 3238.
184. W. Schön and A. Rieche, *Chem. Ber.*, 1967, **100**, 4052.
185. A.R. Doumaux, Jr, J.E. McKeon and D.J. Trecker, *J. Am. Chem. Soc.*, 1969, **91**, 3992.
186. J.-C. Gramain, R. Remuson and Y. Troin, *Tetrahedron*, 1979, **35**, 753, 759.
187. M. Mitzlaff, K. Warning and H. Rehling, *Synthesis*, 1980, 315.
188. T. Shono, Y. Matsumura, K. Uchida and H. Kobayashi, *J. Org. Chem.*, 1985, **50**, 3243.
189. M. Okita, T. Wakamatsu and Y. Ban, *J. Chem. Soc., Chem. Commun.*, 1979, 749.
190. S. Yoshifuji, Y. Arakawa and Y. Nitta, *Chem. Pharm. Bull.*, 1987, **35**, 357.
191. K. Eger, M. Jalalian, E.J. Verspohl and N.-P. Lüpke, *Arzneim.-Forsch./Drug Res.*, 1990, **40**, 1073.

192. A.A. Belyaev and E.V. Krasko, *Izv. Akad. Nauk, Ser. Khim.*, 1992, 1692; *Chem. Abs.*, 1993, **119**, 9111.
193. A. Archelas, R. Furstoss, D. Srairi and G. Maury, *Bull. Soc. Chim. Fr.*, 1986, 234.
194. R. Lukeš and J. Malek, *Chem. Listy*, 1951, **45**, 72; *Chem. Abs.*, 1951, **45**, 9523c.
195. S. Shiotani and T. Kometani, *Chem. Pharm. Bull.*, 1973, **21**, 1053.
196. G.R. Clemo, R. Raper and H.J. Vipond, *J. Chem. Soc.*, 1949, 2095.
197. F.F. Blicke and N.J. Doorenbos, *J. Am. Chem. Soc.*, 1954, **76**, 2317.
198. T. Nagasaka, H. Tamano and F. Hamaguchi, *Heterocycles*, 1986, **24**, 1231.
199. T. Nagasaka, Y. Koseki, H. Hayashi, Y. Yasuda and F. Hamaguchi, *Yakugaku Zasshi*, 1989, **109**, 823; *Chem. Abs.*, 1990, **112**, 216668.
200. H.R. Rickenbacher and M. Brenner, *Chimia*, 1957, **11**, 293.
201. M. Brenner and H.R. Rickenbacher, *Helv. Chim. Acta*, 1958, **41**, 181.
202. W.C. Francis, J.R. Thornton, J.C. Werner and T.R. Hopkins, *J. Am. Chem. Soc.*, 1958, **80**, 6238.
203. R.J. Wineman, E.-P.T. Hsu and C.E. Anagnostopoulos, *J. Am. Chem. Soc.*, 1958, **80**, 6233.
204. W.J. Brouillette and H.M. Einspahr, *J. Org. Chem.*, 1984, **49**, 5113.
205. R.G. Glushkov, V.G. Smirnova, K.A. Zaitseva, N.A. Novitskaya, M.D. Mashkovskii and G.N. Pershin, *Khim.-Farm. Zh.*, 1974, **8**, 14; *Chem. Abs.*, 1974, **81**, 13368y.
206. T. Largman, S. Sifniades and L.J. Schmehl, *Synth. Commun.*, 1979, **9**, 255.
207. H. Yanigasawa, S. Ishimara, A. Ando, T. Kanazaki, S. Miyamoto, H. Koike, Y. Iijima, K. Oizumi, Y. Matsushita and T. Hata, *J. Med. Chem.*, 1988, **31**, 428.
208. S. Sifniades, W.J. Boyle, Jr and J.F. Van Peppen, *J. Am. Chem. Soc.*, 1976, **98**, 3738.
209. J. Crosby, in A.N. Collins, G.N. Sheldrake and J. Crosby (Eds), *Chirality in Industry*, Wiley, New York, 1994, pp. 34–35.
210. M.A. Ogliaruso and J.F. Wolfe, in S. Patai and Z. Rappoport (Eds), *Synthesis of Lactones and Lactams*, Wiley, Chichester, 1993.
211. U. Kraatz, W. Hasenbrink, H. Wamhoff and F. Korte, *Chem. Ber.*, 1971, **104**, 2458.
212. Von G. Reinisch, K. Dietrich and F. Dargazanli, *J. Prakt. Chem.*, 1969, **311**, 455.
213. H.K. Reimschuessel, *J. Heterocycl. Chem.*, 1964, **1**, 193.
214. C. Lambert, B. Caillaux and H.G. Viehe, *Tetrahedron*, 1985, **41**, 3331.
215. H.K. Reimschuessel, J.P. Sibilia and J.V. Pascale, *J. Org. Chem.*, 1969, **34**, 959.
216. H.K. Reimschuessel, *Macromol. Synth.*, 1982, **8**, 37.
217. H.K. Reimschuessel, *J. Org. Chem.*, 1973, **38**, 169.
218. C.G. Overberger and H. Jabloner, *J. Am. Chem. Soc.*, 1963, **85**, 3431.
219. X.T. Phan and P.J. Shannon, *J. Org. Chem.*, 1983, **48**, 5164.
220. J.P. Celerier, M.G. Richaud and G. Lhommet, *Synthesis*, 1983, 195.
221. J.-C. Pommelet, H. Dhimane, J. Chuche, J.P. Celerier, M. Haddad and G. Lhommet, *J. Org. Chem.*, 1988, **53**, 5680.
222. M. Haddad, J.P. Celerier, G. Haviari, G. Lhommet, H. Dhimane, J.-C. Pommelet and J. Chuche, *Heterocycles*, 1990, **31**, 1251.
223. R. Lukeš, V. Dudek, O. Sedláková and J. Korán, *Collect. Czech. Chem. Commun.*, 1961, **26**, 1105; *Chem. Abs.*, 1961, **55**, 18756g.
224. M. Yamaguchi and I. Hirao, *Tetrahedron Lett.*, 1983, **24**, 1719.
225. R.E. Benson and T.L. Cairns, *J. Am. Chem. Soc.*, 1948, **70**, 2115.
226. J.P. Celerier, E. Deloisy-Marchalant, G. Lhommet and P. Maitte, *Org. Synth.*, 1989, **67**, 170.
227. O.M. Friedman, H. Sommer and E. Boger, *J. Am. Chem. Soc.*, 1960, **82**, 5202.
228. V.G. Granik, I.V. Persianova, N.P. Kostyuchenko, R.G. Glushkov and Yu.N. Sheinker, *Zh. Org. Khim.*, 1972, **8**, 181; *Chem. Abs.*, 1972, **76**, 139797c.

229. D.L. Lee, C.J. Morrow and H. Rapoport, *J. Org. Chem.*, 1974, **39**, 893.
230. P. Duhamel and M. Kotera, *J. Org. Chem.*, 1982, **47**, 1688.
231. A.K. Bose, J.L. Fahey and M.S. Manhas, *Tetrahedron*, 1974, **30**, 3.
232. Y. Yamamoto and Y. Morita, *Chem. Pharm. Bull.*, 1984, **32**, 2555.
233. P. Hullot, T. Cuvigny, M. Larchevêque and H. Normant, *Can. J. Chem.*, 1977, **55**, 266.
234. G.P. Tokmakov and I.I. Grandberg, *Izv. Timiryazevsk. S.-Kh. Akad.*, 1979, 151; *Chem. Abs.*, 1980, **92**, 94219c.
235. G.L. Bachman and M.J. Sabacky, *Eur. Pat.*, 83 322; *Chem. Abs.*, 1983, **99**, 158860.
236. E.I. Klabunovskii, D.D. Gogoladze, E.S. Levitina, E.I. Karpeiskaya, L.F. Godunova, L.N. Kaigorodova and G.O. Chivadze, *Izv. Akad. Nauk SSSR, Ser. Khim.*, 1987, 1597; *Chem. Abs.*, 1988, **108**, 167910.
237. D.D. Gogoladze, E.S. Levitina, L.F. Godunova, E.I. Karpeiskaya, E.I. Klabunovskii and G.O. Chivadze, *Izv. Akad. Nauk SSSR, Ser. Khim.*, 1988, 1371; *Chem. Abs.*, 1989, **110**, 38439.
238. J.-C. Meslin and G. Duguay, *Bull. Soc. Chim. Fr.*, 1976, 1200.
239. T. Takeshima, M. Ikeda, M. Yoyoyama, N. Fukada and M. Muraoka, *J. Chem. Soc., Perkin Trans. 1*, 1978, 692.
240. E.P. Kramarova, A.G. Shipov, O.B. Artamkina and Yu.I. Baukov, *Zh. Obshch. Khim.*, 1984, **54**, 1921; *Chem. Abs.*, 1985, **102**, 62311.
241. V.G. Granik, N.S. Kuryatov, V.P. Pakhomov, E.M. Granik, I.V. Persianova and R. G. Glushkov, *Zh. Org. Khim.*, 1972, **8**, 1521; *Chem. Abs.*, 1972, **77**, 139782q.
242. L. Ouazzani-Chadi, J.-C. Quirion, Y. Troin and J.-C. Gramain, *Tetrahedron*, 1990, **46**, 7751.
243. H.C. Wormser and H.N. Abramson, *J. Pharm. Sci.*, 1977, **66**, 1208.
244. M.J. Fisher and L.E. Overman, *J. Org. Chem.*, 1990, **55**, 1447.
245. A. Padwa, D.L. Hertzog, W.R. Nadler, M.H. Osterhout and A.T. Price, *J. Org. Chem.*, 1994, **59**, 1418.
246. R. Grigg, G. Donegan, H.Q.N. Gunaratne, D.A. Kennedy, J.F. Malone, V. Sridharan and S. Thianpatanagul, *Tetrahedron*, 1989, **45**, 1723.
247. K. Suzuki, T. Ohkuma and G. Tsuchihashi, *J. Org. Chem.*, 1988, **53**, 4160.
248. K. Suzuki, T. Ohkuma and G. Tsuchihashi, *J. Org. Chem.*, 1987, **52**, 2929.
249. H.T. Nagasawa and J.A. Elberling, *Tetrahedron Lett.*, 1966, 5393.
250. H.T. Nagasawa, J.A. Elberling, P.S. Fraser and N.S. Mizuno, *J. Med. Chem.*, 1971, **14**, 501.
251. E.E. Sugg, J.F. Griffin and P.S. Portoghese, *J. Org. Chem.*, 1985, **50**, 5032.
252. J.A. Moore and E. Mitchell, in R.C. Elderfield (Ed.), *Heterocyclic Compounds*, Vol. 9, Wiley, Chichester, 1967, p. 224.
253. A.P. Swain, D.F. Braun and S.K. Naegele, *J. Org. Chem.*, 1953, **18**, 1087.
254. Yu.V. Tanchuk amd S.I. Kotenko, *Zh. Org. Khim.*, 1981, **17**, 758; *Chem. Abs.*, 1981, **95**, 132246.
255. L. Birkofer and C.-D. Barnikel, *Chem. Ber.*, 1958, **91**, 1996.
256. A.A. Belyaev and E.V. Krasko, *Synthesis*, 1991, 417.
257. T. Fukumura, *Yukagaku*, 1974, **23**, 456; *Chem. Abs.*, 1974, **81**, 134541.
258. J.T. Edward and H. Stollar, *Can. J. Chem.*, 1963, **41**, 721.
259. C.M. Lee and W.D. Kumler, *J. Org. Chem.*, 1962, **27**, 2052.
260. W. Walter, J. Curts and H. Pawelzik, *Liebigs Ann. Chem.*, 1961, **643**, 29.
261. V.N. Sergeev, S.A. Artamkin, S.V. Pestunovich, A.I. Albanov, M.G. Voronkov and Yu.I. Baukov, *Zh. Obshch. Khim.*, 1992, **62**, 1813; *Chem. Abs.*, 1993, **118**, 234122y.
262. P. Duhamel, M. Kotera and T. Monteil, *Bull. Chem. Soc. Jpn.*, 1986, **59**, 2353.

263. P. Duhamel, M. Kotera, T. Monteil, B. Marabout and D. Davoust, *J. Org. Chem.*, 1989, **54**, 4419.
264. H. Takahata, T. Suzuki, M. Maruyama, K. Moriyama, M. Mozumi, T. Takamatsu and T. Yamazaki, *Tetrahedron*, 1988, **44**, 4777.
265. H. Takahata, K. Takahashi, E.-C. Wang and T. Yamazaki, *J. Chem. Soc., Perkin Trans. 1*, 1989, 1211.
266. F.G. Fang, M. Prato, G. Kim and S.J. Danishefsky, *Tetrahedron Lett.*, 1989, **30**, 3625.
267. R. Gompper and B. Kohl, *Tetrahedron Lett.*, 1980, **21**, 907.
268. Y. Tamaru, M. Kagotani and Z. Yoshida, *J. Org. Chem.*, 1980, **45**, 5221.
269. H. Takahata, Y. Banba, M. Mozumi and T. Yamazaki, *Heterocycles*, 1986, **24**, 947.
270. L.G. Donaruma, R.P. Scelia and S.E. Schonfeld, *J. Heterocycl. Chem.*, 1964, **1**, 48.
271. H.K. Hall, Jr, M.K. Brandt and R.M. Mason, *J. Am. Chem. Soc.*, 1958, **80**, 6420.
272. L. Panizzi, G. Dimaio, P.A. Tardella and L. D'Abbiero, *Ric. Sci.*, 1961, **1**, 312; *Chem. Abs.*, 1962, **57**, 9658i.
273. R. Lukeš and K. Smolek, *Collect. Czech. Chem. Commun.*, 1939, **11**, 506; *Chem. Abs.*, 1940, **34**, 7868[9].
274. J.W. Ralls, *J. Org. Chem.*, 1961, **26**, 66.
275. H. Lettau, A. Buege, P. Harenberg, S. Haertel, K. Jarmer, K. Kock, W. Poeppel, A. Schikora, R. Schneider, *et al.*, *Pharmazie*, 1993, **48**, 410; *Chem. Abs.*, 1993, **119**, 271086f.
276. G.L. Isele and A. Lüttringhaus, *Synthesis*, 1971, 266.
277. S. Sugasawa and T. Fujii, *Chem. Pharm. Bull.*, 1958, **6**, 587; *Chem. Abs.*, 1960, **54**, 14242f.
278. A. Koziara, S. Zawadski and A. Zwierzak, *Synthesis*, 1979, 527.
279. M.F. Shostakovskii, F.P. Sidel'kovskaya and F.L. Kolodkin, *Vysokomol. Soedin.*, 1960, **2**, 1794; *Chem. Abs.*, 1961, **55**, 26516b.
280. H. Krimm, *Chem. Ber.*, 1958, **91**, 1057.
281. H.K. Hall, Jr, *J. Am. Chem. Soc.*, 1958, **80**, 6404.
282. H.S.I. Chao, T.W. Hovatter, B.C. Johnson and S.T. Rice, *J. Polym. Sci., Part A*, 1989, **27**, 3371.
283. E. Stephanou, A. Guggisberg and M. Hesse. *Helv. Chim. Acta*, 1979, **62**, 1932.
284. N. Ogata and H. Tanaka, *Kobunshi Kagaku*, 1971, **28**, 738; *Chem. Abs.*, 1972, **76**, 72829.
285. M.S. Akhtar, W.J. Brouillette and D.V. Waterhous, *J. Org. Chem.*, 1990, **55**, 5222.
286. H. Bara, *Faserforsch. Textiltech.*, 1963, **14**, 368; *Chem. Abs.*, 1964, **60**, 12015c.
287. R.G. Glushkov, V.G. Smirnova, I.M. Zasosova and I.M. Ovcharova, *Khim. Geterotsikl. Soedin.*, 1975, 798; *Chem. Abs.*, 1975, **83**, 178254.
288. B.J. Hoek, J.P.H. Von den Hoff and J.W.M. Steeman, *US Pat.*, 3 093 635; *Chem. Abs.*, 1963, **59**, 11272d.
289. B. Weiss, K.S. Hui, M. Hui, I. Manigault, E. Toth and A. Lajtha, *Res. Commun. Psychol. Psychiatry Behav.*, 1992, **17**, 153; *Chem. Abs.*, 1994, **120**, 95413.
290. B.R. DeCosta, C. Dominguez, X.S. He, W. Williams, L. Radesca and W. Bowen, *J. Med. Chem.*, 1992, **35**, 4334.
291. D.S. Kemp, N.G. Galakatos, S. Drancinis, C. Ashton, N. Fotouhi and T. P Curran, *J. Org. Chem.*, 1986, **51**, 3320.
292. U. Sreenivasan, R.K. Mishra and R.L. Johnson, *J. Med. Chem.*, 1993, **36**, 256.
293. L. Angelucci, P. Calvisi, R. Catini, H. Cosentino, R. Cozzolino, P. Dewitt, O. Ghirardi, F. Giannessi, A. Giuliani, *et al.*, *J. Med. Chem.*, 1993, **36**, 1511.
294. C. Farina, R. Pellegata, M. Pinza and G. Pifferi, *Arch. Pharm.*, 1981, **314**, 108.

295. T. Yokohama, M. Kanesato, T. Kimura and T.M. Suzuki, *Chem. Lett.*, 1990, 693; *Chem. Abs.*, 1990, **113**, 115820.
296. H. Sato, S. Imamura, Y. Kitano, T. Kanda and T. Ashidi, *Bull Chem. Soc. Jpn.*, 1984, **57**, 2162.
297. W.A.W. Cummings and A.C. Davis, *J. Chem. Soc.*, 1964, 4591.
298. I. Saito, H. Sugiyama, N. Furukawa and T. Matsuura, *Tetrahedron Lett.*, 1981, **22**, 3265.
299. R. Pleixats, M. Figueredo, J. Marquet, M. Moreno-Manas and A. Cantos, *Tetrahedron*, 1989, **45**, 7817.
300. C. Parkanyi, H.L. Yuan, B.H.E. Stroemberg and A. Evenzahav, *J. Heterocycl. Chem.*, 1992, **29**, 749.
301. Z.A. Rogovin, E. Khait, I.L. Knunyants and Yu.A. Rymashevskaya, *J. Gen. Chem. USSR*, 1946, **17**, 1316; *Chem. Abs.*, 1948, **42**, 4939h.
302. M. Imoto, M. Sakurai and T. Kono, *J. Polym. Sci.*, 1961, **50**, 467.
303. W. Ziegenbein, A. Schäfler and R. Kaufhold, *Chem. Ber.*, 1955, **88**, 1906.
304. L.E. Wolinski and H.R. Mighton, *J. Polym. Sci.*, 1961, **49**, 217.
305. St. Landa and J. Procházka, *Chem. Listy*, 1943, **37**, 158; *Chem. Abs.*, 1951, **45**, 552c.
306. I.M. Zalesskaya, A.N. Blakitnyi, E.P. Saenko, Yu.A. Fialkov and L.M. Yagupol'skii, *Zh. Org. Khim.*, 1980, **16**, 1194; *Chem. Abs.*, 1980, **93**, 167704.
307. C.G. Overberger and J.H. Kozolwski, *Org. Prep. Proced. Int.*, 1973, **5**, 199; *Chem. Abs.*, 1973, **79**, 115424.
308. R.E. Lyle and G.G. Lyle, *J. Org. Chem.*, 1959, **24**, 1679.
309. G. Boffa, *Gazz. Chim. Ital.*, 1956, **86**, 646.
310. J.A. Davies, C.H. Hassall and I.H. Rogers, *J. Chem. Soc. C*, 1969, 1358.
311. H.E. Holmquist, H.S. Rothrock, C.W. Theobald and B.E. Englund, *J. Am. Chem. Soc.*, 1956, **78**, 5339.
312. Y. Terada and J. Okada, *Yakugaku Zasshi*, 1977, **97**, 207; *Chem. Abs.*, 1977, **87**, 39246.
313. O.E. Edwards, J.M. Paton, M.H. Benn, R.E. Mitchell, C. Watanatada and K.N. Vohra, *Can. J. Chem.*, 1971, **49**, 1648.
314. W. Flitsch, *Chem. Ber.*, 1964, **97**, 1542.
315. J. Kralicek, J. Kondelikova, V. Kubanek and H. Smrckova, *Sb. Vys. Sk. Chem.-Technol. Praze, Org. Chem. Technol.*, 1973, **C18**, 17; *Chem. Abs.*, 1974, **80**, 27495.
316. Z.I. Korshunova, E.R. Zakhs and O.F. Ginzburg, *Zh. Org. Khim.*, 1970, **6**, 504; *Chem. Abs.*, 1970, **72**, 133158.
317. E. Wenkert and B.F. Barnett, *J. Am. Chem. Soc.*, 1960, **82**, 4671.
318. T. Nishitani, H. Horikawa, T. Iwasaki, K. Matsumoto, I. Inoue and M. Miyoshi, *J. Org. Chem.*, 1982, **47**, 1706.
319. W.S. Hamama, M. Hammouda, E.M. Kandeel and E.M. Afsah, *Chin. Pharm. J.*, 1992, **44**, 25; *Chem. Abs.*, 1992, **117**, 26178.
320. P. Hullot, M.L. Cuvigny and H. Normant, *Can. J. Chem.*, 1976, **54**, 1098.
321. R.M. Freidinger, D.S. Perlow and D.F. Veber, *J. Org. Chem.*, 1982, **47**, 104.
322. F.P. Sidel'kovskaya, M.G. Zelenskaya and M.F. Shostakovskii, *Izv. Akad. Nauk SSSR, Otd. Khim. Nauk*, 1959, 901; *Chem. Abs.*, 1960, **54**, 1286d.
323. H. Zahn and L. Zürn, *Chem. Ber.*, 1958, **91**, 1359.
324. H. Zahn and L. Zürn, *Chem. Ber.*, 1961, **94**, 843.
325. Yu.D. Smirnov and A.P. Tomilov, *Zh. Org. Khim.*, 1969, **5**, 864; *Chem. Abs.*, 1969, **71**, 38299.
326. B.A. Mooney, R.H. Prager and A.D. Ward, *Aust. J. Chem.*, 1981, **34**, 2695.
327. V.G. Smirnova, N.A. Novitskaya and R.G. Glushkov, *Khim.-Farm. Zh.*, 1972, **6**, 14; *Chem. Abs.*, 1972, **77**, 109356m.

328. M.W. Majchrzak, A. Kotelko, R. Guryn, J.B. Lambert and S.M. Wharry, *Tetrahedron*, 1981, **37**, 1075.
329. F.F. Blicke and E.-P. Tsao, *J. Am. Chem. Soc.*, 1953, **75**, 3999.
330. A. Gossauer, R.-P. Hinze and H. Zilch, *Angew. Chem.*, 1977, **89**, 429.
331. R.K. Hill and R.T. Conley, *J. Am. Chem. Soc.*, 1960, **82**, 645.
332. J. Kondelikova, J. Kralicek and Z. Smrckova, *Sb. Vys. Sk. Chem.-Technol. Praze, Org. Chem. Technol.*, 1976, **C24**, 5; *Chem. Abs.*, 1977, **86**, 121816.
333. T.E. Stevens, *J. Org. Chem.*, 1969, **34**. 245.
334. L.A. Paquette, *J. Am. Chem. Soc.*, 1962, **84**, 4987; 1963, **85**, 3288.
335. W. Theilacker, K. Ebke, L. Seidl and S. Schwerin, *Angew. Chem.*, 1963, **75**, 208.
336. R.G. Glushkov, V.G. Smirnova, I.M. Zasosova, T.V. Stezhko, I.M. Ovcharova and T.F. Vlasova, *Khim. Geterotsikl. Soedin.*, 1978, 374; *Chem. Abs.*, 1978, **89**, 43306.
337. B. Kasum, R.H. Prager and C. Tsopelas, *Aust. J. Chem.*, 1990, **43**, 355.
338. K. Maruyama and Y. Kubo, *J. Org. Chem.*, 1977, **42**, 3215.
339. I. Panfil and M. Chmielewski, *Tetrahedron*, 1985, **41**, 4713.
340. H. Weidmann and E. Fauland, *Liebigs Ann. Chem.*, 1964, **679**, 192.
341. F.P. Sidel'kovskaya and A.A. Avetisyan, *Izv. Akad. Nauk SSSR, Ser. Khim.*, 1964, 2064; *Chem. Abs.*, 1965, **62**, 7715a.
342. A.A. Avetisyan, F.P. Sidel'kovskaya and R.M. Ispiryan, *Izv. Akad. Nauk SSSR, Ser. Khim.*, 1964, 1303; *Chem. Abs.*, 1964, **61**, 14621b.
343. (a) F.P. Sidel'kovskaya and A.A. Avetisyan, *Dokl. Akad. Nauk SSSR*, 1964, **157**, 632; *Chem. Abs.*, 1964, **61**, 11957e; (b) F.P. Sidel'kovskaya, A.A. Avetisyan and M.F. Shostakovskii, *Izv. Akad. Nauk SSSR, Ser. Khim.*, 1965, 702; *Chem. Abs.*, 1965, **63**, 2895d.
344. Y. Tamaru, M. Mizutani, Y. Furukawa, O. Kitao and Z. Yoshida, *Tetrahedron Lett.*, 1982, **23**, 5319.
345. P. Duhamel, M. Kotera and B. Marabout, *Tetrahedron: Asymmetry*, 1991, **2**, 203.
346. H. Bredereck, G. Simchen and B. Funke, *Chem. Ber.*, 1971, **104**, 2709.
347. M.F. Shostakovskii, F.P. Sidel'kovskaya and A.A. Avetisyan, *USSR Pat.*, 168 704; *Chem. Abs.*, 1965, **63**, 1709b.
348. I. Hermecz, L. Vasvari-Debreczy, A. Horvath, M. Balogh, J. Kokosi, C. Devos and L. Rodriguez, *J. Med. Chem.*, 1987, **30**, 1543.
349. F. Campagna, A. Carotti and G. Casini, *J. Heterocycl. Chem.*, 1990, **27**, 1973.
350. A. Leberre, C. Renault and P. Giraudeau, *Bull. Soc. Chim. Fr.*, 1971, 3245.
351. E.W. Dozorova, S.I. Grisik, I.V. Persianova, R.D. Syubaev, G. Ya. Shvarts and V. G. Granik, *Khim.-Farm. Zh.*, 1985, **19**, 154; *Chem. Abs.*, 1985, **103**, 71272w.
352. H.J. Nitschke and G. Faerber, *Chem. Ber.*, 1954, **87**, 1635.
353. U. Radics, J. Liebscher and M. Pätzel, *Synthesis*, 1992, 673.
354. M. Pätzel, J. Böhrisch and J. Liebscher, *Liebigs Ann. Chem.*, 1991, 975.
355. J.M. Grisar, G.P. Claxton, A.A. Carr amd N.L. Weich, *J. Med. Chem.*, 1973, **16**, 679.
356. J. Cen and G. Gen, *Zhongguo Yiyao Gongye Zazhi*, 1990, **21**, 218; *Chem. Abs.*, 1991, **114**, 122016c.
357. V.G. Granik and R.G. Gluschkov, *Zh. Org. Khim.*, 1971, **7**, 1146; *Chem. Abs.*, 1971, **75**, 110176t.
358. V.G. Granik and R.G. Gluschkov, *Khim.-Farm. Zh.*, 1967, 21; *Chem. Abs.*, 1968, **68**, 12942.
359. K.L. Rinehart and A.D. Patil, *US Pat.*, 4 908 445; *Chem. Abs.*, 1990, **113**, 237817m.

360. K.B. Killday, R. Longley, P.J. McCarthy, S.A. Pomponi, A.E. Wright, R.F. Neale and M.A. Sills., *J. Nat. Prod.*, 1993, **56**, 500.
361. E.A. Prill and S.M. McElvain, *J. Am. Chem. Soc.*, 1933, **55**, 1233.
362. N.J. Leonard and E. Barthel, Jr, *J. Am. Chem. Soc.*, 1949, **71**, 3098.
363. N.J. Leonard and E. Barthel, Jr, *J. Am. Chem. Soc.*, 1950, **72**, 3632.
364. P. Krogsgaard-Larsen and H. Hjeds, *Acta Chem. Scand.*, Ser. B, 1976, **30**, 884.
365. L.E. Overman and I.M. Rodriguez-Campos, *Synlett*, 1992, 995.
366. F.A. Fraser, G.R. Proctor and J. Redpath, *J. Chem. Soc., Perkin Trans. 1*, 1992, 445.
367. P. Dowd and S.-C. Choi, *Tetrahedron*, 1991, **47**, 4847.
368. R.A. Johnson, M.E. Herr, H.C. Murray and G.S. Fonken, *J. Org. Chem.*, 1968, **33**, 3187.
369. K. Bowden, I.M. Heilbron, E.R.H. Jones and B.C.L. Weedon, *J. Chem. Soc.*, 1946, 39.
370. G.M. Singer and W.A. Macintosh, *IARC Sci. Publ.*, 1984, **57**, 459; *Chem. Abs.*, 1985, **103**, 83179j.
371. Y. Imai and S. Kuwatsuka, *Nippon Noyaku Gakkaishi* (*J. Pestic. Sci.*), 1985, **11**, 245; *Chem. Abs.*, 1986, **105**, 166785w.
372. A.L. Klysheva, L.A. Golovleva and A.N. Ilyaletdinov, *Izv. Akad. Nauk Kaz. SSR, Ser. Biol.*, 1980, 29; *Chem. Abs.*, 1981, **94**, 986s.
373. L.I. Hecker and J.E. Saavedra, *Carcinogenesis*, 1980, **1**, 1017; *Chem. Abs.*, 1981, **94**, 169124v.
374. N.J. Leonard and S. Gelfand, *J. Am. Chem. Soc.*, 1955, **77**, 3269.
375. L. Brehm, P. Krogsgaard-Larsen, K. Schaumberg, J.S. Johansen, E. Falch and D.R. Curtis, *J. Med. Chem.*, 1986, **29**, 224.
376. P. Krogsgaard-Larsen, *Acta Chem. Scand.*, Ser. B, 1977, **31**, 584.
377. R. Sauter, G. Griss, W. Grell, R. Hurnaus, R. Reichl and M. Leitold, *Ger. Pat.*, 2 638 828; *Chem. Abs.*, 1978, **89**, 24281f.
378. W. Grell, R. Hurnaus, G. Griss, R. Sauter, M. Leitold and R. Reichl, *Ger. Pat.*, 2 722 416; *Chem. Abs.*, 1979, **90**, 103941v.
379. A. Yokoo and S. Morosawa, *Bull. Chem. Soc. Jpn.*, 1956, **29**, 631.
380. S. Morosawa, *Bull. Chem. Soc. Jpn.*, 1958, **31**, 418.
381. Z. Polívka, J. Mety and M. Protiva, *Collect. Czech. Chem. Commun.*, 1986, **51**, 2034.
382. M. Prost, M. Urbain and R. Charlier, *Helv. Chim. Acta*, 1969, **52**, 1134.
383. A. Yokoo and S. Morosawa, *J. Chem. Soc. Jpn.*, 1956, **77**, 599.
384. R.H. Martin, J. Pecher, J. Peeters and C. van Malder, *Bull. Soc. Chim. Belg.*, 1958, **67**, 256.
385. J. Diamond, W.F. Bruce and F.T. Tyson, *J. Org. Chem.*, 1965, **30**, 1840.
386. J. Diamond, W.F. Bruce and F.T. Tyson, *J. Med. Chem.*, 1964, **7**, 57.
387. A.F. Casy and H. Birnbaum, *J. Chem. Soc.*, 1964, 4130.
388. H. Favre, Z. Hamlet, R. Lanthier and M. Ménard, *Can. J. Chem.*, 1971, **49**, 3075.
389. J.E. Saavedra, *Org. Prep. Proceed. Int.,* 1981, **13**, 129; *Chem. Abs.*, 1981, **95**, 42847y.
390. N.M. Sharkova, N.F. Kucherova, S.L. Portnova and V.A. Zagorevskii, *Khim. Geterotsikl. Soedin.*, 1968, 131; *Chem. Abs.*, 1968, **69**, 86857u.
391. T. Moriya, T. Oki, S. Yamaguchi, S. Morosawa and A. Yokoo, *Bull. Chem. Soc. Jpn.*, 1968, **41**, 230.
392. Z.G. Finney and T.N. Riley, *J. Med. Chem.*, 1980, **23**, 895.
393. J. Deruiter, S. Andurkar, T.N. Riley, D.E. Walters and F.T. Noggle, Jr, *J. Heterocycl. Chem.*, 1992, **29**, 779.

394. J. Lee and A. Ziering, *US Pat.*, 2 802 822; *Chem. Abs.*, 1958, **52**, 2940q
395. M. Kimura and S. Morosawa, *Bull. Chem. Soc. Jpn.*, 1979, **52**, 1437.
396. A. Roglans, J. Marquet and M. Moreno-Manas, *Synth. Commun.*, 1992, **22**, 1249.
397. H. Yamamoto, H. Kawamoto, S. Morosawa and A. Yokoo, *Heterocycles*, 1978, **11**, 267.
398. J.R. Heys and S.G. Senderoff, *J. Org. Chem.*, 1989, **54**, 4702.
399. (a) R.E. Lyle and G.G. Lyle, *J. Am. Chem. Soc.*, 1954, **76**, 3536; (b) R.E. Lyle and G.G. Lyle, *US Pat.*, 2 683 145; *Chem. Abs.*, 1955, **49**, 9047c.
400. N.P. Shulaev, E.P. Badosov, Yu.F. Malina and B.V. Unkovskii, *Sb. Nauch. Tr., Ivanov. Energ. Inst.*, 1972, 100; *Chem. Abs.*, 1974, **80**, 82579j.
401. C.G. Overberger, J. Reichenthal and J.-P. Anselme, *J. Org. Chem.*, 1970, **35**, 138.
402. N.J. Leonard, R.C. Fox and M. Oki, *J. Am. Chem. Soc.*, 1954, **76**, 5708.
403. N. Finch, L. Blanchard and L.H. Werner, *J. Chem. Soc.*, 1977, **42**, 3933.
404. K. Rühlmann, H. Seefluth and H. Becker, *Chem. Ber.*, 1967, **100**, 3820.
405. P.Y. Johnson and D.J. Kerkman, *J. Org. Chem.*, 1976, **41**, 1768.
406. P.Y. Johnson, I. Jacobs and D.J. Kerkman, *J. Org. Chem.*, 1975, **40**, 2710.
407. K. Fukuyama, S. Shimizu, S. Kashino and M. Haisa, *Bull. Chem. Soc. Jpn.*, 1974, **47**, 1117.
408. F.M. Cordero, A. Goti, F. De Sarlo, A. Guarna and A. Brandi, *Tetrahedron*, 1989, **45**, 5917.
409. A. Goti, A. Brandi, F. De Sarlo and A. Guarna, *Tetrahedron*, 1992, **48**, 5283.
410. M.E. Garst, J.N. Bonfiglio and J. Marks, *J. Org. Chem.*, 1982, **47**, 1494.
411. J.Y. Laronze, S. Dridi and J. Sapi, *Synth. Commun.*, 1991, **21**, 881.
412. R.A. Johnson, M.E. Herr, H.C. Murray and L.M. Reineke, *J. Am. Chem. Soc.*, 1971, **93**, 4880.
413. V.M. Thomas and C.L. Holt, *J. Environ. Sci. Health*, 1980, **B15**, 475; *Chem. Abs.*, 1981, **94**, 1058c.
414. W.E. Pereira and F.D. Hostettler, *Environ. Sci. Technol.*, 1993, **27**, 1542; *Chem. Abs.*, 1993, **119**, 55483r.
415. R.S. Tjeerdema and D.G. Crosby, *Xenobiotica*, 1988, **18**, 831; *Chem. Abs.*, 1988, **109**, 206470f.
416. R.S. Tjeerdema and D.G. Crosby, *Aquat. Toxicol.*, 1987, **9**, 305: *Chem. Abs.*, 1987, **106**, 209086j.
417. N.J. Leonard, R.C. Fox, M. Oki and S. Chiavarelli, *J. Am. Chem. Soc.*, 1954, **76**, 630.
418. K.C. Rice and R.E. Wasylishen, *Org. Magn. Reson.*, 1976, **8**, 449; *Chem. Abs.*, 1977, **86**, 170140
419. P. Krogsgaard-Larsen, K. Thyssen and K. Schaumberg, *Acta Chem. Scand., Ser. B*, 1978, **32**, 327.
420. P. Krogsgaard-Larsen and T. Roldskov-Christiansen, *Eur. J. Med. Chem. Chim. Ther.*, 1979, **14**, 157.
421. S. Sakanoue, S. Harusawa, N. Yamazaki, R. Yoneda and T. Kurihara, *Chem. Pharm. Bull.*, 1990, **38**, 2981.
422. M.I. Iorio and M. Miraglia, *Chem. Ther.*, 1973, **8**, 85; *Chem. Abs.*, 1973, **79**, 78581v.
423. R. Hurnaus, G. Griss, R. Sauter, W. Grell, W. Kobinger and L. Pichler, *US Pat.*, 4 409 220; *Eur. Pat.*, 47 412; *Chem. Abs.*, 1982, **97**, 23820j.
424. S. Morosawa, *Bull. Chem. Soc. Jpn.*, 1960, **33**, 1113; *Chem. Abs.*, 1961, **55**, 27363f.
425. A. Yokoo and S. Morosawa, *Bull. Chem. Soc. Jpn.*, 1967, **40**, 1954; *Chem. Abs.*, 1968, **68**, 21873f.
426. S. Morosawa, *Bull. Chem. Soc. Jpn.*, 1960, **33**, 1118; *Chem. Abs.*, 1961, **55**, 27364c.
427. S. Morosawa, *Bull. Chem. Soc. Jpn.*, 1960, **33**, 1108; *Chem. Abs.*, 1961, **55**, 27363b.

428. G. Griss, M. Kleeman, W. Grell and H. Ballhause, *US Pat.*, 3 804 849; *Chem. Abs.*, 1973, **79**, 78777.
429. H. Yamamoto, M. Nakata, S. Morosawa and A. Yokoo, *Bull. Chem. Soc. Jpn.*, 1971, **44**, 153; *Chem. Abs.*, 1971, **74**, 141677.
430. P. Krogsgaard-Larsen, H. Hjeds, S.B. Christensen and L. Brehm, *Acta Chem. Scand.*, 1973, **27**, 3251.
431. P. Krogsgaard-Larsen, H. Mikkelsen, P. Jacobsen, E. Falch, D.R. Curtis, M.J. Peet and J.D. Leah, *J. Med. Chem.*, 1983, **26**, 895.
432. B.E. Maryanoff and M.C. Rebarchak, *J. Org. Chem.*, 1991, **56**, 5203.
433. J. Lévy, J.-Y. Laronze and J. Sapi, *Tetrahedron Lett.*, 1988, **29**, 3303.
434. J. Diamond and W.F. Bruce, *US Pat.*, 2 775 589; *Chem. Abs.*, 1957, **51**, 7445b.
435. A.D. Yanina and M.V. Rubstov, *Zh. Obshch. Khim.*, 1962, **32**, 3693; *Chem. Abs.*, 1963, **58**, 12512c.
436. G. Griss, W. Grell, R. Hurnaus, R. Sauter, M. Leitold, B. Eisele, W. Kaubisch and R. Reichl, *Ger. Pat.*, 2 617 101; *Chem. Abs.*, 1978, **88**, 50833m.
437. H. Yamamoto, T. Komazawa, K. Nakaue and A. Yokoo, *Heterocycles*, 1978, **11**, 275.
438. H. Yamamoto, H. Kawamoto, S. Morosawa and A. Yokoo, *Bull. Chem. Soc. Jpn.*, 1977, **50**, 453.
439. S. Hauptmann and K. Hirschberg, *J. Prakt. Chem.*, 1966, **34**, 272; *Chem. Abs.*, 1967, **66**, 55360k.
440. R. Isaksson and T. Liljefors, *J. Chem. Soc., Perkin Trans. 2*, 1983, 1351.
441. Y. Matsumura, Y. Takeshima and H. Okita, *Bull. Chem. Soc. Jpn.*, 1994, **67**, 304.
442. J.S. Clark and P.B. Hodgson, *J. Chem. Soc., Chem. Commun.*, 1994, 2701.
443. F.G. West, B.N. Naidu and R.W. Tester, *J. Org. Chem.*, 1994, **59**, 6892.
444. K.T. Potts, T. Rochanapruk, A. Padwa, S.J. Coats and L. Hadjiarapoglou, *J. Org. Chem.*, 1995, **60**, 3795.

CHAPTER 4

Azepanediones

Of the nine possible combinations in this group, only azepane-3,6-diones are not mentioned in the literature. However, azepane-2,6-diones, azepane-3,4-diones and azepane-3,5-diones are not properly authenticated. The other five members are well known.

I. AZEPANE-2,3-DIONES

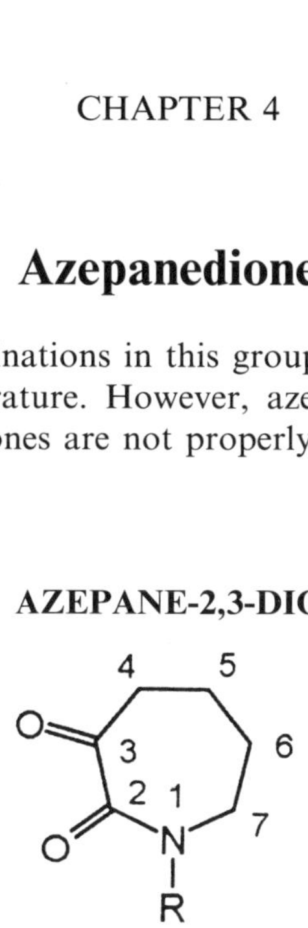

These diones are generally obtained from caprolactams. When 3,3-dihalocaprolactams (e.g. **1**, $R^1 = H$, $R^2 = Cl$) react with secondary amines such as piperidine or morpholine, enamines (**2**, R = H, X = CH_2, O) are obtained.[1]

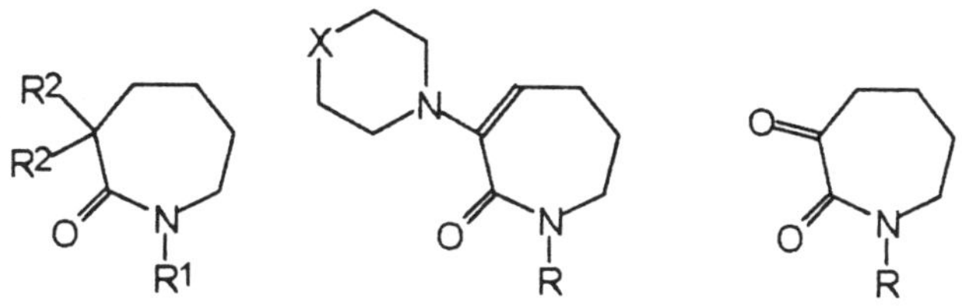

(1)   (2)   (3)

Hydrolysis of the latter with dilute acid provides the required 2,3-diones (**3**; R = H[2] and CH$_3$).[3] In the case of enamine **2** (R = H, X = CH$_2$), it was

(**4**) (**5**)

discovered[4] that passage through a column of silica converted the enamine into the diones **3** (R = H).

The only other method for obtaining azepane-2,3-diones involves a 5 + 2 ring expansion scheme in which the enolic dioxopyrrolidine **4** underwent photochemical [2 + 2] cycloaddition with 2,3-dimethylbut-2-ene yielding adduct **5**. The latter was converted into the azepanedione **6** by heating with sodium hydrogencarbonate in ethanol.[5]

(**6**) (**7**) (**8**)

The azepane-2,3-diones behave in a predictable manner. Infrared spectroscopy revealed carbonyl stretching frequencies at 1630 and 1720 cm^{-1} for the amide carbonyl group and the 3-keto group, respectively[2] in **3** (R = H). In the case of **3** (R = CH$_3$), the corresponding frequencies were 1650 and 1715 cm^{-1}.[3] Addition of arylmagnesium bromide to **3** (R = H) took place at C-3 giving **7**,[4] while sodium borohydride as expected gave **8** from **3** (R = CH$_3$),[3] whereas LiAlH$_4$ caused conversion into 3-hydroxy-N-methylazepane.[3] Hydrogen cyanide also added at C-3.[2] The dione **3** (R = H) reacted with bromine in water to yield a 4-bromo hydrate (**9**).[6] As expected, hydrazines reacted with the ketone carbonyl group (C-3);[7–10] the phenyl hydrazone of dione **3** (R = H) was isolated in both *syn* (**10**) and *anti* (**11**) forms,[8] as were the corresponding isomers from N-methyldione **3** (R = CH$_3$).[8] N-Arylimines **12** (R = H, Cl, OCH$_3$, OH) are also known.[11]

The main impetus for work in this area has come from the desire to obtain polycyclic ring systems involving azepane nuclei, principally carried out by Glushkov and co-workers in the former USSR. In some cases diones[7] were used, in others[1,6–8] the enamines[2] were the starting materials. Thus a variety of multi-ring heterocyclic compounds were constructed. These include indoloazepines (**13**, R = H, OCH$_3$),[1] azepinopyrrolopyrimidines (**14**, R = CH$_3$,

(9) (10) (11) (12)

CH$_2$Ph),[7] pyridopyrroloazepines (**15**, **16**),[10] pyridoazepines (e.g. **17**)[12] and benzofuroazepines (**18**).[13] In some cases,[13] the azepine ring is contracted during reaction. For example, treatment of the O-phenyloxime of dione **3**

(13) (14)

(15) (16)

(17) (18)

(R = H) with gaseous HCl in propan-2-ol at 0 °C gave the pyridocoumarin **19** in good yield along with **18** as a minor product.[13] The furoazepine **20**, the thiazoloazepine **21** and the benzothiazinoazepine **22** have also been reported.[6]

The reactivity at C-4 of azepan-2,3-diones has been exploited. Enamines (**2**) react with aryldiazonium chloride to give arylhydrazones (**23**, R = H, OMe), which are formally derivatives of azepane-2,3,4-triones.[1] By further reaction with various hydrazines, **23** (R = H) was converted into indoloazepines (**24**,

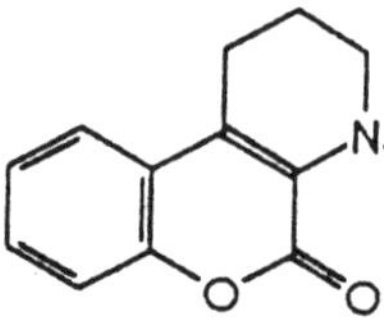

(19) 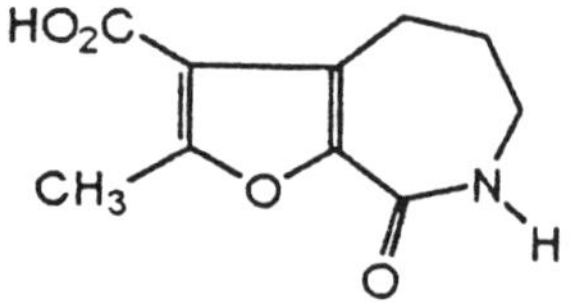 (20) 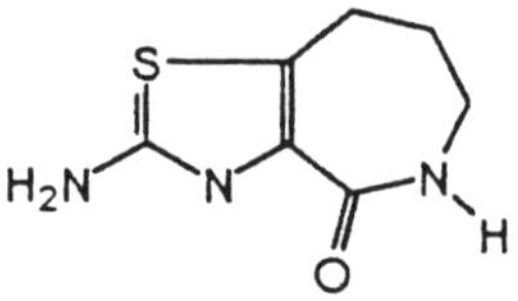 (21)

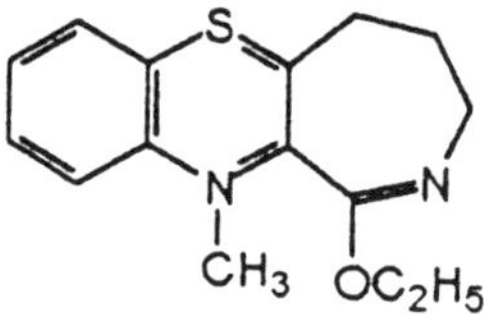

(22) 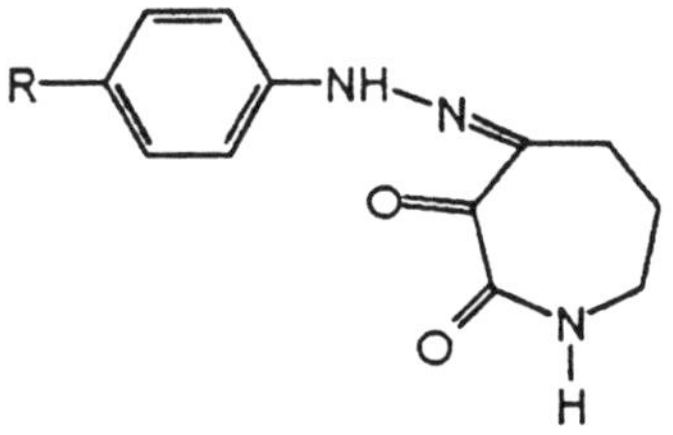 (23)

R = COCH$_3$, 2,4-di-NO$_2$C$_6$H$_3$)[14] and a triazoloazepinone (**25**).[15] Reaction of azepane-2,3-dione (**3**, R = H) with dimethylformamide acetal gave the aminomethylene compound **26**,[16] which, as is usual in such cases, underwent displacement reactions involving the $-$N(CH$_3$)$_2$ group. This led to the synthesis

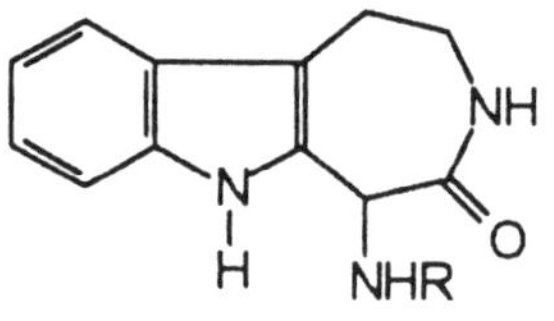

(24) 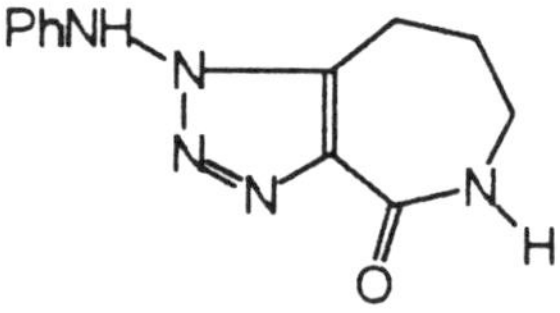 (25) 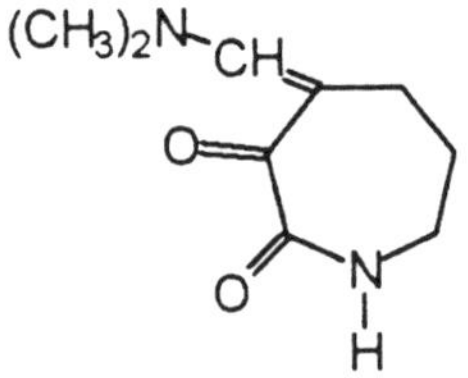 (26)

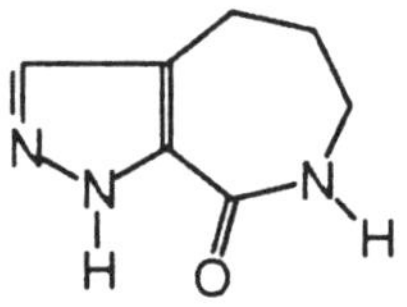

(27) 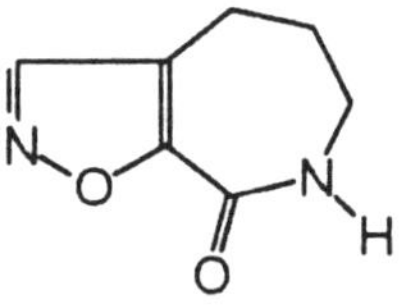 (28)

of pyrazoloazepines (e.g. **27**)[17] using hydrazines and to an isoxazoloazepine (**28**) with hydroxylamine.[18]

Table 1. Azepane-2,3-diones and related imines

Structure	M.p. (°C)/b.p. (°C/mmHg)	Derivatives and m.p. (°C)	Spectroscopic evidence	Ref.
(azepane-2,3-dione, N-H)	71–72		^{13}C NMR, IR	2, 4
(N-CH$_3$ derivative)	126–127/2		IR	3
(tetramethyl, CO$_2$C$_2$H$_5$, N-C$_2$H$_5$ derivative)	55–57		IR, ^{1}H NMR	5
((CH$_3$)$_3$N–CH= derivative)	175–176			16
(HO–CH= derivative)	144.5–145			16
(H$_2$N–C(=S)–NH–N=CH– derivative)	169–170		IR, UV, ^{1}H NMR	17

Table 1 (continued)

Structure	M.p. (°C)/b.p. (°C/mmHg)	Derivatives and m.p. (°C)	Spectroscopic evidence	Ref.
(H₂N-C(=O)-NH-N=CH- azepine-2,3-dione)	215–217		IR, UV, ¹H NMR	17
(Ph-NH-N=CH- azepine-2,3-dione)	144–147		IR, UV, ¹H NMR	17
(Ph-NH-N= azepine-2,3-dione)	205–207			1
(HON=CH- azepine-2,3-dione)	185–186		IR, UV, ¹H NMR	18
(CH₃O-C₆H₄-NH-N= azepine-2,3-dione)	198–200			1
(HO-N= azepine-2,3-dione)	201–202 206–208			8 71

Table 1 (continued)

Structure	M.p. (°C)/b.p. (°C/mmHg)	Derivatives and m.p. (°C)	Spectroscopic evidence	Ref.
(3-oxo-2-hydroxy-2-cyano azepane)	132–134	N,O-di-OAc 132–134		2
(3-(2-phenylhydrazono) azepan-2-one, enol H-bonded)	137–138		IR, UV	8
(3-(phenylhydrazono) azepan-2-one)	213		IR, UV	8
(N-CH3, 3-(2-phenylhydrazono) azepan-2-one, enol)	96–96.5		IR, UV	8
(N-CH3, 3-(phenylhydrazono) azepan-2-one)	214–214.5		IR, UV	8
(3-hydrazono azepan-2-one)	214		IR	8
(3-(thiosemicarbazono) azepan-2-one)	239–241		IR, UV	8

Table 1 (continued)

Structure	M.p. (°C)/b.p. (°C/mmHg)	Derivatives and m.p. (°C)	Spectroscopic evidence	Ref.
PhO-N=...azepanone	125		IR, MS, UV	13
4-pyridyl-HN-N=...azepanone	190–190.5			10
2-pyridyl-HN-N=...azepanone	129–131			10
2-pyridyl-HN-N=...azepanone (isomer)	181–182			10
1,3-dimethyluracil-HN-N=...azepanone	233		IR, UV	7
4-O$_2$N-C$_6$H$_4$-N=...azepanone	168–170			2
4-HO$_2$C-C$_6$H$_4$-N=...azepanone	190–190.5			2

# Azepanediones 259

Table 1 (continued)

Structure	M.p. (°C)/b.p. (°C/mmHg)	Derivatives and m.p. (°C)	Spectroscopic evidence	Ref.
	184–186		IR, ^{1}H NMR	9
	137–138		IR	15
	225–226		IR, UV	14
	205–207		IR, UV	14
	177–178		IR, UV	14
	228 (d)		IR, UV	14

II. AZEPANE-2,4-DIONES

Ring expansion of cyclohexane-1,3-diones has been the most popular method for obtaining azepane-2,4-diones. Thus the enolic diones **29** ($R^1 = R^2 = H$ or CH_3 and $R^1 = CH_3$, $R^2 = H$) or their enol ethers were oximated and subjected to Beckmann rearrangements in PPA to give the azepanediones **30**

($R^1 = R^2 = H$ or CH_3 and $R^1 = CH_3$, $R^2 = H$).[19] A more subtle example involved Beckmann reaction on the bicyclic oxime **31** which, after hydrolysis of **32**, gave the dione **33**.[20] Photolysis of azido ketone **34** also gave dione **30** ($R^1 = R^2 = CH_3$),[21] but thermal versions of this approach appear preferable.[22]

Azepane-2,4-diones have also been made by intramolecular cyclisation of amido esters (36, R = Ph, CH$_2$Ph, X = H, Cl). Deprotonation and cyclisation of 36 only proceeds in the presence of the appropriate crown ether. In this way the parent (35, R = Ph, CH$_2$Ph, X = H)[23] and chloro-substituted systems (35, R = Ph, CH$_2$Ph, X = Cl)[24] have been obtained.

The chemistry of azepane-2,4-dione is unexceptional. Infrared stretching frequencies in the region of 1720 and 1670 cm^{-1} appear typical for the C-4 and C-2 carbonyl groups, respectively.[20,22,24]. The C-4 carbonyl group, as expected, is more easily reduced (NaBH$_4$) than the C-2 carbonyl[25] and is involved in enamine formation (e.g. 37).[26] Condensations with aldehydes takes place at C-3 (e.g. 38, R = H, Cl, NO$_2$),[25] while enol ethers (39, X = OR) are easily obtained from the chloro compound (39, X = Cl).[27] Allyl enol ethers undergo Claisen rearrangement giving a range (40, R^1, R^2, R^3 = H and CH$_3$) of C-3 substituted products.[27]

Azepane-2,4-diones have been employed in conversions to fused systems such as benzofuroazepinones (41, R = H, CH$_3$) by reaction with O-phenylhydroxylamine.[28] It has been claimed[24] that bromination of chloro diones (35, R = Ph, CH$_2$Ph, X = Cl) followed by reaction with S-methylisothiourea gave halosubstituted imidazoazepines (42, R = Ph, CH$_2$Ph).

Table 2. Azepane-2,4-diones

Structure	M.p. (°C)/b.p. (°C/mmHg)	Derivatives and m.p. (°C)	Spectroscopic evidence	Ref.
	94.5		^{1}H NMR	19
	150–153		IR, ^{1}H NMR, MS	20
	98.5–99		^{1}H NMR	19
	98–100		IR, ^{1}H NMR	27
	115–120		IR, ^{1}H NMR	27
	155–165		IR, ^{1}H NMR	27

Table 2 (continued)

Structure	M.p. (°C)/b.p. (°C/mmHg)	Derivatives and m.p. (°C)	Spectroscopic evidence	Ref.
(Ph-allyl azepanedione)	128–132		IR, ^{1}H NMR	27
(CH$_3$-allyl azepanedione)	97–107		IR, ^{1}H NMR	27
(dimethylallyl azepanedione)	130–132		IR, ^{1}H NMR	27
(methallyl azepanedione)	153–155		IR, ^{1}H NMR	27
(CO$_2$CH$_3$-ethyl azepanedione)	121–124 150–200/0.1		IR, ^{1}H NMR, MS	20
(6,6-dimethyl azepanedione)	146–147 144–144.5 145.5–146.5		^{1}H NMR	26 21, 22 19

Table 2 (continued)

Structure	M.p. (°C)/b.p. (°C/mmHg)	Derivatives and m.p. (°C)	Spectroscopic evidence	Ref.
[3-CH$_3$, 6,6-di-CH$_3$ azepane-2,4-dione, NH]	124–125		IR, ^{1}H NMR	25
[3-C$_2$H$_5$, 6,6-di-CH$_3$ azepane-2,4-dione, NH]	132–133		IR, ^{1}H NMR	25
[3-(CH$_3$)$_2$CH, 6,6-di-CH$_3$ azepane-2,4-dione, NH]	150–151		IR, ^{1}H NMR	25
[3-Cl, N-Ph azepane-2,4-dione]	130/0.05	5-Bromo 50	IR, ^{1}H NMR	24
[3-Cl, N-CH$_2$Ph azepane-2,4-dione]	120/0.05	5-Bromo 140/0.05	IR, ^{1}H NMR	24
[3-(1-methylallyl), 6,6-di-CH$_3$ azepane-2,4-dione, NH]	118–120		IR, ^{1}H NMR	27

Table 2 (continued)

Structure	M.p. (°C)/b.p. (°C/mmHg)	Derivatives and m.p. (°C)	Spectroscopic evidence	Ref.
(3,3-diallyl-7,7-dimethyl azepane-2,4-dione)	115–116		IR, ^{1}H NMR	27
(3,3-bis(cinnamyl) azepane-2,4-dione)	155–157		IR, ^{1}H NMR	27
(3,3-bis(but-2-enyl) azepane-2,4-dione)	68.5–70		IR, ^{1}H NMR	27
(3,3-bis(methallyl) azepane-2,4-dione)	77–80		IR, ^{1}H NMR	27
(4-phenylhydrazono azepane-2,4-dione)	185–188		IR, UV	14
(3-(4-chlorobenzyl)-7,7-dimethyl azepane-2,4-dione)	189–190			25

Table 2 (continued)

Structure	M.p. (°C)/b.p. (°C/mmHg)	Derivatives and m.p. (°C)	Spectroscopic evidence	Ref.
	217–218			25
	193–194			25
	187–188			25

III. AZEPANE-2,5-DIONES

Several methods are recorded for production of azepane-2,5-diones. However, the most popular, developed by Kanaoka and co-workers,[29–35] involves photolysis of N-substituted succinimides. Others have contributed to our knowledge of this useful methodology.[36–40]

It is widely accepted that when succinimides are irradiated (~240 nm), hydrogen atom transfer takes place (**43a** → **43b**) so that the reaction proceeds via an azetidine (**43c**), leading to azepane-2,5-diones substituted as indicated in **44**.

Various alkyl-substituted succinimides were initially used,[29] later extended to alkenyl-substituted succinimides[34] and also those carrying oxygen or sulphur atoms in the side-chain[32] and nitrogen protected by acyl groups[33,35] (see Table 3). It is unfortunate that some products were wrongly formulated[33] and later republished[35] correctly without further comment. In the case of **44** ($R^1 = R^2 = CH_3$, $R^3 = R^4 = H$) obtained from **43** ($R^1 = R^2 = CH_3$, $R^3 = R^4 = H$), an X-ray analysis[37] has settled beyond doubt the regiochemistry of the photolysis. Unsymmetrically ring-substituted succinimides (e.g. **45**) gave mixtures of regioisomeric products **46** and **47**[31] as one would expect, given the above mechanism of reaction.

The photochemical conversion of *N*-substituted succinimides is tolerant of a variety (e.g. Ph, OH, CN, $CO_2C_2H_5$) of attachments on Cγ in **43**,[38] but absence of a hydrogen atom at Cγ (**43**) causes the reaction to proceed in other directions, leading to useful syntheses of bi- and tricyclic systems.[32,33,35,39,40]

Beckmann rearrangment has played a part in obtaining certain azepane-2,5-diones. Thus the oximes **48** and **49** have been expanded to azepane-2,5-dione[41] and 5-hydroxycaprolactam,[42,43] respectively. In the latter case (Chapter 3), oxidation gave the dione **44** ($R^1 = R^2 = R^3 = R^4 = H$).

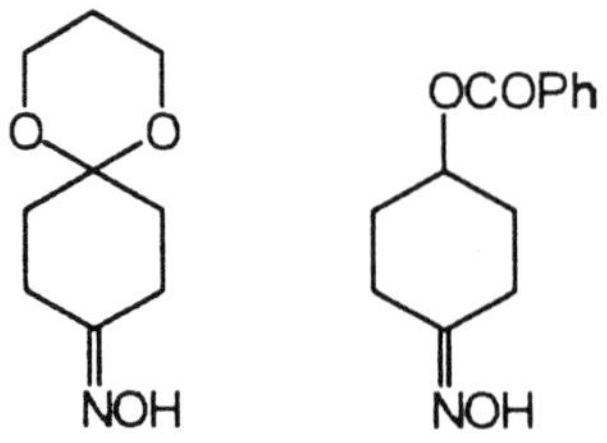

(48)   (49)

Schmidt reaction on benzoquinones gives 2,5-dihydro[1*H*]azepine-2,5-diones (Chapter 10), which can be hydrogenated catalytically (5% Pd/C) to yield azepane-2,5-diones,[44] e.g. **50 → 51**.

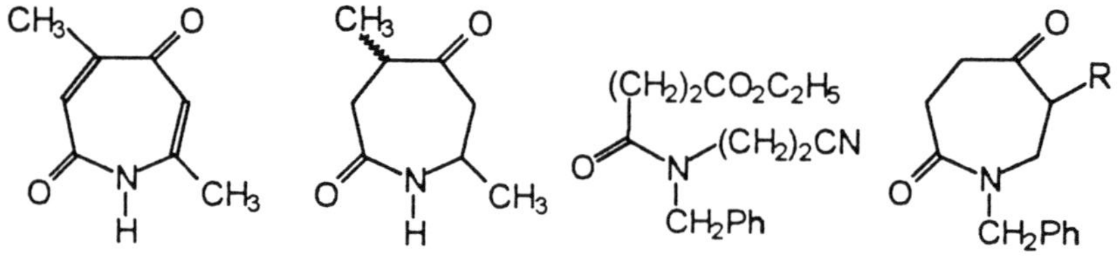

(50)   (51)   (52)   (53)

Intramolecular Dieckmann-type cyclisation of the cyano ester **52** is said to take place using KO^{t}Bu in toluene; the product (**53**, R = CN) appears to be enolic (^{1}H NMR), but on vigorous acid hydrolysis was claimed to yield an *N*-benzyldione (**53**, R = H).[24] This matter seems worthy of further examination.

Azepane-2,5-diones have not been much further studied.

Table 3. Azepane-2,5-diones

Structure	M.p. (°C)/b.p. (°C/mmHg)	Derivatives and m.p. (°C)	Spectroscopic evidence	Ref.
(azepane-2,5-dione)	Oil 139–140 140–141 137–140 138–138.5	2,4-DNPH 234 4,4-Ethylene-dithiane 165–166	IR IR, UV, ^{1}H NMR IR, ^{1}H NMR	41 29, 30 42 38 43
(3-CH$_3$)	120–121 117–120		IR, UV, ^{1}H NMR	29 38
(7-CH$_3$)	139.5–140.5		IR, UV, ^{1}H NMR	29
(3-C$_2$H$_5$)	84–86 85–87		 IR, UV, ^{1}H NMR	38 29
(3-^{n}Bu)	57–60		IR, ^{1}H NMR	38
(3,7-(CH$_3$)$_2$)	109–110		IR, UV, ^{1}H NMR	29

Table 3 (continued)

Structure	M.p. (°C)/b.p. (°C/mmHg)	Derivatives and m.p. (°C)	Spectroscopic evidence	Ref.
7,7-dimethyl-2,5-dioxoazepane (NH)	175–176		IR, UV, ^{1}H NMR X-ray	29 37
3,7-dimethyl-2,5-dioxoazepane (NH)	166–168		IR, ^{1}H NMR	44
6-phenyl-2,5-dioxoazepane (NH)	154–156		IR, MS, ^{1}H NMR	38
6-hydroxy-2,5-dioxoazepane (NH)	159–163		IR, MS, ^{1}H NMR	38
N-benzyl-2,5-dioxoazepane	57		IR, ^{1}H NMR	24
6-(pyrrolidine-1-carbonyl)-2,5-dioxoazepane (NH)	136–140		IR, MS, ^{1}H, ^{13}C NMR	35

Table 3 (continued)

Structure	M.p. (°C)/b.p. (°C/mmHg)	Derivatives and m.p. (°C)	Spectroscopic evidence	Ref.
(azepanedione with pyrrolidinone substituent)	154–156		IR, MS, ^{1}H, ^{13}C NMR	33, 35
(azepanedione with –CH$_2$N(CH$_3$)CO$_2$C$_2$H$_5$)	78–81		IR, MS, ^{1}H, ^{13}C NMR	33, 35
(azepanedione with –N(CH$_3$)COCH$_3$)	115.5–117.5		IR, MS, ^{1}H, ^{13}C NMR	33, 35
(azepanedione with –CH$_2$N(CH$_3$)COCH$_3$)	115.5–117		IR, MS, ^{1}H, ^{13}C NMR	33, 35
(azepanedione with –N(CH$_3$)$_2$)			IR, MS, ^{1}H, ^{13}C NMR	40
(azepanedione with morpholino substituent)	176–178		^{1}H, ^{13}C NMR	39

Table 3 (continued)

Structure	M.p. (°C)/b.p. (°C/mmHg)	Derivatives and m.p. (°C)	Spectroscopic evidence	Ref.
	121–124		IR, MS, ^{1}H NMR	34
	118–120		IR, MS, ^{1}H NMR	34
	158–161		IR, MS, ^{1}H NMR	34
	169–171		IR, MS, ^{1}H NMR	34
	145–150		IR, MS, ^{1}H NMR	34
	142–144		IR, MS, ^{1}H NMR	34

Table 3 (continued)

Structure	M.p. (°C)/b.p. (°C/mmHg)	Derivatives and m.p. (°C)	Spectroscopic evidence	Ref.
azepane-2,5-dione with -(CH₂)₂-morpholine			IR, MS, ^{1}H, ^{13}C NMR	40
azepane-2,5-dione with -SCH₃	122–123			32
azepane-2,5-dione with -CH₂SCH₃	99–100			32
azepane-2,5-dione with -CH₂SCH₃ and -CO₂CH₃	156–157			32
azepane-2,5-dione with -CH₃, -CH₃, -CON(pyrrolidine)	203–204.5		IR, MS, ^{1}H, ^{13}C NMR	35
azepane-2,5-dione with -CH₃, -CH₂OH	Oil	O-Acetate 111–112.5 Di-Me acetal 153–154	IR, ^{1}H NMR	36

Table 3 (continued)

Structure	M.p. (°C)/b.p. (°C/mmHg)	Derivatives and m.p. (°C)	Spectroscopic evidence	Ref.
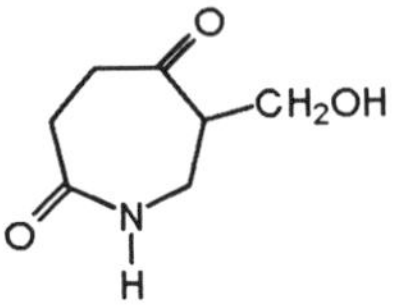 (azepanedione with CH₂OH, NH)		Di-Et acetal 132.5–133.5 Di-Me acetal 153–154		36
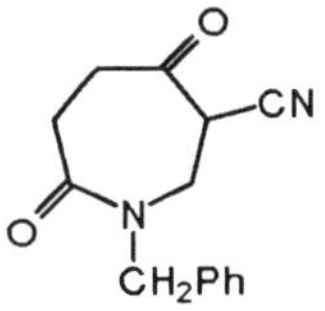 (azepanedione with CN, N-CH₂Ph)	99		IR, ^{1}H NMR	24
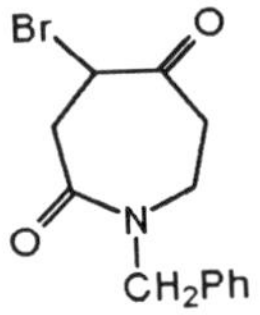 (azepanedione with Br, N-CH₂Ph)	70		IR, ^{1}H NMR,	24

IV. AZEPANE-2,6-DIONES

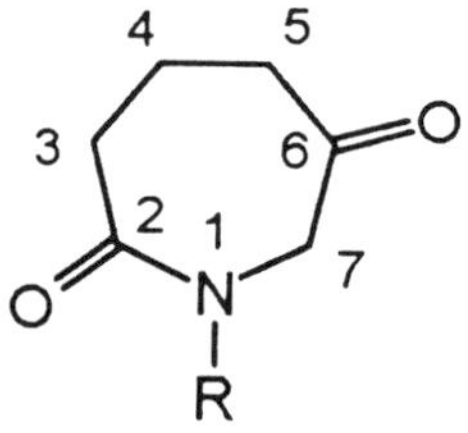

Surprisingly, there is only one reference to a member of this series in the literature.[45] It was reported[45] that refluxing the diester **54** with trimethylsilyl iodide in

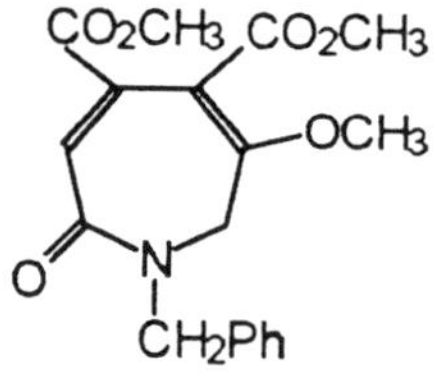

(54) 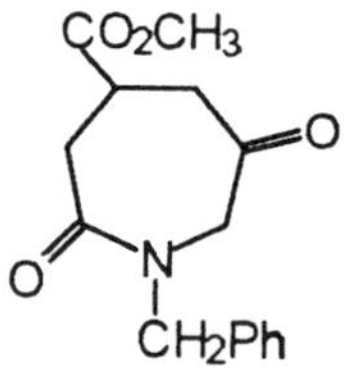   (55)

chloroform for 2.5 days gave a 25% yield of the dione **55**, the structure of which seemed reasonably well supported by spectroscopic, analytical and other data. Unfortunately, this reaction could not be repeated and, in any case, it is not entirely clear how reduction could have occurred. One has to conclude that azepane-2,6-diones remain an interesting synthetic challenge.

V. AZEPANE-2,7-DIONES

Azepane-2,7-diones (**56**), otherwise known as adipimides, have been well studied. The most useful methods for their preparation involve either cyclisation of adipic acid derivatives or oxidation of caprolactams (Chapter 3, Section I).

(**56**)

Adipimide (**56**, R = H) was obtained by heating adipamide at 250–260 °C,[46] whilst *N*-methyladipimide (*N*-methylazepane-2,7-dione) (**56**, R = CH$_3$) arose from *N*-methyladipamic acid (**57**) either by heating at 250–260 °C[47] or by treatment with thionyl chloride at 0–4 °C.[48] 3,3-Disubstituted azepane-2,7-diones

(**57**)　(**58**)　(**59**)

have been made by cyclisation of the appropriate nitrile esters or dinitriles (**58**, R^3 = CO$_2$C$_2$H$_5$ or CN) by heating with acetic and sulphuric acids[49] (see Table 4). The parent system (**56**, R = H) was obtained from the amido ester **59** by heating with potassium methoxide in toluene, but the yield was only 35%.[50] This tendency of adipic acid derivatives was unexpectedly noted in the case of the cyano acid **60**, which gave the imide **61** when heated with polyphosphoric acid at 90 °C instead of cyclising to a benzosuberone.[51] Adipic anhydrides have been claimed as precursors for *N*-substituted azepane-2,7-diones by heating

(60) (61)

with aliphatic primary amines.[52] Adipoyl chloride has also been a source of azepane-2,7-diones by reaction with an aliphatic amine in the presence of triethylamine[53] or with the iminophosphorane $PhCH_2N=PPh_3$. In the latter case,[54] the yield was a mere 2% of **56** (R = CH_2Ph), but the principal product

(62) (63) (64)

(65) (66)

(31%) was the chloro compound **62** (R = CH_2Ph), from which, presumably, more of the azepanedione **56** (R = CH_2Ph) could be obtained. Other analogues of **62** (R = Ph and $CH_2CO_2C_2H_5$) were obtained similarly in acceptable yields[54] (Chapter 6), so it seems possible that this approach could be developed into a method for obtaining *N*-substituted azepane-2,7-diones.

Turning to the oxidation of caprolactams, several oxidant systems have been used. First, peracetic acid at −10 °C (in the presence of manganic acetylacetonate) was shown to give an 84% yield of adipimide;[55] more recently,[56] the combination of ruthenium tetraoxide and sodium periodate was demonstrated to be useful in some cases. For example, *N*-isopropylcaprolactam gave 89% of

N-isopropylazepane-2,7-dione (**56**, R = ^{i}Pr).[56] It should be noted that the otherwise promising anodic oxidation of cyclic amides (Chapter 3, Section I), when applied to N-substituted caprolactams, gave several products none of which was an azepane-2,7-dione.[57] Oxidation of 3-N-phthaloylcaprolactam **63** (X = Y = H) to **63** (X + Y = O) has been brought about by potassium permanganate in aqueous acetic acid (anhydrous acetic acid was not effective)[58] and also by the agency of benzeneseleninic anhydride [(PhSeO$_2$)$_2$O].[59] Hydroboration of the tetrahydroazepinone **64** provided a product which, when treated with chromic oxide, led to the azepane-2,7-dione **65**.[60]

There is an interesting report[61] of a photochemically induced ring expansion which converted 2-nitrocyclohexanone into the N-hydroxyimide **56** (R = OH) in moderate yield. The report[62] concerning ring expansion of the dioxime **66** with bromine in liquid SO$_2$ is, however, open to doubt. The initial product claimed was a monooxime of adipimide (**67**) having a melting point of

(**67**) (**68**)

155 °C; it may or may not be so. However, the product (m.p. 166.5 °C) of reacting **67** with thionyl chloride cannot be adipimide (**56**, R = H) as stated, because the latter substance has a well authenticated melting point of 98 °C (see Table 4).

A recipe for converting azepane-2,7-diones into the monothio derivative **68** has been published.[63] The dione was first converted into the O-ethyl lactim ether (via silver salt and ethyl iodide), which was then treated with dry H$_2$S; a yield of 51% was thus obtained.

Table 4. Azepane-2,7-diones

Structure	M.p. (°C)/b.p. (°C/mmHg)	Derivatives and m.p. (°C)	Spectroscopic evidence/data	Ref.
(N-H)	98 166.5 (?) 93–94	Oxime 155	IR	46, 72 62 50, 55, 73
(N-CH₃)	119/18 96–100/6		UV, IR, ^{1}H NMR, MS	47 48
(N-C₂H₅)	103/3		IR, ^{1}H NMR	56
(N-CH(CH₃)₂)	84/2		IR, MS, ^{1}H NMR	56
(N-CH₂Ph)	76		IR, MS, ^{1}H NMR	54
(N-OH)	169–171		IR	61
(N-(CH₂)₂OCH₂Ph)			IR	53

Table 4 (continued)

Structure	M.p. (°C)/b.p. (°C/mmHg)	Derivatives and m.p. (°C)	Spectroscopic evidence/data	Ref.
(phthalimido-azepanedione)	265–270 (d)		$[\alpha]_D^{21}$ −31.3	59
(phthalimido-azepanedione, enantiomer)	265–268		$[\alpha]_D^{21}$ +30.2	59
(C$_2$H$_5$)$_2$N(CH$_2$)$_2$–, Ph-substituted azepanedione	109–110			49
Ph, C$_2$H$_5$-substituted azepanedione	198–204/0.05			49
2-pyridyl, Ph-substituted azepanedione	146–147			49
(CH$_3$)$_2$N(CH$_2$)$_2$–, Ph-substituted azepanedione	180/0.05		IR, MS, ^{1}H NMR	51

Table 4 (continued)

Structure	M.p. (°C)/b.p. (°C/mmHg)	Derivatives and m.p. (°C)	Spectroscopic evidence/data	Ref.
	106–107		IR, MS, ^{1}H NMR	60
	84/0.05		n_D^{20} 1.4779	52
	91/0.05		n_D^{20} 1.4795	52
	55/0.025		n_D^{20} 1.4789	52
	167/760		n_D^{20} 1.3620, d_4^{20} 1.8047	74

VI. AZEPANE-3,4-DIONES

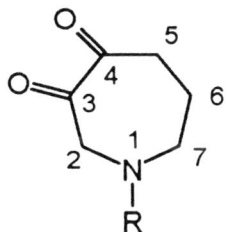

The only mention of azepane-3,4-diones (**69**) in the literature indicated that they were inferred constituents of mixtures obtained when azepan-4-ones were

(**69**)   (**70**)

reacted with selenium dioxide (Chapter 3, Section III).[64,65] The principal products were azepan-3,4-diones (not isolated; Section VIII), which reacted with aminoacetamidine to give 2-amino-6,7,8,9-tetrahydro-5H-pyrazino[2,3-d]azepines (**70**). The latter were adequately authenticated and were accompanied by isomeric by-products which in one case (**69**, R = $CO_2C_2H_5$) appeared to arise from the 3,4-dione.[66]

VII. AZEPANE-3,5-DIONES

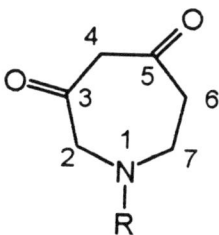

It has been claimed that the dione **71** (R = $COSC_2H_5$) could be detected when certain microorganisms transformed the 4-ketoazepane Ordram after application to rice paddies.[67] No data in support of this contention was offered

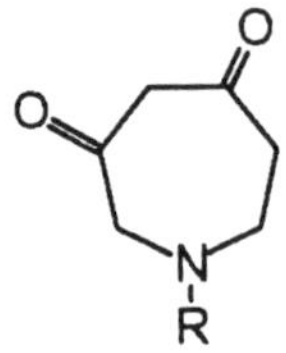

(71)

and, perhaps surprisingly, no other references to this ring system have been found.

VIII. AZEPANE-3,6-DIONES

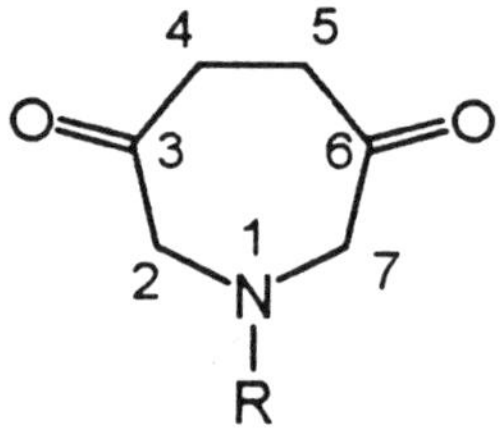

There have been no compounds based on the azepane-3,6-dione structure reported in the literature.

IX. AZEPANE-4,5-DIONES

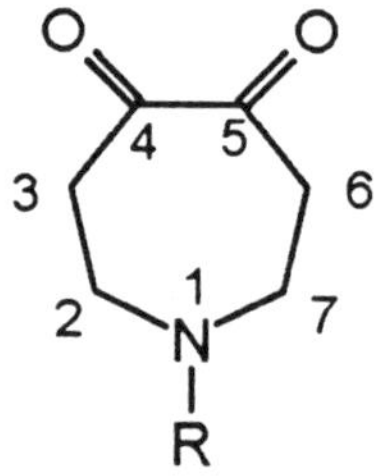

As recounted in Section VI, oxidation of a series of azepan-4-ones with SeO_2 in aqueous dioxane gave N-substituted azepan-4,5-diones (e.g. **72**, R = COC_6H_4Cl,

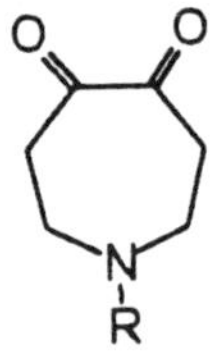

(72)

Azepanediones

CH₃CO, CO₂C₂H₅), although they were not purified and contained some quantities of the isomeric 3,4-diones.[64,65] It is now established that the addition of water to the dioxane was unnecessary.[66] Authentic diones (**74**) in this series were isolated from acyloins (**73**, R = ^{t}Bu, CH₃) by oxidation with lead tetra acetate;[68,69] **74** (R = ^{t}Bu)[68] was fully characterised, but **74** (R = CH₃) was less

(73) (74) (75)

(76)

well described. It was pointed out that the corresponding protected acyloins (**75**) also gave diones after treatment with bromine in carbon tetrachloride.[69] There seems no reason why other azaacyloins (Chapter 3, Section III) should not be suitable precursors for azepane-4,5-diones, but this area has not been of interest.

Finally, there has been a report[70] that a polyketone (**76**) underwent a Mannich reaction yielding the hydrochloride of the azepane-4,5-dione (**77**). It is possible that this claim is correct but it is based on scanty information and is worthy of more detailed study.

(77)

Table 5. Azepane-4,5-diones

Structure	M.p. (°C)/b.p. (°C/mmHg)	Derivatives and m.p. (°C)	Spectroscopic evidence	Ref.
(3,3,6,6-tetramethyl-1-methyl-azepane-4,5-dione)			^{1}H NMR	68, 69
(3,3,6,6-tetramethyl-1-tBu-azepane-4,5-dione)	73–74/0.1		IR, ^{1}H NMR, UV	68
(bis-indanedione substituted N-methyl azepanedione)		HCl >300	IR	70
(N-CO$_2$C$_2$H$_5$ azepane-4,5-dione)	Not purified			65
(N-COCH$_3$ azepane-4,5-dione)	Not purified			65

REFERENCES

1. R.G. Glushkov, V.A. Volskova, V.G. Smirnova and O.Yu. Magidson, *Dokl. Akad. Nauk SSSR*, 1969, **187**, 327; *Chem. Abs.*, 1969, **71**, 112753.

2. V.G. Smirnova, N.A. Novitskaya and R.G. Glushkov, *Khim.-Farm. Zh.*, 1972, **6**, 14; *Chem. Abs.*, 1972, **77**, 109356m.
3. R.G. Glushkov, V.G. Smirnova, K.A. Zaitseva, N.A. Novitskaya, M.D. Mashkovskii and G.N. Pershin, *Khim.-Farm. Zh.*, 1974, **8**, 14; *Chem. Abs.*, 1974, **81**, 13368.
4. W.J. Brouillette and H.M. Einspahr, *J. Org. Chem.*, 1984, **49**, 5113.
5. S.T. Reid and D. De Silva, *Tetrahedron Lett.*, 1983, **24**, 1949.
6. R.G. Glushkov, V.G. Smirnova, I.M. Zasosova, T.V. Stezhko, I.M. Ovcharova and T.F. Vlasova, *Khim. Geterotsikl. Soedin.*, 1978, 374; *Chem. Abs.*, 1978, **89**, 43306.
7. R.G. Glushkov, I.M. Zasosova and I.M. Ovcharova, *Khim. Geterotsikl. Soedin.*, 1977, 1398; *Chem. Abs.*, 1978, **88**, 50781.
8. R.G. Glushkov, V.G. Smirnova, I.M. Zasosova and I.M. Ovcharova, *Khim. Geterotsikl. Soedin.*, 1975, 798; *Chem. Abs.*, 1975, **83**, 178254.
9. R.G. Glushkov, I.M. Zasosova, I.M. Ovcharova, E.A. Rudzit, A.S. Saratikov, N. S. Livshits and N.P. Kostyuchenko, *Khim.-Farm. Zh.*, 1978, **12**, 48; *Chem. Abs.*, 1978, **89**, 43178.
10. L.N. Yakhontov, R.G. Glushkov, E.V. Pronina, and V.G. Smirnova, *Dokl. Akad. Nauk SSSR*, 1973, **212**, 389; *Chem. Abs.*, 1973, **79**, 137004.
11. V.G. Smirnova and R.G. Glushkov, *Khim.-Farm. Zh.*, 1973, **7**, 3; *Chem. Abs.*, 1974, **80**, 47816n.
12. E.W. Dozorova, S.I. Grisik, I.V. Persianova, R.D. Syubaev, G.Ya. Shvarts and V.G. Granik, *Khim.-Farm. Zh.*, 1985, **19**, 154; *Chem. Abs.*, 1985, **103**, 71272w.
13. R.G. Glushkov, I.M. Zasosova, I.M. Ovcharova, N.P. Solove'va, O.S. Anisimova and Yu.N. Sheinker, *Khim. Geterotsikl. Soedin.*, 1978, 1504; *Chem. Abs.*, 1979, **90**, 87318.
14. R.G. Glushkov, I.M. Zasosova, I.M. Ovcharova, N.P. Solov'eva and Yu.N. Sheinker, *Khim. Geterotsikl. Soedin.*, 1979, 954; *Chem. Abs.*, 1979, **91**, 193210.
15. R.G. Glushkov, I.M. Zasosova and I.M. Ovcharova, *Khim. Geterotsikl. Soedin.*, 1978, 1429; *Chem. Abs.*, 1979, **90**, 38848.
16. R.G. Glushkov, O. Ya. Belyaeva, V.G. Granik, M.K. Polievktov, A.B. Grigor'ev, V.E. Serokhvostova and T.F. Vlasova, *Khim. Geterotsikl. Soedin.*, 1976, 1640; *Chem. Abs.*, 1977, **86**, 121287f.
17. R.G. Glushkov, T.F. Stezhko, T.F. Vlasova and O.S. Anisimova, *Khim. Geterotsikl. Soedin.*, 1978, 1248; *Chem. Abs.*, 1979, **90**, 87346.
18. R.G. Glushkov and T.F. Stezhko, *Khim. Geterotsikl. Soedin.*, 1978, 1252; *Chem. Abs.*, 1979, **90**, 22784.
19. Y. Tamura, Y. Kita and M. Terashima, *Chem. Pharm. Bull.*, 1971, **19**, 529.
20. Y. Tamura, Y. Kita and J. Uraoka, *Chem. Pharm. Bull.*, 1972, **20**, 876.
21. S. Sato, *Bull. Chem. Soc. Jpn.*, 1968, **41**, 2524.
22. Y. Tamura, Y. Yoshimura and Y. Kita, *Chem. Pharm. Bull.*, 1972, **20**, 871.
23. G.R. Proctor and M. Waly, unpublished data.
24. M. Waly, *J. Prakt. Chem.*, 1994, **336**, 86.
25. T. Duong, R.H. Prager, A.D. Ward and D.I. B. Kerr, *Aust.J. Chem.*, 1976, **29**, 2651.
26. T. Duong, R.M. Prager, J.M. Tippett, A.D. Ward and D.I. B. Kerr, *Aust. J. Chem.*, 1976, **29**, 2667.
27. B.A. Mooney, R.H. Prager and A.D. Ward, *Aust. J. Chem.*, 1980, **33**, 2717.
28. N.F. Kucherova, L.A. Aksanova, L.M. Sharkova and V.A. Zagorevskii, *Khim. Geterotsikl. Soedin.*, 1973, 149; *Chem. Abs.*, 1973, **78**, 136128.
29. Y. Kanaoka and Y. Hatanaka, *J. Org. Chem.*, 1976, **41**, 400.
30. Y. Kanaoka, Y. Hatanaka and H. Okajima, *J. Photochem.*, 1985, **28**, 569.
31. Y. Kanaoka, H. Okajima and Y. Hatanaka, *Heterocycles*, 1977, **8**, 339.

32. H. Nakai, Y. Sato, T. Mizoguchi, M. Yamazaki and Y. Kanaoka, *Heterocycles*, 1977, **8**, 345.
33. M. Machida, S. Oyadomari, H. Takechi, K. Ohno and Y. Kanaoka, *Heterocycles*, 1982, **19**, 2057.
34. M. Machida, K. Oda and Y. Kanaoka, *Chem. Pharm. Bull.*, 1984, **32**, 950.
35. H. Takechi, S. Tateuchi, M. Machida, Y. Nishibata, K. Aoe, Y. Sato and Y. Kanaoka., *Chem. Pharm. Bull.*, 1986, **34**, 3142.
36. K. Maruyama and Y. Kubo, *J. Org. Chem.*, 1977, **42**, 3215.
37. T.Y. Fu, J.R. Scheffer and J. Trotter, *Can. J. Chem.*, 1994, **72**, 1952.
38. B.A. Mooney, R.H. Prager and A.D. Ward, *Aust.J. Chem.*, 1981, **34**, 2695.
39. L.R. B. Bryant and J.D. Coyle, *J. Chem. Res. (S)*, 1982, 164.
40. J.D. Coyle and L.R.B. Bryant, *J. Chem. Soc., Perkin Trans 1*, 1983, 2857.
41. A.H. Rees, *J. Chem. Soc.*, 1962, 3097.
42. Z. Majer, M. Kajtar, M. Tichy and K. Blaha, *Collect. Czech. Chem. Commun.*, 1982, **47**, 950.
43. R.C. Hider and D.I. John, *J. Chem. Soc., Perkin Trans. 1*, 1972, 1825.
44. E.J. Moriconi and I.A. Maniscalco, *J. Org. Chem.*, 1972, **37**, 208.
45. G.W. Heinicke, A.M. Morella, J. Orban, R.H. Prager and A.D. Ward, *Aust. J. Chem.*, 1985, **38**, 1847.
46. H.K. Hall, Jr, and A.K. Schneider, *J. Am. Chem. Soc.*, 1958, **80**, 6409.
47. W. Flitsch, *Chem. Ber.*, 1964, **79**, 1542.
48. R. Shapiro and S. Nesnow, *J. Org. Chem.*, 1969, **34**, 1695.
49. E. Tagmann, E. Sury and K. Hoffmann, *Helv. Chim. Acta*, 1952, **35**, 1235, 1541.
50. I.V. Micovic, M.D. Ivanovic and D.M. Piatak, *J. Serb. Chem. Soc.*, 1988, **53**, 419; *Chem. Abs.*, 1989, **111**, 232056.
51. M.T. Omar, G.R. Proctor and D.I. C. Scopes, *J. Chem. Res. (S)*, 1988, 383.
52. E. Agouri, *Br. Pat.*, 1 094 273; *Chem. Abs.*, 1968, **69**, 10106c.
53. V.K. Antonov, V.S. Morgulyan and M.M. Shemyakin, *Zh. Obshch. Khim.*, 1967, **37**, 1597; *Chem. Abs.*, 1968, **68**, 12434.
54. T. Aubert, M. Farnier and R. Guilard, *Synthesis*, 1990, 149.
55. (a) A.R. Doumaux, Jr, J.E. McKeon and D.J. Trecker, *J. Am. Chem. Soc.*, 1969, **91**, 3992; (b) A.R. Doumaux, Jr, and D.J. Trecker, *J. Org. Chem.*, 1970, **35**, 2121.
56. S. Yoshifuji, Y. Arakawa and Y. Nitta, *Chem. Pharm. Bull.*, 1987, **35**, 357.
57. M. Okita, T. Wakamatsu and Y. Ban, *J. Chem. Soc., Chem. Commun.*, 1979, 749.
58. A.A. Belyaev and E.V. Krasko, *Izv. Akad. Nauk Ser. Khim.*, 1992, 1692; *Chem. Abs.*, 1993, **119**, 9111.
59. K. Eger, M. Jalalian, E.J. Verspohl and N.-P. Lupke, *Arzneim.-Forsch./Drug Res.*, 1990, **40**, 1073.
60. M.R. Hatswell, R.H. Prager and A.D. Ward, *Aust.J. Chem.*, 1993, **46**, 135.
61. S.T. Reid and J.N. Tucker, *J. Chem. Soc., Chem. Commun.*, 1971, 1609.
62. N. Tokura, R.Tada and K. Yokoyama, *Bull. Chem. Soc. Jpn.*, 1961, **34**, 1812.
63. A. Gossauer, R.-P. Hinze and H. Zilch, *Angew. Chem., Int. Ed. Engl.*, 1977, **16**, 418.
64. R. Hurnaus, G. Griss, R. Sauter, W. Grell, W. Kobinger and L. Pichler, *US Pat.*, 4 409 220; *Chem. Abs.*, 1982, **97**, 23820j.
65. R. Hurnaus, G. Griss, R. Sauter, W. Grell, W. Kobinger and L. Pichler, *Eur. Pat.*, 47412.A1; *Chem. Abs.*, 1982, **97**, 23820j.
66. R. Hurnaus, personal communication.
67. A.L. Klysheva, L.A. Golovleva and A.N. Ilyaletdinov, *Izv. Akad. Nauk Kaz. SSR, Ser. Biol.*, 1980, 19; *Chem. Abs.*, 1981, **94**, 9865.
68. P.Y. Johnson, I. Jacobs and D.J. Kerkman, *J. Org. Chem.*, 1975, **40**, 2710.
69. R. Isaksson and T. Liljefors, *J. Chem. Soc., Perkin Trans. 2*, 1983, 1351.

70. W.S. Hamama, M. Hammouda and E.M. Afsah, *Z. Naturforsch., Teil B*, 1988, **43**, 897; *Chem. Abs.*, 1989, **110,** 74997s.
71. M. Brenner and H.R. Rickenbacher, *Helv. Chim. Acta*, 1958, **41**, 181; *US Pat.*, 2 938 029; *Chem. Abs.*, 1961, **55**, 14549.
72. E.N. Zil'berman, *Zh. Obshch. Khim.*, 1955, **25**, 2127; *Chem. Abs.*, 1955, **50**, 8458f.
73. B. Lanska and J. Sebenda, *Angew. Makromol. Chem.*, 1988, **164**, 181; *Chem. Abs.*, 1989, **111**, 96444.
74. G.B. Fedorova, I.M. Dolgopol'skii, M.A. Suvorova, V.A. Gubanov and P. E. Gracheva, *Zh. Org. Khim.*, 1973, **9**, 2211; *Chem. Abs.*, 1974, **80**, 36675.

CHAPTER 5

Tetrahydroazepines

I. 2,3,4,5-TETRAHYDRO[1H]AZEPINES

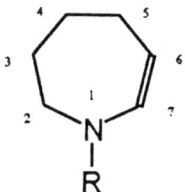

PREPARATION

Given the ready availability of caprolactam (**1**, R = H), it is perhaps not surprising that many synthetic approaches in the literature to 2,3,4,5-tetrahydro[1H]azepines have been based on either the lactam itself (**1**, R = H), or one of its N-substituted derivatives (**1**, R = alkyl or acyl). For example, **1** (R = CH$_3$) has been converted using an appropriate Grignard reagent into the corresponding alkyl (**2**, R = C$_5$H$_{11}$, C$_6$H$_{13}$, C$_7$H$_{15}$, C$_{10}$H$_{21}$)[1] and aryl (**2**, R = Ph, 1-naphthyl)[2,3] derivatives. A useful modification of this approach

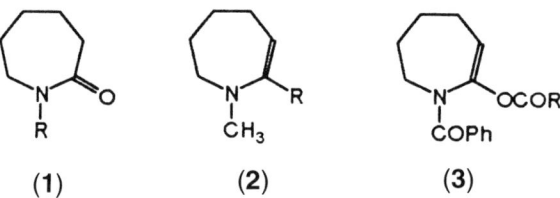

(1)   (2)   (3)

involved treatment of the lithio derivative of **1** (R = COPh) with an acyl chloride to give **3** (R = Ph, Et).[4]

Treatment of caprolactams (**4**, R = H, Cl, CH$_3$) with COCl$_2$ also proved a useful method of achieving substituted 2,3,4,5-tetrahydro[1H]azepines, giving chloro derivatives (**5**, R = H, Cl, CH$_3$).[5-9] The product **5** (R = H), could be further derivatised on the N atom to give the amides **6** (R = NH$_2$, NHPh)[6] or the ester **6** (R = OEt);[10] alternatively brominated on the double bond, **5** (R = H) gave **7** (R = Cl),[6] which could be further converted into the amide **7** (R = NH$_2$) or the ester **7** (R = OEt).[6]

(4)   (5)   (6)   (7)

The action of P$_2$S$_5$ on N-methylcaprolactam **1** (R = CH$_3$), followed by treatment with methyl iodide and then base, has been reported to give the methylthio derivative of 2,3,4,5-tetrahydro[1H]azepine **8**,[11] whereas the same sequence of reactions with caprolactam (**1**, R = H) gave the 3,4,5,6-tetrahydro[2H]azepine (**9**).[12] Reaction of **8** with aryl isocyanates occurred exclusively at the double bond giving products such as **10** (R = H, CH$_3$, Cl, Br),[13,14] whereas **9** reacted to give the N-substituted 2,3,4,5-tetrahydro[1H]azepine **11** (R = NHPh).[15] N-Substitution also occurred on acylation of **9**, giving **11** (R = CH$_3$, Ph),[16] and on treatment with ethyl chloroformate, giving **11** (R = OCH$_3$).[17]

(8)   (9)   (10)   (11)

Removal of the methylthio group in the analogous derivatives **12** (R = H, n-C$_5$H$_{11}$), **13** (R = H, n-C$_5$H$_{11}$) and **14** was achieved smoothly with Raney nickel to give **15** (R = H, n-C$_5$H$_{11}$),[18] **16** (R = H, n-C$_5$H$_{11}$)[19] and **17**,[19a] respectively.

(12)   (13)   (14)

In contrast to the well documented use of the thioenol ethers **8–14**[11-19] for the preparation of the 2,3,4,5-tetrahydro[1H]azepine ring system, little use has been

(15) (16) (17)

made of the corresponding enol ethers 2 (R = OSiMe$_3$, and its 1-trifluoroacetyl derivative),[20,21] mainly because they readily re-form the lactam system (1) on acid treatment. The saturated ethers 18 (R = CO$_2$Et, CHO), however, have been utilised, yielding the 2,3,4,5-tetrahydro[1H]azepines 19 (R = CO$_2$Et, CHO)[22,23] on demethoxylation with NH$_4$Br. In a similar fashion, treatment of the phenyl selenide 20 with hydrogen carbonate introduced a double bond into the ring to give 19 (R = CH=NCMe$_3$).[24]

(18) (19) (20)

Ring expansion reactions for the preparation of 2,3,4,5-tetrahydro[1H]azepines have been reported, but in general they cannot be considered important methods of preparation because the yields are poor and often a wide range of by products is formed. For example, a Demjanov reaction of 1-methyl-3-aminomethylpiperidine (21) gave nine products, including 1-methyl-2,3,4,5-tetrahydro[1H]azepine (2, R = H),[25a] and a photolytic rearrangement of the bicyclic system 22 produced three products including 23, isolated in 9% yield.[26]

(21) (22) (23)

An interesting ring expansion reaction involving the insertion of a two-carbon chain into 2-alkyl-Δ^1-pyrrolinium salts such as 24 (R^1 = H, R^2 = CH$_3$ and R^1 = Ph, R^2 = H) using dimethyl acetylenedicarboxylate (DMAD) in the presence of triethylamine gave a mixture of the tetrahydro[1H]azepines (25, R^1 = H, R^2 = CH$_3$ and R^1 = Ph, R^2 = H) along with the corresponding pyrrolidines (**26**, $R^1 = H$, $R^2 = CH_3$ and $R^1 = Ph$, $R^2 = H$). The products (**25** and **26**) could be separated by chromatography, although only modest yields (34% and 13%, respectively) of the azepines **25** were claimed.[27]

(**24**) (**25**) (**26**)

Finally, a novel approach[28] involved treatment of cinnamalaniline (**27**) with KCN in dimethyl sulphoxide or dimethylformamide, to give the intermediate ion **28**, which then underwent a 1:4 addition to a second mole of **27**, to give the dimeric product **29**, which cyclised with loss of cyanide ion to give the 4,5-*trans*-diphenyltetrahydroazepine **30**.

PROPERTIES

The 2,3,4,5-tetrahydro[1*H*]azepines are low-melting solids or oils. Being enamines, they are fairly reactive compounds, particularly the members of the series which are unsubstituted on the N atom. Alkylation or acylation on the N atom produces more stable compounds, which nevertheless still undergo typical enamine reactions. Substitution on the double bond, at position 6, has already been mentioned, producing derivatives such as **10**.[13] Many 6-substituted derivatives can react further at position 7 to form fused systems. For example, the lactam acetal **31** on treatment with acrylonitrile in benzene gave three products (**1**; R = CH$_3$, **32** and **33**)[29] in a ratio of 1:5:4. It is probable that

(31)   (32)   (33)

the fused-ring product **33** was formed from **32** as an intermediate, since it was also formed from **32** directly on heating in ethylene glycol.

A fused-ring system was also obtained when the amides **10** (R = H, CH$_3$, Cl, Br), formed as described above from **8** with aryl isocyanates in diethyl ether at room temperature, underwent further reaction at reflux temperature in toluene to give the bicyclic compounds **34** (R = H, CH$_3$, Cl, Br).[13,14]

(34)

2,3,4,5-Tetrahydro[1H]azepines have also been reported to undergo ring-contraction reactions. For example, the enaminium bromides **35** and **36** derived from the alkyl derivatives **15** (R = n-C$_5$H$_{11}$),[18b] and **17**,[19a] on treatment with aqueous triethylamine, gave the corresponding piperidines **37** and **38**.

(35)   (36)   (37)

In the case of the N-methyl derivative (**37**), the authors were uncertain of the relative configuration of the n-C$_5$H$_{11}$ group and the ester,[18b] but in the case of the N-benzyl derivative (**38**) they presented strong evidence in favour of a *trans*

(38) (39)

relationship.[19a] A similar sequence of reactions resulting in a ring contraction had previously been reported from the aldehyde 16 (R = H)[19b] giving the piperidine 39 as the product.

Table 1 lists those 2,3,4,5-tetrahydro[1H]azepines whose identities are generally supported by chemical analyses and, in most cases, by spectroscopic measurements. Blanks in the table indicate that the compound has been claimed but not characterised.

Table 1. 2,3,4,5-Tetrahydro[1H]azepines

Structure	M.p. (°C)/b.p. (°C/mmHg)	Spectroscopic evidence	Ref.
N-CH₃		^{1}H NMR	25
N-CHO	108–111/15	MS, ^{1}H NMR	22a
N-CO₂CH₃	94–95/25	IR, ^{1}H NMR	149
N-CO₂C₂H₅	No data given		23

Table 1 (continued)

Structure	M.p. (°C)/b.p. (°C/mmHg)	Spectroscopic evidence	Ref.
[azepine with N–CH=N–C(CH₃)₃]	45/115	IR, ¹H NMR	24
[4-OH azepine, N-CO₂C₂H₅]	75–77/2.5	IR, ¹H NMR	23
[4-ethyl azepine, N-CO₂CH₃]			32
[3-CO₂CH₃ azepine, N-CH₃]	Oil	IR, ¹H NMR	18a
[3-(3,4-dimethoxyphenyl) azepine, N-CH₃]	47–49	IR, MS, ¹H NMR	148
[3-COCH₃ azepine, N-CHO]	134–137/1	MS, ¹H NMR	22b

Table 1 (continued)

Structure	M.p. (°C)/b.p. (°C/mmHg)	Spectroscopic evidence	Ref.
(azepine-azepine bis-CHO)	160–210/0.2	MS, ^{1}H NMR	22b
(azepine-tetrahydroisoquinoline bis-CHO)	200–250/0.2	MS, ^{1}H NMR	22b
(3-CHO azepine, N-CH$_2$Ph)	69–70	IR, ^{1}H, ^{13}C NMR	19b
(3-(3,4-dimethoxyphenyl) azepine, N-CO$_2$CH$_3$)	71–73	IR, MS, ^{1}H NMR	148
(3-CHO, 2-Ph azepine, N-H)	129–129.5	IR, ^{1}H NMR, UV, MS	26a, 26b
(2-Ph azepine, N-CH$_3$)	152–155/18 150–160/20	IR	2b, 3

Table 1 (continued)

Structure	M.p. (°C)/b.p. (°C/mmHg)	Spectroscopic evidence	Ref.
1-methyl-2-C$_5$H$_{11}$-azepine	106–112/12		1
1-methyl-2-C$_6$H$_{13}$-azepine	118–120.5/6		1
1-methyl-2-C$_7$H$_{15}$-azepine	138–140/7		1
1-methyl-2-C$_{10}$H$_{21}$-azepine	162–165/11		1
1-methyl-2-(1-naphthyl)-azepine	62–63 145–155/0.5	IR, UV	2a, 2b
1-methyl-2-CF$_3$-azepine	152.1/760	IR, MS, ^{1}H NMR	151
1-methyl-2-OSi(CH$_3$)$_3$-azepine	62–65/2	IR, ^{1}H NMR	20

Tetrahydroazepines 297

Table 1 (continued)

Structure	M.p. (°C)/b.p. (°C/mmHg)	Spectroscopic evidence	Ref.
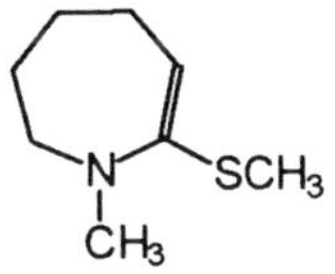 N-CH₃, SCH₃ azepine	86–87.5/10		11
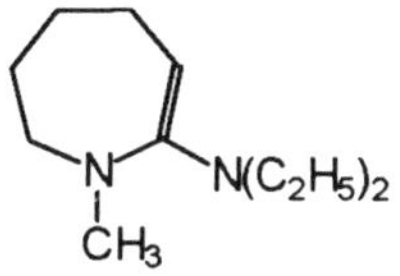 N-CH₃, N(C₂H₅)₂ azepine	91/11	IR, UV	30
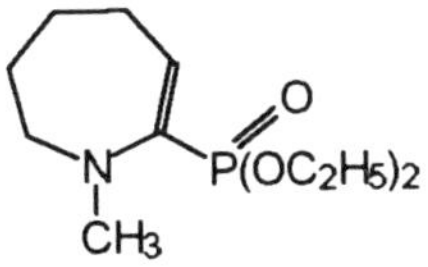 N-CH₃, P(O)(OC₂H₅)₂ azepine	88/0.2	IR, ^{1}H NMR	31
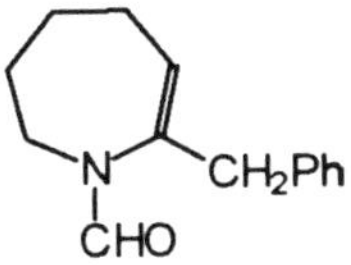 N-CHO, CH₂Ph azepine	100/0.05	^{1}H NMR	24
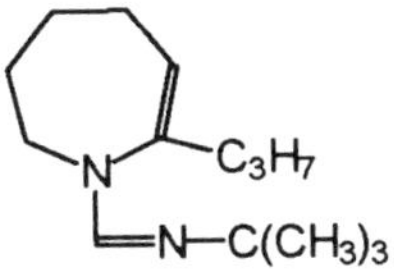 N=N-C(CH₃)₃, C₃H₇ azepine	Oil	IR, ^{1}H NMR	24
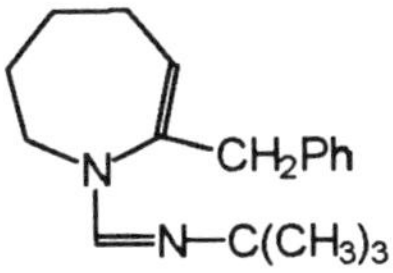 N=N-C(CH₃)₃, CH₂Ph azepine	Oil	IR, ^{1}H NMR	24
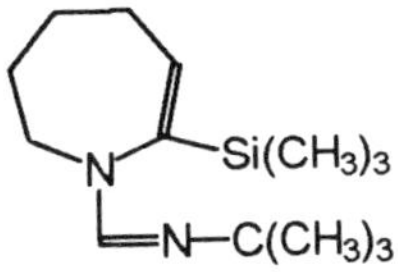 N=N-C(CH₃)₃, Si(CH₃)₃ azepine	Oil	IR, ^{1}H NMR	24

Table 1 (continued)

Structure	M.p. (°C)/b.p. (°C/mmHg)	Spectroscopic evidence	Ref.
2-Cl, N-CO$_2$C$_2$H$_5$ azepine	69/0.1	IR	10
2-SCH$_3$, N-CO$_2$CH$_3$ azepine	143–146/0.1	MS, ^{1}H NMR	17
2-OCH$_3$, N-COCH$_3$ azepine	79–82/0.4	IR, ^{1}H NMR	37
2-SCH$_3$, N-COCH$_3$ azepine	103–104/0.4	IR, MS, ^{1}H NMR	16
2-SCH$_3$, N-COPh azepine	135/0.0004	IR, MS, ^{1}H NMR	16
2-OCOC$_2$H$_5$, N-COPh azepine	98–99	IR, ^{1}H NMR	4
2-OCOPh, N-COPh azepine	85–86	IR, MS, ^{1}H NMR	4

Tetrahydroazepines

Table 1 (continued)

Structure	M.p. (°C)/b.p. (°C/mmHg)	Spectroscopic evidence	Ref.
N-CO$_2$Ph, OSO$_2$CF$_3$		IR, ^{1}H, ^{13}C NMR	150
N-CONHPh, SCH$_3$	Oil	IR, MS, ^{1}H NMR	15
N-COCl, OCH$_3$	76–78/0.1	IR, ^{1}H NMR	37
N-COCl, Cl	112–114/10, 125/12	IR	5–7
N-CONH$_2$, Cl	89	IR	6
N-CONHPh, Cl	88–89		6
N-COCF$_3$, OSi(CH$_3$)$_3$	117–118/7	IR, ^{1}H NMR	21

Table 1 (continued)

Structure	M.p. (°C)/b.p. (°C/mmHg)	Spectroscopic evidence	Ref.
(azepine-NH linked to azepine-N with Cl)	125/2		36
(azepine with N-COCH$_2$Cl and NPh-COCH$_2$Cl substituent)	138–139	^{1}H NMR	33
(N-CH$_3$ azepine with CH$_3$ and Cl)			9
(N-CH$_3$ azepine with CO$_2$CH$_3$ and SCH$_3$)	Oil	IR, ^{1}H NMR	18a
(N-CH$_3$ azepine with CONH-Ph and SCH$_3$)	147–149	IR, MS, ^{1}H NMR	13, 14
(N-CH$_3$ azepine with CONH-C$_6$H$_4$-CH$_3$ and SCH$_3$)	144–146	IR, MS, ^{1}H NMR	13, 14
(N-CH$_3$ azepine with CONH-C$_6$H$_4$-Cl and SCH$_3$)	152–154	IR, MS, ^{1}H NMR	13, 14

Table 1 (continued)

Structure	M.p. (°C)/b.p. (°C/mmHg)	Spectroscopic evidence	Ref.
(N-CH₃ tetrahydroazepine with SCH₃ and CONH-C₆H₄-Br substituents)	157–159	IR, MS, ^{1}H NMR	13, 14
(N-CH₃ tetrahydroazepine with OC₂H₅ and CH₂CH₂CN substituents)	129–130/1	IR, UV	29
(N-CH₃ tetrahydroazepine with N(CH₃)₂ and dimedone-ylidene substituent)	159–162	MS, ^{1}H NMR	34
(N-CH₃ tetrahydroazepine with C₅H₁₁ and CO₂CH₃ substituents)	175–179/11	IR, MS, ^{1}H NMR	18b
(N-CH₂Ph tetrahydroazepine with C₅H₁₁ and CO₂CH₃ substituents)	Oil	IR, ^{1}H NMR	19a
(N-CH₂Ph tetrahydroazepine with C₅H₁₁ and CHO substituents)	Oil	IR, ^{1}H, ^{13}C NMR	19a
(N-CH₂Ph tetrahydroazepine with SCH₃ and CHO substituents)	88–89	IR, ^{1}H, ^{13}C NMR	19b

Table 1 (continued)

Structure	M.p. (°C)/b.p. (°C/mmHg)	Spectroscopic evidence	Ref.
3-Cl, 2-Cl azepane, N-COCl	42 110–112/2		8
3-Br, 2-Cl azepane, N-COCl	38–40 130/5	IR	6
3-Br, 2-Cl azepane, N-CO$_2$C$_2$H$_5$	120–125/3	IR	6
3-Br, 2-Cl azepane, N-CONH$_2$	131	IR	6
3-CO$_2$CH$_3$, 2-SCH$_3$, 7-C$_5$H$_{11}$, N-CH$_3$	Oil	IR, MS, ^{1}H NMR	18b
3-CO$_2$CH$_3$, 2-SCH$_3$, 7-C$_5$H$_{11}$, N-CH$_2$Ph	Oil	IR, ^{1}H NMR	19a
3-CHO, 2-SCH$_3$, 7-C$_5$H$_{11}$, N-CH$_2$Ph	Oil	IR, ^{1}H NMR	19a

Table 1 (continued)

Structure	M.p. (°C)/b.p. (°C/mmHg)	Spectroscopic evidence	Ref.
	145/0.03	IR, MS, ^{1}H, ^{13}C NMR	35
	172	IR, MS, ^{1}H NMR	28
	134–135	IR, UV, ^{1}H, ^{13}C NMR, MS	27
	164	IR, UV, ^{1}H NMR, MS	27

II. 2,3,4,7-TETRAHYDRO[1H]AZEPINES

PREPARATION

Substituted 2,3,4,7-tetrahydro[1H]azepines are fairly well represented in the literature, although the parent compound itself does not appear to have been reported. A frequently used method of preparation involves dehydration of an appropriate alcohol, itself usually prepared from the corresponding ketone. The disadvantage of this approach is that the product is often a mixture of double bond isomers which can be difficult to separate.

For example, the alcohol **40** on dehydration with aqueous HCl gave a mixture of the isomers **41** and **42** in a ratio of 2.7:1 after 24 h, reversing to a 1:3 ratio after 48 h.[38] Formation of a single isomer could only be guaranteed when the dehydration resulted in a conjugated system, as in the acids **45** and **46**, formed from the alcohols **43** and **44**, respectively, on treatment with aqueous mineral acid.[39,40]

A claim[41] that the 2,3,4,7-tetrahydro[1H]azepine **48** was the sole product of dehydration of the alcohol **47** was not fully supported by the evidence presented.

Tetrahydroazepines

(47)   (48)

Dehydrohalogenation of azepane halides (**49**, X = Cl, Br, I)[23] and reduction of the amide **50**[42,43] appear to offer more reliable means of synthesising 2,3,4,7-tetrahydro[1H]azepines specifically, giving **51** and **52**, respectively, free from double-bond isomers.

(49)   (50)   (51)   (52)

An alternative strategy which proved useful was ring formation from an appropriately substituted non-cyclic amine. This has been utilised both for fairly simple systems such as the formation of **54** from **53** [R = H, Si(CH$_3$)$_3$] in the presence of HCHO[44] and for more highly substituted products such as

(53)   (54)   (55)   (56)

56 (R = CH$_3$, Ph) from **55** (R = CH$_3$, Ph).[45,46] An interesting, novel variation on this theme[47] is the cyclisation of the tertiary amine **57** with loss of ethylene, to give **58**, using a complexed molybdenum catalyst to effect the ring closure.

Ring expansion reactions for the preparation of 2,3,4,7-tetrahydro[1H]azepines have also been reported. For example, the 1-methyl derivative **59** was also reported as a minor product from the Demjanov rearrangement of the aminomethylpiperidine **21**[25] as described above, but again, because of the

(57)   (58)

low yields and the complexity of the mixture formed, this cannot be considered as an important synthetic procedure.

A more satisfactory ring-expansion method utilised 2-phenyl-2-vinylaziridine (**60**) which could be shown to undergo thermal cycloaddition with electron-poor olefins, either via a concerted 4 + 2 addition, or a dipolar intermediate. With these reagents (**61**, R = CN, CO$_2$CH$_3$, SO$_2$Ph), the 6-phenyl-2,3,4,7-tetrahydro[1H]azepines (**62**, R = CN, CO$_2$CH$_3$, SO$_2$Ph) were formed in good yields.[48]

(59)   (60)   (61)   (62)

The more highly substituted aziridine **63** (R = H, CH$_3$) had previously been shown to react with acetylenes such as 2-propynyltriphenylphosphonium bromide (**64**) to give phosphonium bromide salts of 2,3,4,7-tetrahydro[1H]azepines (**65**, R = H, CH$_3$).[49]

(63)   (64)   (65)

A further possibility involving ring expansion for synthesising 2,3,4,7-tetrahydro[1H]azepines was the reported isomerisation of the 2-vinylpyrrolidines **66a–c**[50] using methyl iodide, which gave the corresponding quaternised tetrahydroazepines **67a–c**. Thermal dealkylation of one of the derivatives (**67b**) gave the corresponding tetrahydroazepine (**68b**), which was unstable in air, but could be readily re-quaternised with methyl iodide.

(66)   (67)   (68)   (69)

(a) R¹ = R² = H
(b) R¹ = H, R² = OCH₃
(c) R¹ = R² = OCH₃

A similar sequence of reactions using cinnamyl bromide with the 2-vinylpyrrolidine **66a** gave the corresponding quaternary salt (**69**), which was readily dealkylated with Na/Hg to give **68a**. Direct thermal ring expansion, without forming the intermediate quaternary salts, of the vinylpyrrolidines **66** in the presence of a variety of Lewis acids proved unsuccessful.[50]

PROPERTIES

The 2,3,4,7-tetrahydro[1H]azepines are low-melting solids or oils, frequently characterised as crystalline salts such as picrates or hydrochlorides. They are fairly stable compounds and demonstrate the expected chemical properties of secondary or tertiary amines depending on whether or not they are substituted on the N atom. The more commonly reported reactions, alkylation or acylation on the N atom, occur readily using traditional methods such as Eschweiler Clark alkylation[42] or acyl halides.[43]

Table 2 lists those 2,3,4,7-tetrahydro[1H]azepines whose identities are supported in most cases by chemical analyses and by spectroscopic measurements. Blanks in the table indicate that the compound has been claimed but not characterised.

Table 2. 2,3,4,7-Tetrahydro[1H]azepines

Structure	M.p. (°C)/b.p. (°C/mmHg)	Derivatives and m.p. (°C)	Spectroscopic evidence	Ref.
N-CH₃ azepine				25
N-nBu azepine	100/1		MS, IR, ^{1}H, ^{13}C NMR	44
N-H azepine, 3-CO₂H		HBr, 224.5–225.5	IR, UV, ^{1}H NMR	39
N-H azepine, 4-CO₂H		HBr, 183–184.5	UV, ^{1}H NMR	40
N-CO₂C₂H₅, 2-OCH₃ azepine	68–71/1.5		IR, ^{1}H NMR	23
N-CO₂CH₃, 2-C₂H₅ azepine				32
N-CH₂Ph, 3-CH₃ azepine	Oil		MS, IR, ^{1}H, ^{13}C NMR	47

Table 2 (continued)

Structure	M.p. (°C)/b.p. (°C/mmHg)	Derivatives and m.p. (°C)	Spectroscopic evidence	Ref.
1-benzyl-3-methyl-tetrahydroazepine	Oil		MS, IR, ^{1}H, ^{13}C NMR	47
4-ethyl-1-methyl-tetrahydroazepine	112–115/25	Picrate 117–119		41
1-methyl-4-phenyl-tetrahydroazepine	Oil		UV	50
1-methyl-4-(3,4-dimethoxyphenyl)-tetrahydroazepine	70–75/0.03			50
3-iodo-1-tosyl-tetrahydroazepine	Colourless oil		IR, MS, ^{1}H NMR	152
3-(tert-butyldimethylsilylmethyl)-1-tosyl-tetrahydroazepine	No data given			153

Table 2 (continued)

Structure	M.p. (°C)/b.p. (°C/mmHg)	Derivatives and m.p. (°C)	Spectroscopic evidence	Ref.
(CH$_3$O$_2$C, Ph-substituted azepine, NH)	56–57			48
(NC, Ph-substituted azepine, NH)	70–71			48
(PhSO$_2$, Ph-substituted azepine, NH)	105–106			48
(Ph, CH$_3$-substituted azepine, N-CH$_3$)	73–75/0.05	HCl, 172–174	^{1}H NMR	38
(CH$_3$, CH$_3$, CH$_3$-substituted azepine, NH)	32–35/0.5	HBr, 149–151	MS, ^{1}H NMR	43
(Ph, CH$_3$, P$^+$Ph$_3$Br$^-$ substituted azepine, NH)	229–234		MS, IR, ^{1}H, ^{13}C NMR	49

Table 2 (continued)

Structure	M.p. (°C)/b.p. (°C/mmHg)	Derivatives and m.p. (°C)	Spectroscopic evidence	Ref.
	140–155		MS, IR, ^{1}H, ^{13}C NMR	49
	184–185		MS, IR, UV, ^{1}H NMR	45
	199–201		MS, IR, UV, ^{1}H NMR	45
	259–267		MS, IR, UV, ^{1}H NMR	46
	213–214		UV, ^{1}H NMR	50

Table 2 (continued)

Structure	M.p. (°C)/b.p. (°C/mmHg)	Derivatives and m.p. (°C)	Spectroscopic evidence	Ref.
4-(4-methoxyphenyl)-1,1-dimethyl azepinium iodide	186–188		UV, ^{1}H NMR	50
4-(3,4-dimethoxyphenyl)-1,1-dimethyl azepinium iodide	196–197		UV, ^{1}H NMR	50
1-cinnamyl-1-methyl-4-phenyl azepinium bromide	Oil			50
1-cinnamyl-4-(3,4-dimethoxyphenyl)-1-methyl azepinium bromide	164–167		UV, ^{1}H NMR	50

Table 2 (continued)

Structure	M.p. (°C)/b.p. (°C/mmHg)	Derivatives and m.p. (°C)	Spectroscopic evidence	Ref.
(4-methoxyphenyl-substituted tetrahydroazepinium bromide, N-benzyl, N-methyl)	167–170		UV, ^{1}H NMR	50
(3,4-dimethoxyphenyl-substituted tetrahydroazepinium bromide, N-benzyl, N-methyl)	154–159		UV	50
(methyl, phenyl, methyl-substituted azepine with phosphonium ylide, Ph$_3$Br)	232–234		MS, IR, ^{1}H, ^{13}C NMR	49
(tetramethyl-N-methyl tetrahydroazepine)	No data presented			42
(trimethyl tetrahydroazepine N-acyl with 2-iodophenyl)	Oil		MS, IR, ^{1}H NMR	43

III. 2,3,6,7-TETRAHYDRO[1H]AZEPINES

PREPARATION

2,3,6,7-Tetrahydro[1H]azepines are fairly well documented and have been synthesised by a number of approaches. A synthetic method which was particularly useful for preparing symmetrically substituted derivatives utilised intramolecular reductive coupling of Mannich base hydrochlorides in the presence of low-valency titanium, prepared *in situ* from $TiCl_4$ and zinc powder.[51] In this way the amines **70** (R^1 = H, R^2 = CH_3, C_2H_5, ^nBu; and R^1 = R^2 = CH_3) were converted into the corresponding 2,3,6,7-tetrahydro[1H]azepines **71** (R^1 = H, R^2 = CH_3, C_2H_5, ^nBu; and R^1 = R^2 = CH_3) in yields around 60%.[51]

(70)   (71)

A symmetrically substituted 2,3,6,7-tetrahydro[1H]azepine (**73**) was also formed from the diester **72** with trimethylsilyl chloride and sodium in toluene.[52]

(72)   (73)

Intramolecular aminomethylation–desilylation of ω-monoalkylaminoallyltrimethylsilanes such as **74** (R = ^iPr, ^nBu, CH_2Ph) in the presence of formaldehyde and trifluoroacetic acid readily gave the corresponding

Tetrahydroazepines 315

2,3,6,7-tetrahydro[1H]azepines (**75**, R = ^{i}Pr, ^{n}Bu, CH$_2$Ph) in yields of 60–96%.[53] Reaction of **74** (R = ^{n}Bu) with ^{n}PrCHO in place of formaldehyde gave the corresponding 4-*n*-propyl analogue (**76**) and using the disilylated derivative **77** with formaldehyde gave the 4-trimethylsilyl derivative **78**,[53] although in both of these cases the yields were substantially lower (18 and 40%, respectively).

(CH$_3$)$_3$SiCHCH=CH$_2$
|
CH$_2$CH$_2$NHR

(**74**)   (**75**)

Traditional chemical methods, such as reduction or Grignard alkylation of a carbonyl group followed by dehydration of the alcohol formed, as discussed in the previous section, have also been used for the preparation of 2,3,6,7-tetrahydro[1H]azepines. This approach has been applied to the preparation of the acid **79**[40] and the 4-phenyl derivative **80**,[38,54] although formation of isomeric mixtures of products sometimes occurs.[38] Simultaneous direct reduction of the amide carbonyl group and the imine double bond of **81** gave the tetramethyl derivative **82**.[55]

(**76**)   (CH$_3$)$_3$SiCHCH=CHSi(CH$_3$)$_3$
           |
           CH$_2$CH$_2$NH^{n}Bu

(**77**)   (**78**)

Formation of an enamine system from a carbonyl group has also been used to give products which are formally 2,3,6,7-tetrahydro[1H]azepines, as in the preparation of the amines **84** (R = COCH$_3$, CO$_2$Et, CH$_2$Ph) and **85**

(**79**)   (**80**)   (**81**)   (**82**)

(R = CH$_2$Ph, CO$_2$Et) from the corresponding ketones (**83**, R = COCH$_3$, CO$_2$Et, CH$_2$Ph[56] and R = CH$_2$Ph, CO$_2$Et).[57] As might be expected, the enamines **84** and **85** are not particularly stable and **85** (R = CH$_2$Ph) has not been isolated in a pure form.

(**83**)   (**84**)   (**85**)

An interesting rearrangement of the bicyclic tosylate **86** using water was claimed to give the parent 2,3,6,7-tetrahydro[1H]azepine **87** (R = H) along with 3-vinylpyrrolidine (**88**) as a 91:9 mixture.[58] It was proposed that the

(**86**)   (**87**)   (**88**)

reaction proceeded via the initially formed immonium salt **89**, which in turn underwent hydrolysis with loss of formaldehyde. The minor amounts of 3-vinylpyrrolidine formed resulted from a sigmatropic [3,3] rearrangement of the immonium salt **89** to its isomer **90** before hydrolysis. The unsubstituted azepine **87** (R = H) was not isolated, but was converted *in situ* into the N-tosyl derivative **87** (R = CH$_3$C$_6$H$_4$SO$_2$).

An alternative procedure involving reaction of the bicyclic tosylate **86** with aqueous sodium hydroxide in the presence of tosyl chloride was claimed to give the tosylate **87** (R = CH$_3$C$_6$H$_4$SO$_2$) as the sole product. Unfortunately, no supporting physical-chemical evidence was quoted to support the assigned structures **87** (R = H, CH$_3$C$_6$H$_4$SO$_2$).

(**89**)   (**90**)

PROPERTIES

In general, the 2,3,6,7-tetrahydro[1H]azepines are low-melting solids or oils, frequently characterised by conversion into the crystalline picrates or hydrochlorides. As an exception to this general rule, the 4,5-diphenyl derivatives such as **71** are readily crystallisable, high-melting solids. Stability is rarely a problem with the exception of the 4-amino derivatives **84** and **85**, which, being enamines, readily underwent hydrolytic decomposition.[57] These are also the most reactive derivatives. For example, the benzylamino derivatives **84** (R = $COCH_3$, CO_2Et, CH_2Ph) readily underwent condensation reactions with hydroxylamine to give the corresponding benzylammonium salts **91** (R = $COCH_3$, CO_2Et, CH_2Ph), which in turn afforded the isoxazolin-5-ones **92** (R = $COCH_3$, CO_2Et, CH_2Ph) with methanolic hydrogen chloride.[56] Similarly, the amine **85** (R = CO_2Et) condensed with diethyl malonate in the presence of sodium ethoxide to give the 6,7,8,9-tetrahydro-5H-pyrido[2,3-d]azepin-2-one **93** (R = CO_2Et). Treatment of this latter compound with hydrochloric acid removed the 3-carbethoxy group and gave the pyridoazepinone **93** (R = H).[57]

(91)   (92)   (93)

Table 3 lists those 2,3,6,7-tetrahydro[1H]azepines whose identities are supported by chemical analyses and by spectroscopic measurements.

Table 3. 2,3,6,7-Tetrahydro[1H]azepines

Structure	M.p. (°C)/b.p. (°C/mmHg)	Derivatives and m.p. (°C)	Spectroscopic evidence	Ref.
N-n-C$_3$H$_7$	Oil		IR, ^{1}H NMR	53
N-n-C$_4$H$_9$	Oil		IR, ^{1}H NMR	53
N-CH$_2$Ph	Oil		IR, MS, ^{1}H NMR	53
4-CO$_2$H, N-H		HBr, 174.5–176.5	IR, ^{1}H NMR	40
4-n-C$_3$H$_7$, N-n-C$_4$H$_9$	Oil		IR, ^{1}H NMR	53
4-Si(CH$_3$)$_3$, N-n-C$_4$H$_9$	Oil		IR, ^{1}H NMR	53

Table 3 (continued)

Structure	M.p. (°C)/b.p. (°C/mmHg)	Derivatives and m.p. (°C)	Spectroscopic evidence	Ref.
4-Ph, 3-CH₃, N-CH₃ tetrahydroazepine	99–100/0.3 73–75/0.05	Picrate 126–127 HCl 96–98	^{1}H NMR ^{1}H NMR	54 38
4,5-diPh, N-CH₃ tetrahydroazepine	103–104		IR, MS, ^{1}H NMR	51
4,5-diPh, N-C₂H₅ tetrahydroazepine	60–61		IR, MS, ^{1}H NMR	51
4,5-diPh, N-nC₄H₉ tetrahydroazepine	55–56		IR, MS, ^{1}H NMR	51
4,5-di(4-CH₃-C₆H₄), N-C₂H₅ tetrahydroazepine	222–224		IR, MS, ^{1}H NMR	51

Table 3 (continued)

Structure	M.p. (°C)/b.p. (°C/mmHg)	Derivatives and m.p. (°C)	Spectroscopic evidence	Ref.
(CH₃)₃SiO, OSi(CH₃)₃ azepine, N-CH₂Ph	151–153/0.2		IR, MS, ¹H NMR	52
NH₂, CO₂C₂H₅ azepine, N-CH₂Ph	Unstable		IR, ¹H NMR	57
NH₂, CO₂C₂H₅ azepine, N-CO₂C₂H₅	75–76		IR, ¹H NMR	57
PhCH₂NH, CO₂C₂H₅ azepine, N-COCH₃	106–108.5		IR, UV, ¹H NMR	56
PhCH₂NH, CO₂C₂H₅ azepine, N-CO₂C₂H₅	36.5–39		IR, UV, ¹H NMR	56

Table 3 (continued)

Structure	M.p. (°C)/b.p. (°C/mmHg)	Derivatives and m.p. (°C)	Spectroscopic evidence	Ref.
PhCH₂NH-[ring]-CO₂C₂H₅, N-CH₂Ph	Oil		IR, UV, ¹H NMR	56
CH₃,CH₃-[ring]-CH₃,CH₃, NH	Oil	Picrate 142–143	IR, ¹H NMR	55

IV. 3,4,5,6-TETRAHYDRO[2H]AZEPINES

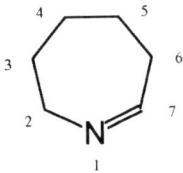

This series of compounds is the best represented of the tetrahydroazepines with over 500 references in the literature, dating back to 1930. A selection had therefore to be made to include the most useful methods of preparation of the different types of derivatives. Table 4 is fairly comprehensive and includes a large number of the structures published, but in order to keep it to a reasonable length, references to polymers have been excluded, and patent references have been kept to a minimum.

PREPARATION

The parent 3,4,5,6-tetrahydro[2H]azepine (**94**) has been mentioned in the literature as a likely intermediate in the catalytic hydrogenation of 7-amino-3,4,5,6-tetrahydro[2H]azepine (**95**),[59] but the main product isolated was the completely reduced material (**96**). Consequently, the existence of a stable parent (**94**) cannot be considered as definitely established.

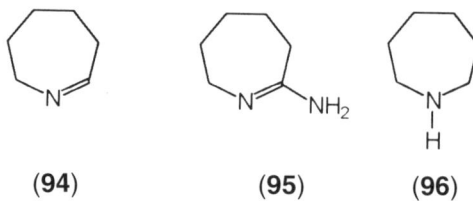

(94)   (95)   (96)

Much of the chemistry associated with the preparation of substituted 3,4,5,6-tetrahydro[2H]azepines is based on caprolactam (1, R = H) or one of its enol ethers (97, R = alkyl). Details of a procedure for the preparation of the simple enol ethers 97 (R = CH$_3$, C$_2$H$_5$) first appeared in 1948,[60] although the compounds were used as starting materials for other derivatives as early as 1930,[61] but no details of preparation were given in the earlier paper. Treatment of caprolactam (1) with dimethyl or diethyl sulphate[60] was used in a number of preparations,[62-69] and various boiling points were quoted for the product (see Table 4). An improved procedure from caprolactam (1, R = H) using diazomethane and a silica catalyst was claimed in 1979.[70] The 7-chloro derivative 98, prepared by the action of phosgene on caprolactam was also used as an intermediate, followed by treatment with the corresponding alcohol[71] and this procedure was extended[72] to give a range of ethers [97, R = CH$_3$, CH(CH$_3$)$_2$, cyclohexyl, PhCH$_2$, Ph]. Although the method was successful, the use of phosgene as a reagent did not gain wide acceptance,

(1)   (97)   (98)   (99)   (100) ROCOCl

and a later improvement[73] from the thiolactam 99 and the half-ester half-acid chloride 100 extended the range of ethers available [97, R = C$_2$H$_5$, CH$_2$CH(CH$_3$)$_2$, CH$_2$C(CH$_3$)$_3$.

The 7-thio ethers 101 (R = alkyl) have been prepared by routes similar to those used for the preparation of the 7-ethers (97, R = alkyl), using the thiolactam 99 as starting material. For example, thiolactam 99 with methanesulphonic acid,[62,66] diazomethane,[70] or methyl iodide,[74,75] gave the 7-methylthio ether 101 (R = CH$_3$), and other analogues (101, R = PhCH$_2$, CH=CHCH$_2$, CH$_2$=CH, NH$_2$COCH$_2$), have been prepared[73,76-79] using a variety of reagents.

Ring expansion of the oximes 102 and 103 with methanesulphonyl chloride followed by treatment with ^{i}BuAlSMe has been used[80] to give the 2-methyl- and 6-methyl-7-thio ethers 104 and 105, respectively, and a ring expansion reaction[81] of the thioacetals **106** (R = H, CH$_3$) with sodium azide or trimethylsilyl azide, followed by treatment of the unisolated intermediate azides **107**

(**101**) (**102**) (**103**) (**104**) (**105**)

(**106**) (**107**)

(R = H, CH$_3$) with NaOH/Na$_2$S$_2$O$_3$ gave the corresponding thio ethers **101** (R = CH$_3$) and **105**.

The preparation of 7-alkyl-3,4,5,6-tetrahydro[2H]azepines has been reported. Direct reaction of Grignard reagent on caprolactam (**1**) gave the 7-n-butyl derivative **108** (R = ^{n}Bu),[82] but required heating to 150°C to effect the reaction. Milder methods have been reported subsequently using the enol ether **97** (R = CH$_3$) rather than caprolactam to give a range of alkyl derivatives [**108**, R = CH$_2$=CH(CH$_2$)$_3$ and R = CH$_3$, C$_2$H$_5$, ^{n}Pr, ^{n}Bu, Ph, PhCH$_2$].[83,84]

(**108**) (**109**) (**110**) (**111**)

Preparation of 7-alkyl derivatives (**108**, R = alkyl) by direct alkylation of caprolactam using alkyllithium reagents does not seem to be possible, but successful reactions have been reported when the nitrogen atom has been protected. For example, the trimethylsilyl derivative **109** and the 1-vinyl derivative **110** with methyllithium both gave the 7-methyl derivative **108** (R = CH$_3$).[85,86] In contrast, no pure product could be isolated from **110** using phenyllithium.[86] More recently, several 7-alkyl derivatives (**108**, R = CH$_3$, ^{n}Pr, ^{n}Bu, ^{t}Bu, Ph) have been prepared under fairly mild conditions from the enol ether **97** (R = CH$_3$) and the corresponding alkyllithium in diethyl ether.[87] The 7-alkyl derivatives **108** (R = CH$_3$, C$_2$H$_5$, ^{n}Pr, CH$_2$CH$_2$Ph) have also been prepared by decarboxylation by boric acid of the esters **111** (R = H, CH$_3$, C$_2$H$_5$, CH$_2$Ph).[88,89]

(112) **(113)** **(114)** **(115)**

Direct alkylation of the enol ether **97** (R = CH$_3$) to give the 6-alkyl enol ethers **112** (R = CH$_3$, C$_2$H$_5$, ^{n}Pr) has been achieved[67] using ^{i}Pr$_2$NLi to give the lithio derivative **113**, followed by treatment with the appropriate alkyl halide. The reaction is presumed to go via an intermediate, probably the isomers **114** (R = CH$_3$, C$_2$H$_5$, ^{n}Pr) which were not isolated, but were converted into **112** by acidification with the weak acid ammonium chloride. Attempted acid hydrolysis[67] of the enol ethers **112** (R = CH$_3$, C$_2$H$_5$, ^{n}Pr) was unsuccessful, whereas the use of mineral acid instead of ammonium chloride on the intermediates **114** (R = CH$_3$, C$_2$H$_5$, ^{n}Pr) gave the lactams **115** (R = CH$_3$, C$_2$H$_5$, ^{n}Pr), albeit in poor yields.

As is the case with the isomeric tetrahydroazepines already described, ring-expansion reactions have also proved useful for the preparation of alkyl-substituted 3,4,5,6-tetrahydro[2H]azepines. In some instances ring expansion from a six-membered ring has been used to obtain derivatives with unusual substitution patterns which are not readily available by direct substitution of the azepine itself. For example, heating the aziridine **116** (R = H) gave the 3,3,7-trimethyl derivative **117** (R = H),[35] although the isomeric product **118** (see Table 1) resulted on similar treatment of the ester **116** (R = CO$_2$CH$_3$).[35]

(116) **(117)** **(118)**

In contrast, ultraviolet irradiation at room temperature of the aziridines **116** (R = H, CO$_2$CH$_3$) gave mixtures,[90] and although the products **117** (R = H) and a rearranged product **117** (R = CO$_2$CH$_3$) originating from **118** were detected, no pure materials could be isolated because of further rearrangement during the reaction.

In other instances, pyrolytic processes have been used to prepare derivatives with relatively simple substitution patterns and since the products formed are probably more readily available from caprolactam (**1**, R = H) or its enol ether

(**97**, R = CH$_3$), it is possible that the primary aim of these efforts was to study the reaction mechanisms. Nevertheless, some useful synthetic methods resulted. For example, pyrolysis of the diamine **119** gave a mixture of cyclohexene and the enamine **120**,[91] isolated as the picrate, the structure of which was confirmed by a separate synthesis from the 7-chloro derivative **98** treated with piperidine.[91]

(**119**)   (**120**)

Ring expansion of hydroxylamines via the amine sulphonate esters has also been investigated as a route to alkyl-substituted 3,4,5,6-tetrahydro[2H]-azepines. The sulphonates **121** (R^1 = CH$_3$, R^2 = NO$_2$, CF$_3$) were shown to rearrange to the 7-methyl product **108** (R = CH$_3$),[92,93] and the reaction was extended to other analogues (**121**, R^1 = CH$_3$, ^{n}Bu, CH$_2$Ph, R^2 = NO$_2$)[94] to give the corresponding 7-alkyl derivatives **108** (R = CH$_3$, ^{n}Bu, CH$_2$Ph). The 2,7-dialkyl product **123** has also been prepared from the corresponding hydroxylamine (**122**).[94]

(**108**)   (**121**)   (**122**)   (**123**)

Acid treatment of the azides **125** (R = CH$_3$, C$_2$H$_5$, cyclohexyl), prepared from the corresponding tertiary alcohols (**124**, R = CH$_3$, C$_2$H$_5$, cyclohexyl), yielded the 7-alkyl derivatives (**108**, R = CH$_3$, C$_2$H$_5$, cyclohexyl)[95] with loss of nitrogen, and similar treatment of the azides **127** and **129** from the secondary

(**124**)   (**125**)   (**126**)   (**127**)

alcohols **126** and **128** gave the 7-methyl and 4,7-dialkyl derivatives (**108**, R = CH$_3$ and **130**), respectively.[96]

(**128**)   (**129**)   (**130**)

The final variation on this theme is the Beckmann rearrangement of the methane sulphonate (**131**) in the presence of the trimethyl silyl ether (**132**), which gave the cyclohexanone derivative (**133**)[97] on treatment with diethylaluminium chloride.

(**131**)   (**132**)   (**133**)

Several procedures involving ring closure of esters,[98] ketones,[99,100] acetylenes[101,102] and azides[103-106] have been utilised successfully to form the 3,4,5,6-tetrahydro[2H]azepine ring system. The amide esters **134** (R = CH$_3$, Ph), derived from ε-aminocaproic acid, on treatment with NaOH/Ca(OH)$_2$, cyclised and decarbethoxylated to give the corresponding 7-methyl- and 7-phenyl-3,4,5,6-tetrahydro[2H]azepines (**108**, R = CH$_3$, Ph),[98] and the latter compound (**108**, R = Ph) was also formed from the phthalimido derivative **135**, which on hydrolysis[99] or hydrazinolysis[100] released the free amine which cyclised. The corresponding 3,4-dimethoxyphenyl derivative (**136**) was prepared in a similar manner.[99]

(**134**)   (**135**)   (**136**)

In an alternative cyclisation procedure, the acetylene **137** on treatment under Leuckart conditions with ammonium formate gave the formamido derivative **138**, which in turn was treated with acid to give an unisolated intermediate (**139**) which cyclised to 2-phenyl-7-methyl-3,4,5,6-tetrahydro[2H]azepine **140**.[101]

PhCO(CH₂)₄C≡CH   PhCH(CH₂)₄C≡CH   PhCH(CH₂)₄C(O)-CH₃
                     NHCHO   NH₂

(137)   (138)   (139)   (140)

The acetylenes **141** [R = H, (CH₃)₃Si] have also been cyclised, after reduction with NaCNBH₃ to the unisolated heptynylhydroxylamines **142** [R = H, (CH₃)₃Si] and refluxing in toluene. In each case the product was the nitrone **143** with the terminal alkyne **142** (R = H) cyclising more efficiently than the silyl derivative **142** [R = (CH₃)₃Si].[102]

R—≡—(CH₂)₃CHCH=NOH   R—≡—(CH₂)₃CHCH₂NHOH
               CH₃   CH₃

(141)   (142)   (143)

The cyclisation of terminal azides was described in a series of papers by Vaultier and co-workers.[103–106] The keto azide **144** (R = CH₃) on treatment with triphenylphosphine underwent an intramolecular aza-Wittig reaction via the transient phosphinimine **145** to give the 7-methyl-3,4,5,6-tetrahydro[2H]azepine **108** (R = CH₃).[103] The reaction was subsequently extended to the cyclopropyl ester **146**, which under the same reaction conditions cyclised to give the 7-cyclopropyl derivative **147**,[104] and the generality of the reaction was demonstrated further by cyclising the azides **144** (R = CH₃, Ph, cyclohexyl[105] and R = CH₃)[106] to give the corresponding 7-substituted analogues **108** (R = CH₃, Ph, cyclohexyl).

N₃(CH₂)₅COR   [(CH₂)₅COCH₃ / N=PPh₃]

(144)   (145)

N₃(CH₂)₅C(O)—(cyclopropyl)—CO₂C₂H₅

(146)   (147)

The parent 7-amino-3,4,5,6-tetrahydro[2H]azepine **148** (R = H) has been described[59] as the product formed by cyclisation of the dinitrile **149**. Its

3-phenyl analogue **150** was prepared by the action of ammonia or ammonium chloride on the 7-ethyl ether **151** in an autoclave at elevated temperature.[107]

(148) (149) (150)

(151) (152)

The 7-substituted-amino-3,4,5,6-tetrahydro[2H]azepines **148** have been prepared in abundance, by the action of an amine on either the 7-methyl ether **97** (R = CH_3) or the 7-chloro derivative **98**. These routes have been used to prepare a range of 7-alkylamino- (**148**, R = CH_3, n-C_6H_{13}, n-C_8H_{17}, n-$C_{12}H_{25}$, n-$C_{14}H_{29}$),[108–111] 7-arylamino (**148**, R = Ph, 2-, 3-, 4- and 5-substituted phenyl, 2-thiazolyl)[110–114] and 7-aralkylamino derivatives [**148**, R = CH_2Ph, $(CH_2)_2Ph$].[108] In addition, a further range of secondary amines **148** with a variety of substituted alkyl or acyl groups have been described,[5,69,111,115–121] as well as a number of tertiary amines (**152**, R^1 = alkyl, R^2 = phenyl, acyl).[32,122,123] Full details of these structures, and a further range of related 7-amino derivatives,[124–129] are given in Table 4.

7-Hydrazino-3,4,5,6-tetrahydro[2H]azepine (**153**, R = H) was first described in 1930,[61] at which time there was some doubt whether the compound existed as structure **153** (R = H) or as its tautomer **154** (R = H). Subsequent investigation has done little to clarify the situation, and while later papers formulate the unsubstituted 7-hydrazino derivative as **153** (R = H),[65,130,131] there is some evidence for the existence of the tautomeric form **154** with the acyl-substituted hydrazino analogues (**154**, R = PhCO, 4-$CH_3OC_6H_4CO$, 4-ClC_6H_4CO and 4-pyridyl-CO)[132] and the thioacyl derivatives (**154**, R = CH_3SCS, C_2H_5NHCS, PhNHCS),[133] particularly when the material is in solution.[133] Most authors, however, favour the tautomeric form **153**.

(153) (154) (155)

The favoured method of preparation of both unsubstituted (**153**, R = H),[65,130] and substituted derivatives (**153**, R = alkyl, acyl)[108,132,134] is by treatment of the enol ether **97** (R = CH$_3$) or the thioether **101** (R = CH$_3$) with the appropriate hydrazine. For example, the alkyl hydrazines **153** (R = n-C$_4$H$_9$, n-C$_6$H$_{13}$, n-C$_8$H$_{17}$, n-C$_{10}$H$_{21}$, n-C$_{12}$H$_{25}$) have been prepared from both starting materials[108] and the acyl hydrazines **153** (R = CH$_3$CO, C$_2$H$_5$CO, 4-NO$_2$C$_6$H$_4$CO, 3- and 4-pyridyl-CO)[134] and (**153**, R = PhCO, 4-CH$_3$OC$_6$H$_4$CO, 4-ClC$_6$H$_4$CO and 4-pyridyl-CO)[132] were prepared from the enol ether. The reaction of hydrazine and phenylhydrazine with the thioether **101** (R = CH$_3$) has proved useful for the preparation of the dimer **155** and the phenyl analogue **153** (R = Ph),[135] respectively. A full list of related 7-substituted hydrazino analogues is included in Table 4.

PROPERTIES

Apart from the parent 3,4,5,6-tetrahydro[2H]azepine (**94**)[59] and the nitrone **156**,[136] the compounds in this class exhibit remarkable stability, even withstanding distillation at temperatures as high as 200 °C in some cases, and often forming stable, high-melting salts. Remarkably also, with a few rare exceptions, including the two compounds mentioned above (**94** and **156**), most derivatives have a substituent at position 7, emphasising again their common origin from caprolactam (**1**, R = H).

(**156**) (**97**) (**101**)

(**108**) (**148**) (**153**)

The chemistry of the enol ethers **97** and the thioether **101** is unremarkable, the compounds finding their main use as synthetic intermediates for other 7-substituted derivatives such as the 7-alkyl (**108**), 7-amino (**148**) and 7-hydrazino (**153**) derivatives. Chemically the two latter series of compounds (**148** and **153**) have proved to be extremely useful starting materials for the preparation of fused bicyclic systems. For example, the fused diazoles **157** and **158** have been prepared by cyclisation of the bromoethylamine **159** and the acetal **160**, respectively,[116,117] and the more highly substituted derivative **161** was prepared from the corresponding ketone **162**.[107]

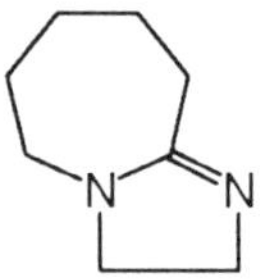

(157) 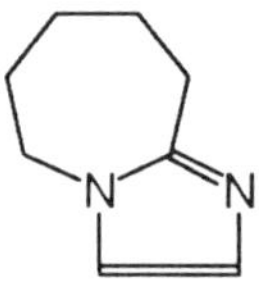 (158) 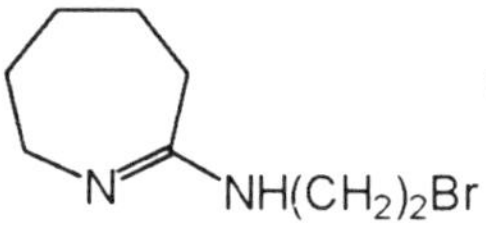 (159) 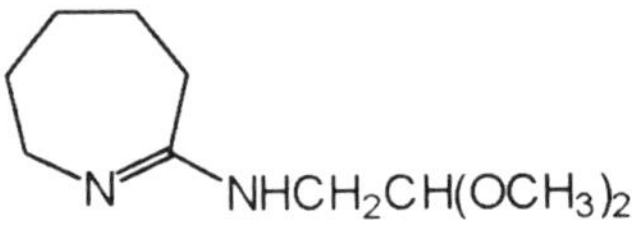 (160)

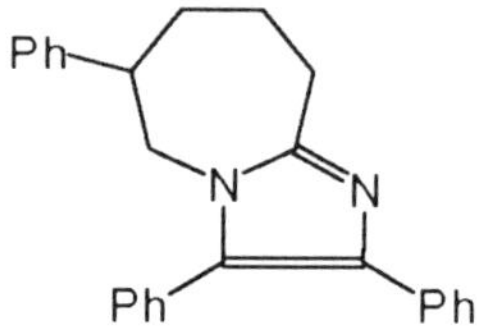

(161) 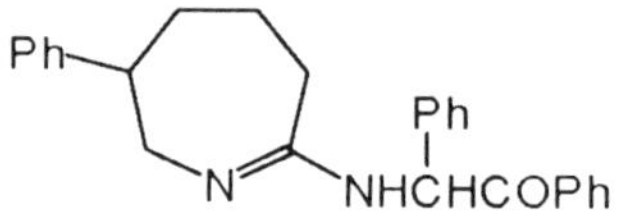 (162)

The 3-keto analogues **163** (R = H, and R = ^{i}Pr, CH$_2$Ph, H) have been prepared also from the corresponding acids **164** (R = H, and R = ^{i}Pr, CH$_2$Ph, H).[115,120]

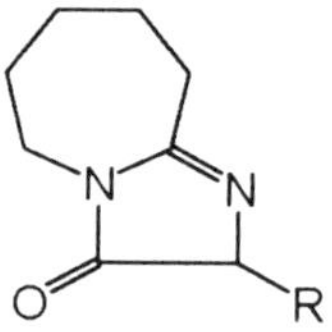

(163) 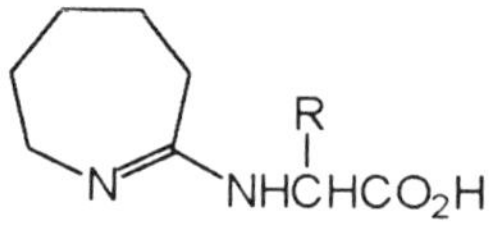 (164)

Following similar procedures, the fused 1,2,4-triazoles **165** (R = H, CH$_3$) were prepared from 7-hydrazino-3,4,5,6-tetrahydro[2*H*]azepine (**153**, R = H) on treatment with formic acid or acetic acid,[130] and the pyridino derivative **166** was formed by cyclisation of the corresponding isonicotinoyl hydrazide (**167**).[130] The fused tetrazole 'corazole' (**168**) has been prepared by cyclisation of the hydrazine **153** (R = H) using sodium nitrite and 20% sulphuric acid.[65]

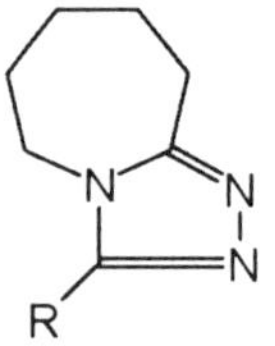

(165) 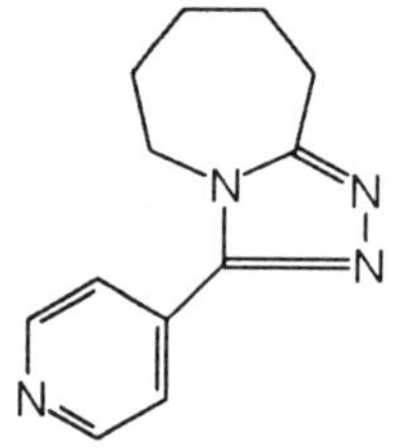 (166) 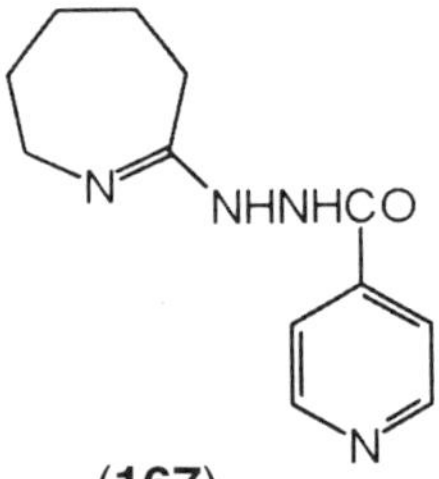 (167) 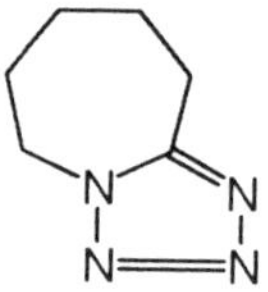 (168)

A 1,3-dipolar cycloaddition has been reported to give epimeric mixtures of the fused products **169** (R = Ph, OC$_2$H$_5$, CH$_2$OH, CN, CO$_2$CH$_3$, CHO) when

Tetrahydroazepines

the nitrone **156** was reacted with the alkenes **170** (R = Ph, OC$_2$H$_5$, CH$_2$OH, CN, CO$_2$CH$_3$, CHO). Separation of the main products from the co-formed stereoisomers, however, was not achieved.[136]

(169)   CH$_2$=CHR (170)   (171) with NHOSO$_2$CH$_3$ (172)

Finally, ring expansion of the methane sulphonate of 7-hydroxylamino-3,4,5,6-tetrahydro[2H]azepine (**171**) has been reported, giving the eight-membered ring **172** on treatment with base.[124] A full list of 3,4,5,6-tetrahydro[2H]azepines is given in Table 4.

Table 4. 3,4,5,6-Tetrahydro[2H]azepines

Structure	M.p. (°C)/b.p. (°C/mmHg)	Derivatives and m.p. (°C)	Spectroscopic evidence	Ref.
(azepine with N=)	Oil			59
(N-oxide)				136
(azepine-OCH$_3$)	65–67/24			60
	82.5/11			62
	64–66/20		IR	63
	66–67/24			
	71.5–72/24			66
	164.5–165.5/760			
	100–102/75			67
	50–52/0.4			69
	60–62/20			71
	48–50/16		^{1}H NMR, IR, MS	72a
				64,
				65,
				70

Table 4 (continued)

Structure	M.p. (°C)/b.p. (°C/mmHg)	Derivatives and m.p. (°C)	Spectroscopic evidence	Ref.
(azepine-OC$_2$H$_5$)	30–31/0.7 81–82/26 88–89/28	HCl 92–95	IR	10 60 61 68 73
(azepine-OCH(CH$_3$)$_2$)	32–33/0.4		^{1}H NMR, IR, MS	72a
(azepine-OCH$_2$CH(CH$_3$)$_2$)	115–116/20	HCl 103–106		73
(azepine-OCH$_2$C(CH$_3$)$_3$)	98–100/15	HCl 165–166		73
(azepine-O-cyclohexyl)	73–74/0.2		^{1}H NMR, IR, MS	72a
(azepine-O-phenyl)	87–90/0.2		^{1}H NMR, IR, MS	72a
(azepine-OCH$_2$-phenyl)	115–119/0.4		^{1}H NMR, IR, MS	72a
(azepine-OCO-C$_6$H$_4$-CH$_3$)	85/6.5		IR	63

Table 4 (continued)

Structure	M.p. (°C)/b.p. (°C/mmHg)	Derivatives and m.p. (°C)	Spectroscopic evidence	Ref.
(azepine)-SCH₃	82.5/12 100.5–101.5/21 87–88/15–20 Oil	 HI 175–177 Methiodide 122–124 HI	 ^{1}H NMR, IR ^{1}H, ^{13}C NMR, IR	62 66 74 74 75 80 81 70
(azepine)-SC₂H₅	Oil		^{1}H NMR, IR	80
(azepine)-SCH=CH₂				78
(azepine)-SCH₂CH=CH₂		HBr		75 77
(azepine)-S^{n}C₄H₉	95/1			146
(azepine)-SCH₂CONH₂		HCl 158–159		79
(azepine)-SCH₂Ph	140–145/7 150–152.5/4	HCl 163–165 HBr 151–153 HCl 102		73 76 75

Table 4 (continued)

Structure	M.p. (°C)/b.p. (°C/mmHg)	Derivatives and m.p. (°C)	Spectroscopic evidence	Ref.
(azepine-CH₃)	43–47/8			84
	Oil		^{1}H NMR, IR, MS	86
	Oil		MS	87
	110/760			88, 89
			IR, ^{1}H NMR	92, 93
	Oil		^{1}H NMR, IR	94
	140–149/760			96
	110–160/21	Picrate 189–190		98
	70–80/18		IR, ^{13}C NMR	103
	44–45/18	Picrate 194–195	^{1}H, ^{13}C NMR, IR, MS	105
				85, 95, 106
(azepine-C₂H₅)	60–63/15			84
	60/20			88, 89
				95
(azepine-C₃H₇)	73–75/11			84
	Oil		MS, ^{1}H NMR	87
	78/20			88, 89
(azepine-nC₄H₉)	99–100/16	Picrolonate 173		84
				82
	Oil			87
			MS, ^{1}H NMR	94
(azepine-tC₄H₉)	Oil		MS, ^{1}H NMR	87
(azepine-(CH₂)₃CH=CH₂)	110–113/14.5			83
(azepine-cyclopropyl)	70–80/5	Picrate 141	^{1}H, ^{13}C NMR, IR, MS	105

Tetrahydroazepines   335

Table 4 (continued)

Structure	M.p. (°C)/b.p. (°C/mmHg)	Derivatives and m.p. (°C)	Spectroscopic evidence	Ref.
(7-membered N-ring with cyclohexyl)				95
(7-membered N-ring with 2-oxocyclohexyl)				139
(7-membered N-ring with 1-methyl-2-oxocyclohexyl)	Oil		IR, ^{1}H NMR	97
(7-membered N-ring with cyclopropyl-CO$_2$C$_2$H$_5$)	110–115/0.05		^{1}H, ^{13}C NMR, IR, MS	104
(bis-azepine with NH)	Oil		^{1}H NMR, IR, MS	138
(7-membered N-ring with Ph)	139–141/9 Oil 103–108/4 142–245/15 108/0.6 85–95/1	Methiodide 176–179 Picrate 175–176 MeClO$_4$ salt	 MS, ^{1}H NMR ^{1}H NMR ^{1}H, ^{13}C NMR, IR, MS	84 87 98 99 100 105 142
(7-membered N-ring with CH$_2$Ph)	143.5–146/11 Oil		 ^{1}H, ^{13}C NMR, IR, MS	84 94

Table 4 (continued)

Structure	M.p. (°C)/b.p. (°C/mmHg)	Derivatives and m.p. (°C)	Spectroscopic evidence	Ref.
azepine-CH₂-C₆H₃(OCH₃)₂	195–200/0.2 64.5–66			99
azepine-CH₂CH₂Ph	105/0.05			88, 89
azepine-(CH₂)₅NH₂	134/0.6 142–146/9			137a 137b 137c
azepine-NH₂	93–94/1.5	HCl 159.5–160.5 Phenyl hydrogen methyl phosphonate 110–111.5		59 60 147
azepine-NH-CH₃	106–107 100–102 109–110	HCl 215–216 2HCl 252–255	 ^{1}H NMR ^{1}H, ^{13}C NMR ^{1}H NMR	33 110 109 111 118 144
azepine-NH-C₂H₅	110–111/10 57–60			144
azepine-NH-^{n}C₃H₇	74–77/0.07 65–67			144
azepine-NH-CH(CH₃)₂	128–131/22 86–87			144

Tetrahydroazepines 337

Table 4 (continued)

Structure	M.p. (°C)/b.p. (°C/mmHg)	Derivatives and m.p. (°C)	Spectroscopic evidence	Ref.
azepine-NH–C_6H_{13}-n	105–109/0.002	HCl 146–147		108
azepine-NH–C_8H_{17}-n	110–117/0.001	HCl 152–154		108
azepine-NH–$C_{12}H_{25}$-n	53–54	HCl 148–149		108
azepine-NH–$C_{14}H_{29}$-n	56–58			108
azepine-NH–$C_{18}H_{37}$-n	73–74	Oxalate 105.5–107		60
azepine-NH–CHPh(CH$_2$CH(CH$_3$)$_2$)	106–107			108
azepine-NH–CH$_2$CH$_2$NHC_6H_{13}-n	113–115/0.002			108
azepine-NH–CH$_2$CH$_2$NHC_8H_{17}-n	130–131/0.003			108
azepine-NH–(CH$_2$)$_3$(OCH$_2$CH$_2$)$_2$OC$_2$H$_5$	142–147/0.01			108

Table 4 (continued)

Structure	M.p. (°C)/b.p. (°C/mmHg)	Derivatives and m.p. (°C)	Spectroscopic evidence	Ref.
azepine-NH-(CH₂)₃NH-(2-Cl-C₆H₄)	150/0.0001			108
azepine-NH-(CH₂)₃NH-(3-Cl-C₆H₄)	160/0.0001			108
azepine-NH-(CH₂)₃NH-(4-Cl-C₆H₄)	150/0.0001	HCl 218–220		108
azepine-NHCH₂CH(OCH₃)₂		HCl 172–173		117
azepine-NHCH₂Ph	76–77 75–76.5			108 144
azepine-NHCH₂CH₂Ph	120–121/0.01			108
azepine-NHCH₂(tetrahydropyrimidine)		2HCl 210–212		118
azepine-NH-cyclohexyl	129–130 129.5–131			60 144
azepine-NH-Ph	126–128/1–2 105 105.6–106 102–104		¹H NMR, MS, IR, UV	33 112 110 113

Table 4 (continued)

Structure	M.p. (°C)/b.p. (°C/mmHg)	Derivatives and m.p. (°C)	Spectroscopic evidence	Ref.
2-Cl-C₆H₄ hexahydroazepine imine	92–94		¹H NMR, MS, IR, UV	113
	126–128			60
2-OCH₃-C₆H₄ hexahydroazepine imine	86–88			60
4-Cl-C₆H₄ hexahydroazepine imine	128–130		¹H NMR, MS, IR, UV	113
4-Br-C₆H₄ hexahydroazepine imine	127–128		¹H NMR, MS, IR, UV	113
4-OCH₃-C₆H₄ hexahydroazepine imine	83–85		¹H NMR, MS, IR, UV	113
4-CH₃-C₆H₄ hexahydroazepine imine			¹H, ¹³C NMR	111
2,4-Cl₂-C₆H₃ hexahydroazepine imine	104–105		¹H NMR, MS, IR, UV	113
3,4-Cl₂-C₆H₃ hexahydroazepine imine	109–110		¹H NMR, MS, IR, UV	113

Table 4 (continued)

Structure	M.p. (°C)/b.p. (°C/mmHg)	Derivatives and m.p. (°C)	Spectroscopic evidence	Ref.
azepine-NH-(2-Cl,4-CH₃-C₆H₃)	110–113		¹H NMR, MS, IR, UV	113
azepine-NH-(2-Cl,6-CF₃-C₆H₃)	113–115		¹H NMR, MS, IR, UV	113
azepine-NH-(CH₂)₆-NH-azepine	160–160.5			60
azepine-N(CH₃)₂	95–96/14	HCl 185–187		145
azepine-N(CH₃)(C₆H₅)	82/0.05	p,p'-Dichloro-diphenyldi-sulphimide 222		122
azepine-N(CONHCH₃)(C₆H₅)	108–110		¹H NMR, MS, IR, UV	113
azepine-N(CONHCH₃)(2-Cl-C₆H₄)	Glassy solid		¹H NMR, MS, IR, UV	113

Table 4 (continued)

Structure	M.p. (°C)/b.p. (°C/mmHg)	Derivatives and m.p. (°C)	Spectroscopic evidence	Ref.
tetrahydroazepine-N(CONHCH₃)-C₆H₄-4-Cl	95–97		¹H NMR, MS, IR, UV	113
tetrahydroazepine-N(CONHCH₃)-C₆H₄-4-Br	121–123		¹H NMR, MS, IR, UV	113
tetrahydroazepine-N(CONHCH₃)-C₆H₄-4-OCH₃	106–108		¹H NMR, MS, IR, UV	113
tetrahydroazepine-N(CONHCH₃)-C₆H₃-2,4-Cl₂	76–79		¹H NMR, MS, IR, UV	113
tetrahydroazepine-N(CONHCH₃)-C₆H₃-3,4-Cl₂	103–105		¹H NMR, MS, IR, UV	113
tetrahydroazepine-N(CONHCH₃)-C₆H₃-3-Cl-4-CH₃	Glassy solid		¹H NMR, MS, IR, UV	113
tetrahydroazepine-N(CONHCH₃)-C₆H₃-2-Cl-6-CF₃	Glassy solid		¹H NMR, MS, IR, UV	113

Table 4 (continued)

Structure	M.p. (°C)/b.p. (°C/mmHg)	Derivatives and m.p. (°C)	Spectroscopic evidence	Ref.
azepine-NHCHP(O)(OC$_2$H$_5$)$_2$ with phenyl	Oil		^{1}H, ^{31}P NMR	111
azepine-NHCHP(O)(OH)(OC$_2$H$_5$) with phenyl	233		^{1}H, ^{31}P NMR	111
azepine-NHCHP(O)(OH)(OC$_2$H$_5$) with 4-methylphenyl	256–258		^{1}H, ^{13}C, ^{31}P NMR MS	111
azepine-NHCHP(O)(OH)(OC$_2$H$_5$) with 4-methoxyphenyl	268–269		^{1}H, ^{31}P NMR, MS	111

Tetrahydroazepines 343

Table 4 (continued)

Structure	M.p. (°C)/b.p. (°C/mmHg)	Derivatives and m.p. (°C)	Spectroscopic evidence	Ref.
[azepine-NHCH(4-ClC6H4)P(O)(OH)(OC2H5)]	282–283		^{1}H, ^{31}P NMR, MS	111
[azepine-NHCH(C6H5)P(O)(OH)(C3H7)]	247–248		^{1}H, ^{31}P NMR, MS	111
[azepine-NHCH2CH2Br]		HBr 182		116
[azepine-NHCH2CO2H]	173–174 176–177	HCl 152–153	MS	115 120
[azepine-NH(CH2)5CO2H]	80	HCl 152–154 p-Toluene-sulphonate 124 HCl 148, 150 Benzene-sulphonate 99	IR (salt) IR, ^{1}H NMR (salt) IR (salt)	5 119 119 119 144
[azepine-NH(CH2)5CO2CH3]	154/0.005			5
[azepine-NHCH(CH(CH3)2)CO2H]	173–175		^{1}H NMR, IR, MS	120

Table 4 (continued)

Structure	M.p. (°C)/b.p. (°C/mmHg)	Derivatives and m.p. (°C)	Spectroscopic evidence	Ref.
azepine-NHCH(CH$_2$Ph)CO$_2$H	139–141		^{1}H NMR, IR, MS	120
azepine-NHCH(Ph)CO$_2$H	167–168		^{1}H NMR, IR, MS	120
azepine-N(CH$_3$)COCH$_2$Cl			^{1}H NMR	33
azepine-N(Ph)COCH$_2$Cl			^{1}H NMR	33
azepine-NH(CH$_2$)$_2$SO$_3$H	279–280		IR, ^{1}H NMR	154
azepine-NH(CH$_2$)$_3$SO$_3$H	307–308		IR, ^{1}H NMR	154
azepine-azetidinone(CH$_3$,CH$_3$)	82–85/0.1 87–88			125

Tetrahydroazepines 345

Table 4 (continued)

Structure	M.p. (°C)/b.p. (°C/mmHg)	Derivatives and m.p. (°C)	Spectroscopic evidence	Ref.
(azepine-N=C-N-azetidinone with CH₃ groups)	92/0.4, 73–74			125
(azepine-N=C-N-isothiazolidine-1,1-dioxide)	129–131		IR, ¹H NMR	154
(azepine-N=C-N-piperidine)	56–58/0.01 133–134/760	Picrate 118	IR, ¹H NMR	91 114
(azepine-N=C-N-caprolactam)	30.5–32 31–32 113/0.04 30–32		IR, ¹H NMR ¹H NMR, MS	140b 140a 126
(dichloro-azepine-N=C-N-dichlorocaprolactam)	84.5–85		IR, ¹H NMR	126
(azepine-N=C-NH-thiazole)				114
(azepine-N=C-NH-CH₂-NH-C=N-azepine)	120–121			110

Table 4 (continued)

Structure	M.p. (°C)/b.p. (°C/mmHg)	Derivatives and m.p. (°C)	Spectroscopic evidence	Ref.
azepine–NH–C₆H₄–NH–azepine	213–214			110
azepine–N(piperazine)N–azepine				114
azepine–NHOSO₂CH₃	89–90			124
azepine–NHCONHSO₂–C₆H₅	190		IR, ^{1}H NMR	121
azepine–NHCONHSO₂–C₆H₄–CH₃	172		IR, ^{1}H NMR	121
azepine–NHCONHSO₂–C₆H₄–Cl	189		IR, ^{1}H NMR	121
azepine–N(CH₃)–C(=S)–NH–Ph	106		IR, ^{1}H NMR	123
azepine–NH₂NH₂	103.5–104.5 112	Picrate 136		60, 61, 65, 130, 132
azepine–NH–NH–$C_4H_9^n$	144–145/12	HCl 153–157		108

Table 4 (continued)

Structure	M.p. (°C)/b.p. (°C/mmHg)	Derivatives and m.p. (°C)	Spectroscopic evidence	Ref.
azepine-NH-NH-nC_6H_{13}	105–109/0.01	HCl 155–158		108
azepine-NH-NH-nC_8H_{17}	128–131/0.03	HCl 146–150		108
azepine-NH-NH-$nC_{10}H_{21}$	145–149/0.01			108
azepine-NH-NH-$nC_{12}H_{25}$	159–163/0.01			108
azepine-NH-NH-CH$_2$Ph	140–145/0.01	Oxalate 156–157		108
azepine-NHNHPh	109–111 113–115 108		IR	110 130 135
azepine-NHNH-tetrazole	225			134
azepine-NHNH-azepine	125.5–126.5 125.5		IR	60 135
azepine-NHNH-benzisothiazole-SO$_2$	197–198			134
azepine-NH-triazole	242			134

Table 4 (continued)

Structure	M.p. (°C)/b.p. (°C/mmHg)	Derivatives and m.p. (°C)	Spectroscopic evidence	Ref.
azepine-NHNH.COCH$_3$	176–177	HCl 181		134
azepine-NHNH.COCH$_3$	169			134
azepine-NHNH.CO-C$_6$H$_5$	173–179			132
azepine-NHNH.CO-C$_6$H$_4$-OCH$_3$	179–185			132
azepine-NHNH.CO-C$_6$H$_4$-Cl		HCl 199–201		132
azepine-NHNH.CO-C$_6$H$_4$-NO$_2$	188			134
azepine-NHNH.CO-pyridyl	108			134
azepine-NHNH.CO-pyridyl	208–210 210			130 134 132
azepine-NHNH-CO$_2$C$_2$H$_5$	122			134

Tetrahydroazepines 349

Table 4 (continued)

Structure	M.p. (°C)/b.p. (°C/mmHg)	Derivatives and m.p. (°C)	Spectroscopic evidence	Ref.
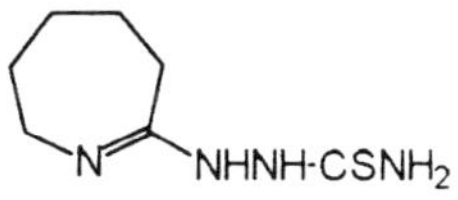 azepine-NHNH·CSNH₂	240–250			134
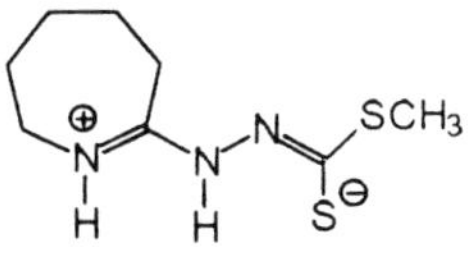 azepinium-NH-N(H)-C(S⁻)-SCH₃	174–176		^{1}H, ^{13}C NMR, MS, UV	133
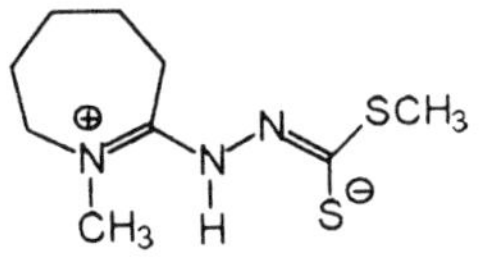 N-methyl azepinium-NH-N(H)-C(S⁻)-SCH₃	158–160		^{1}H, ^{13}C NMR, MS, UV	133
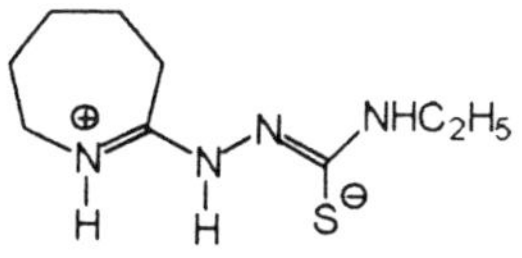 azepinium-NH-N(H)-C(S⁻)-NHC₂H₅	125		^{1}H NMR, MS	133
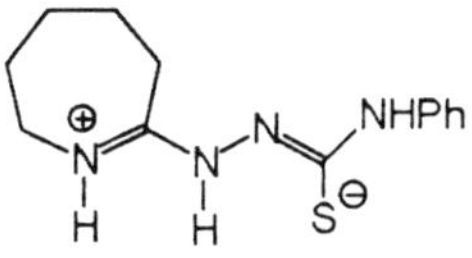 azepinium-NH-N(H)-C(S⁻)-NHPh	166–167		^{1}H NMR, MS	133
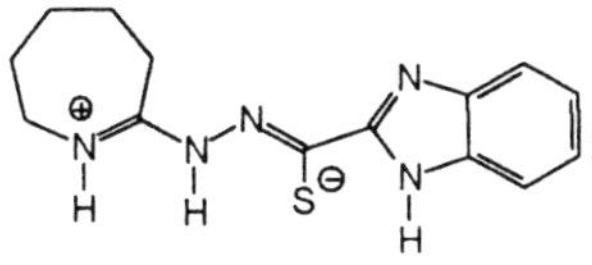 azepinium-NH-N(H)-C(S⁻)-(benzimidazole)	241–243		^{1}H NMR, MS	133
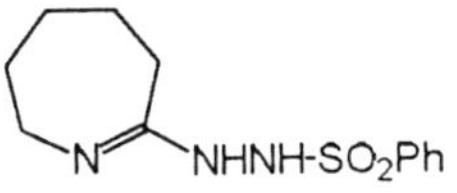 azepine-NHNH-SO₂Ph	190–191			134
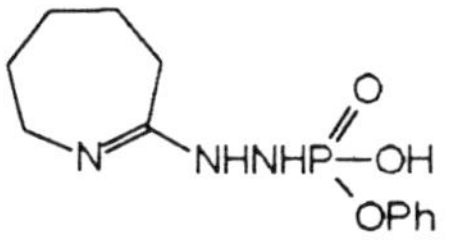 azepine-NHNHP(O)(OH)(OPh)			IR, ^{1}H NMR	131

Table 4 (continued)

Structure	M.p. (°C)/b.p. (°C/mmHg)	Derivatives and m.p. (°C)	Spectroscopic evidence	Ref.
azepine-NHNHP(=S)(OH)(CH₃)			IR, ¹H NMR	131
azepine-NHNHP(=S)(OH)(OC₂H₅)			IR, ¹H NMR	131
azepine-NHNHP(=S)(OH)(OCH₃)			IR, ¹H NMR	131
azepine-Cl		HCl 66–70		71, 72a,b, 112
N-methyl azepinium-NHPh, N-SCH₃, I⁻	192–193			128
N-methyl azepinium-N(C₆H₄Cl)SCH₃, I⁻	168–170			128
azepine-N=C(NHC₂H₅)S-CH₂-C₆H₄-CN	109–111		¹H NMR, MS	127

Tetrahydroazepines

Table 4 (continued)

Structure	M.p. (°C)/b.p. (°C/mmHg)	Derivatives and m.p. (°C)	Spectroscopic evidence	Ref.
azepine-Ge(C₂H₅)₃			^{1}H, ^{13}C NMR, MS	141
CH₃-azepine-C₄H₉	Oil		^{1}H NMR, IR, MS	94
CH₃-azepine-OCH₃	110/25 70–73/8			67 129
CH₃-azepine-SCH₃	45/0.1 90–92/20 130–135/5	HI 120–125	^{1}H NMR, IR	67 74 80
CH₃-azepine-NH₂		HCl 94–97		129
Ph-azepine-CH₃	105–106/0.5		^{1}H, ^{13}C NMR, IR	101
Ph-azepine-OC₂H₅	107–110/0.2			107
Ph-azepine-NH₂		HCl 258		107
CH₃-azepine-CH(CH₃)₂	198–201/766		IR	96

Table 4 (continued)

Structure	M.p. (°C)/b.p. (°C/mmHg)	Derivatives and m.p. (°C)	Spectroscopic evidence	Ref.
3-methyl-2-methoxy azepine	60–62/12 70–73/8			67 129
3-ethyl-2-methoxy azepine	70–72/11			67
3-propyl-2-methoxy azepine	89–90/10			67
3-methyl-2-methylthio azepine	Oil		^{1}H NMR, IR ^{1}H, ^{13}C NMR, IR	80 81
3-methyl-2-amino azepine		HCl 94–97		129
N-oxide azepinium	Oil		IR, MS, ^{1}H NMR	102a 102b
dimethyl azepine with acetonyl			^{1}H, ^{13}C NMR	35 90
dimethyl azepine with ethoxycarbonyl ketone			^{1}H, ^{13}C NMR	90

Table 4 (continued)

Structure	M.p. (°C)/b.p. (°C/mmHg)	Derivatives and m.p. (°C)	Spectroscopic evidence	Ref.
(structure)	40/0.1			67
(structure)	Oil		^{1}H NMR, IR, MS	143

REFERENCES

1. R. Lukeš, V. Dudek, O. Sedláková and J. Korán, *Collect. Czech. Chem. Commun.*, 1961, **26**, 1105; *Chem. Abs.*, 1961, **55**, 18757c.
2. (a) A. Cervinka and L. Hub, *Collect. Czech. Chem. Commun.*, 1964, **30**, 3111; *Chem. Abs.*, 1965, **63**, 13212b; (b) A. Cervinka and L. Hub, *Tetrahedron Lett.*, 1964, 463.
3. W. Carruthers and R.A. Johnstone, *J. Chem. Soc.*, 1965, 1653.
4. J. Stehlicek, B. Valter and J. Sebenda, *Makromol. Chem.*, 1986, **187**, 513; *Chem. Abs.*, 1986, **104**, 207748m.
5. H.R. Meyer, *Kunststoffe–Plastics*, 1956, **3**, 160; *Chem. Abs.*, 1958, **52**, 11781q.
6. T. Fukumoto and M. Murakami, *Nippon Kagaku Sasshi*, 1963, **84**, 736; *Chem. Abs.*, 1964, **60**, 653q.
7. Stamicarbon NV, *Br. Pat.*, 901 169, 1962; *Chem. Abs.*, 1963, **58**, 6810e.
8. J.H. Ottenheym and J.W. Garritsen, *Ger. Pat.*, 1 157 210, 1963; *Chem. Abs.*, 1964, **60**, p6756e.
9. B. Haveaux, A. Dekoker, M. Rens, A.R. Sidani, J. Toye and L. Ghosez, *Org. Synth.*, 1979, **59**, 26.
10. M.T. Tetenbaum, *J. Org. Chem.*, 1966, **31**, 4298.
11. R. Gompper and W. Elser, *Tetrahedron Lett.*, 1964, 1971.
12. (a) H. Takahata, A. Tomiguchi and T. Yamazaki, *Chem. Pharm. Bull.*, 1980, **28**, 1000; (b) J. Körösi, *J. Prakt. Chem.*, 1964, **23**, 212.
13. H. Takahata, M. Nakamo, A. Tomiguchi and T. Yamazaki, *Heterocycles*, 1982, **17** (Special Edition), 413.
14. H. Takahata, T. Nakajima, M. Nakano, A. Tomiguchi and T. Yamazaki, *Chem. Pharm. Bull.*, 1985, **33**, 4299.
15. U. Kraatz, *Liebigs Ann. Chem.*, 1976, **3**, 412.
16. H. Takahata, T. Yamazaki and K. Aoe, *J. Org. Chem.*, 1985, **50**, 4648.
17. K. Hartke and A. Brutsche, *Arch. Pharm. (Weinheim)*, 1993, **326**, 63; *Chem. Abs.*, 1993, **119**, 28100g.
18. (a) P. Duhamel and M. Kotera, *J. Org. Chem.*, 1982, **47**, 1688; (b) P. Duhamel and M. Kotera, *J. Chem. Res. (S)*, 1982, 276; *(M)*, 1982, 2851.

19. (a) P. Duhamel, M. Kotera, T. Monteil and B. Marabout, *J. Org. Chem.*, 1989, **54**, 4419; (b) P. Duhamel, M. Kotera and T. Montiel, *Bull. Chem. Soc. Jpn.*, 1986, **59**, 2353.
20. E.P. Kramarova, N.A. Anisimova and Yu.I. Baukov, *Zh. Obshch. Khim.*, 1976, **46**, 1658; *Chem. Abs.*, 1976, **85**, 143188h.
21. E.P. Kramarova, A.G. Shipov, O.B. Artamkina and Yu. I. Baukov, *Zh. Obshch. Khim.*, 1984, **54**, 1921; *Chem. Abs.*, 1985, **102**, 62311k.
22. (a) K. Nyberg, *Synthesis*, 1976, **8**, 545; (b) L. Eberson, M. Malmberg and K. Nyberg, *Acta Chem. Scand., Ser. B*, 1984, **38**, 391; *Chem. Abs.*, 1984, **101**, 211067u.
23. S. Torii, T. Inokuchi, F. Akahosi and M. Kubota, *Synthesis*, 1987, **3**, 242.
24. A.I. Meyers, P.D. Edwards, T.R. Bailey and E.G. Jagdmann, Jr, *J. Org. Chem.*, 1985, **50**, 1019.
25. (a) L.B. Dmitriev, I.I. Grandberg and V.A. Moskalenko, *Izv. Timiryazevsk. Skh. Akad.*, 1968, (2) 195; *Chem. Abs.*, 1968, **69**, 43770t; (b) V.V. Vil'yams and L.B. Dmitriev, *Dokl. Timiryazevsk. Skh. Akad.*, 1965, **115**, 247; *Chem. Abs.*, 1966, **65**, 8868b.
26. (a) O. Seshimoto, T. Kumagai, K. Shimizu and T. Mukai, *Chem. Lett.*, 1977, **10**, 1195; (b) T. Kumagai, K. Shimizu, Y. Kawamura and T. Mukai, *Tetrahedron*, 1981, **37**, 3365.
27. G. Dannhardt and R. Obergrusberger, *Arch. Pharm. (Weinheim)*, 1981, **314**, 787; *Chem. Abs.*, 1982, **96**, 6510v.
28. G. Singh and A.K. Mandel, *Heterocycles*, 1982, **18** (Special Edition), 291.
29. A.M. Zhidkova, V.G. Granik, N.S. Kuryatov, V.P. Pakhomov, O.S. Anisimova and R.G. Glushkov, *Khim. Geterotsikl. Soedin.*, 1974, 1089; *Chem. Abs.*, 1975, **82**, 16640n.
30. V.G. Granik and R.G. Glushkov, *Zh. Org. Khim.*, 1971, **7**, 1146; *Chem. Abs.*, 1971, **75**, 110176t.
31. V. Yu. Mavrin, V.V. Moskva and T.N. Apal'kova, *Zh. Obshch. Khim.*, 1988, **58**, 1668; *Chem. Abs.*, 1988, **109**, 170177z..
32. T. Shono, J. Terauchi, Y. Ohki and Y. Matsumura, *Tetrahedron Lett.*, 1990, 6385.
33. T.V. Aleshnikova, A.F. Prokof'eva.V.V. Negrebetskii and N.N. Mel'nikov, *Zh. Obshch. Khim.*, 1989, **59**, 282; *Chem. Abs.*, 1989, **111**, 134330b.
34. A.K. Shanarazov, N.P. Solov'eva, V.V. Chistyakov and V.G. Granik, *Khim. Geterotsikl. Soedin.*, 1991, **2**, 218; *Chem. Abs.*, 1991, **115**, 48779f.
35. E.P. Müller, *Helv. Chim. Acta*, 1985, **68**, 1107.
36. T. Fukumoto and M. Murakami, *Nippon Kagaku. Zasshi*, 1963, **84**, 740; *Chem. Abs.*, 1964, **60**, 654a.
37. H. Keifer, *Synthesis*, 1972, **2**, 81.
38. M. Iorio and M. Miraglia, *Chim. Ther.*, 1973, **8**, 85.
39. P. Krogsgaard-Larsen, K. Thyssen and K. Schaumburg, *Acta Chem. Scand., Ser. B,* 1978, **32**, 327; *Chem. Abs.*, 1979, **90**, 39082m.
40. P. Krosgaard-Larsen and T. Roldskov-Christiansen, *Eur. J. Med. Chem. Chim. Ther.*, 1979, **14**, 157.
41. A.D. Yanina and M.V. Rubtsov, *Zh. Obshch. Khim.*, 1962, **32**, 3693; *Chem. Abs.*, 1963, **58**, 12512e.
42. S.J. Neeson and P.J. Stevenson, *Tetrahedron Lett.*, 1988, **29**, 3993.
43. R. Grigg, V. Santhakumar, V. Sridharan, P. Stevenson, A. Teasdale, M. Thornton-Pett and T. Worakun, *Tetrahedron*, 1992, **47**, 9703.
44. C. Flann, T.C. Malone and L.E. Overman, *J. Am. Chem. Soc.*, 1987, **109**, 6097.

45. D.I. Bishop, I.K. Al-Khawaja and J.A. Joule, *J. Chem Res. (S)*, 1981, 361; *(M)*, 1981, 4279.
46. D.I. Bishop, I.K. Al-Khawaja, F. Heatley and J.A. Joule, *J. Chem Res. (S)*, 1982, 159; *(M)*, 1982, 1766.
47. G.C. Fu and R.M. Grubbs, *J. Am. Chem. Soc.*, 1992, **114**, 7324.
48. (a) A. Hassner, W. Chau and R. D'Costa, *Isr. J. Chem.*, 1982, **72**, 76; *Chem. Abs.*, 1982, **97**, 198048r; (b) A. Hassner, R. D'Costa, A. McPhail and W. Butler, *Tetrahedron Lett.*, 1981, 3691; (c) A. Hassner and W. Chau, *Tetrahedron Lett.*, 1982, 1989.
49. M.A. Calcagno and E.E. Schweizer, *J. Org. Chem.*, 1978, **43**, 4207.
50. H.W. Bersch, R. Rissmann and D. Schon, *Arch. Pharm. (Weinheim)*, 1982, **315**, 749; *Chem. Abs.*, 1982, **97**, 162796u.
51. W. Chen, J. Feng, Z. Zhou and J. Zhang, *Synthesis*, 1989, **3**, .82.
52. N. Finch, L. Blanchard and L.H. Werner, *J. Org. Chem.*, 1977, **42**, 3933.
53. B. Guyot, J. Pornet and L. Miginiac, *J. Organomet. Chem.*, 1990, **386**, 19.
54. J. Diamond, W.F. Bruce and F.T. Tyson, *J. Med. Chem.*, 1964, **7**, 57.
55. (a) T. Sasaki, S. Eguchi and M. Ohno, *J. Am. Chem. Soc.*, 1970, **92**, 3192; (b) T. Sasaki, S. Eguchi and M. Ohno, *J. Org. Chem.*, 1972, **37**, 466.
56. P. Krogsgaard-Larsen, H. Hjeds, S.B. Christensen and L. Brehm, *Acta Chem. Scand.*, 1973, **27**, 3251; *Chem. Abs.*, 1974, **80**, 95798e.
57. H. Yamamoto, H. Kawamoto, S. Morosawa and A. Yokoo, *Heterocycles*, 1978, **11**, 267.
58. (a) C.A. Grob, W. Kunz and P.R. Marbet, *Tetrahedron Lett.*, 1975, 2612; (b) C.A. Grob, M. Bolleter and W. Kunz, *Angew. Chem.*, 1980, **92**, 734; *Chem. Abs.*, 1981, **94**, 14725t.
59. (a) A. Ya. Lazaris, E.N. Zil'berman, E.V. Lunicheva and E.M. Vedin, *Zh. Prikl. Khim.*, 1965, **38**, 1097; *Chem. Abs.*, 1965, **63**, 6857c; (b) H.C. Kozlov, V.A. Tarasevich and S.I. Kozintsev, *Dokl. Akad. Nauk BSSR*, 1984, **28**, 37; *Chem. Abs.*, 1984, **100**, 174237f.
60. R.E. Benson and T.L. Cairns, *J. Am. Chem. Soc.*, 1948, **70**, 2115.
61. R. Stollé, O Schattner and F. Hanush, *Chem. Ber.*, 1930, **63B**, 1032; *Chem. Abs.*, 1930, **24**, 4034.
62. H. Behringer and H. Meier, *Liebigs Ann. Chem.*, 1957, **607**, 67; *Chem. Abs.*, 1958, **52**, 13748b.
63. W.Z. Heldt, *J. Am. Chem. Soc.*, 1958, **80**, 5880.
64. R.G. Glushkov and E.S. Golouchinskaya, *Zh. Prikl. Khim.*, 1959, **32**, 920; *Chem. Abs.*, 1959, **53**, 17111q.
65. R.G. Glushkov and E.S. Golouchinskaya, *Med. Prom. SSSR*, 1960, **14**, 12; *Chem. Abs.*, 1960, **54**, 22605b.
66. J. Körösi, *J. Prakt. Chem.*, 1964, **23**, 212; *Chem Abs.*, 1964, **60**, 13315f.
67. T. Duong, R.H. Prager, A.D. Ward and D.I.B. Kerr, *Aust. J. Chem.*, 1976, **29**, 2651.
68. R.W. Ralls, *J. Org. Chem.*, 1961, **26**, 66.
69. P. Schlack, *US Pat.*, 2356622, 1944; *Chem. Abs.*, 1945, **39**, 1420.
70. H. Nishiyama, H. Nagase and K. Ohno, *Tetrahedron Lett.*, 1979, 4671.
71. H. Ulrich, B. Tucker and R. Richter, *J. Org. Chem.*, 1978, **43**, 1544.
72. (a) J. Jurczak, T. Kozluk, W. Kulicki, M. Pietraszkiewicz and J. Szymanski, *Synthesis*, 1983, **5**, 382; (b) J. Jurczak, T. Kozluk, W. Kulicki, M. Pietraszkiewicz and J. Szymanski, *Pol. J. Chem.*, 1983, **57**, 447; *Chem. Abs.*, 1984, **101**, 110706w.
73. E. Mohacsi and E.M. Gordon, *Synth. Commun.*, 1984, **14**, 1159.
74. D.J. Fry and B.A. Lea, *Br. Pat.*, 795134, 1958; *Chem. Abs.*, 1959, **53**, p3722f.

75. A.A. Avetisyan, Zh.G. Boyadzhyan, A.A. Zazyan and M.T. Dangyan, *Khim. Geterotsikl. Soedin.*, 1968, 102; *Chem. Abs.*, 1968, **69**, 106691r.
76. D.E. Beattie, R. Crossley, K.H. Dickinson and G.M. Dover, *Eur. J. Med. Chem. Chim. Ther.*, 1983, **18**, 277; *Chem. Abs.*, 1984, **100**, 44878e.
77. O.P. Sidel'kovskaya, A.A. Avetisyan, V.I. Seifman and M.T. Dangyan, *Khim. Atsetilena, Tr. Vses. Konf. 3rd*, 1968, 253 (Pub. 1972); *Chem. Abs.*, 1973, **79**, 18521m.
78. F.P. Sidel'kovskaya and A.A. Avetisyan, *USSR Pat.*, 168 704 1965; *Chem. Abs.*, 1965, **63**, p1709b.
79. M. Ohta, K. Yoshida and S. Sato, *Bull. Chem. Soc. Jpn.*, 1966, **39**, 1269; *Chem. Abs.*, 1966, **65**, 13682d.
80. K. Maruoka, T. Miyazak, M. Ando, Y. Matsumura, S. Sakani, K. Hattori and H. Yamamoto, *J. Am. Chem. Soc.*, 1983, **105**, 2831.
81. B.M. Trost, M. Vaultier and M.L. Santiago, *J. Am. Chem. Soc.*, 1980, **102**, 7929.
82. K. Heyns and W. Pyrus, *Chem. Ber.*, 1955, **88**, 678; *Chem. Abs.*, 1956, **50**, 3300e.
83. O. Cervinka. *Chem Listy*, 1958, **52**, 1145; *Chem. Abs.*, 1958, **52**, 18406d.
84. V. Dudek and H.-O. Li, *Collect. Czech. Chem. Commun.*, 1965, **30**, 2472; *Chem. Abs.*, 1965, **63**, 11497f
85. D.H. Hua, W. Shou, S.N. Bharathi, T. Katsuhira and A.A. Bravo, *J. Org. Chem.*, 1990, **55**, 3682.
86. J. Bielawski, S. Brandange and L. Lindblom, *J. Heterocycl. Chem.*, 1978, **15**, 97.
87. C.A. Zezza, M.B. Smith, B.A. Ross, A. Arhin and P.L.E. Cronin, *J. Org. Chem.*, 1984, **49**, 4397.
88. D. Bacos, J.P. Celerier and C. Lhommet, *Tetrahedron Lett.*, 1987, 2353.
89. P. Delbecq, D. Bacos, J.P. Celerier and C. Lhommet, *Can. J. Chem.*, 1991, **69**, 1201.
90. E.P. Müller, *Helv. Chim. Acta*, 1986, **69**, 692.
91. E. Vilsmaier, G. Kristen and C. Tetzlaff, *J. Org. Chem.*, 1988, **53**, 1806.
92. R.V. Hoffman, A. Kumar and G.A. Buntain, *J. Am. Chem. Soc.*, 1985, **107**, 4731.
93. R.V. Hoffman and A. Kumar, *J. Org. Chem.*, 1985, **50**, 1859.
94. R.V. Hoffman and G.A. Buntain, *J. Org. Chem.*, 1988, **53**, 3316.
95. Y. Yukawa and K. Tanaka, *Mem. Inst. Sci. Ind. Res., Osaka Univ.*, 1957, **14**, 199; *Chem. Abs.*, 1958, **52**, 5413h
96. J.H. Boyer and F.C. Cantor, *J. Am. Chem. Soc.*, 1955, **77**, 3287.
97. Y. Matsumura, J. Fujiwara, K. Maruoka and H. Yamamoto, *J. Am. Chem. Soc.*, 1983, **105**, 6312.
98. I. Murakoshi, Y. Shikakura and J. Haginiwa, *Yakagaku Zasshi*, 1964, **84**, 671; *Chem. Abs.*, 1964, **61**, 9465a.
99. K.A. Maier and O. Hromatka, *Monatsh. Chem.*, 1971, **102**, 513; *Chem. Abs.*, 1971, **75**, 5669h.
100. S. Wawzonek and J.M. Shradel, *Org. Prep. Proced. Int.*, 1977, **9**, 281; *Chem. Abs.*, 1978, **88**, 22489k.
101. M. Miocque, O. Lafont and L. Maserier-Demagny, *Bull. Soc. Chim. Fr.*, 1977, **11–12**, Part 2, 1237.
102. (a) M.E. Fox, A.B. Holmes, I.T. Forbes and M. Thompson, *Tetrahedron Lett.*, 1992, 7421; (b) M.E. Fox, A.B. Holmes, I.T. Forbes and M. Thompson, *J. Chem. Soc., Perkin Trans. 1*, 1994, 3379.
103. P.H. Lambert, M. Vaultier and R. Carrie, *J. Chem. Soc., Chem. Commun.*, 1982, 1224.
104. M. Vaultier, P.H. Lambert and R. Carrie, *Bull. Soc. Chim. Belg.*, 1985, **94**, 449.
105. M. Vaultier, P.H. Lambert and R. Carrie, *Bull. Soc. Chim. Fr.*, 1986, 83.
106. A. Benalil, A. Guerin, B. Carboni and M. Vaultier, *J. Chem. Soc., Perkin Trans. 1*, 1993, 1061.

107. M. Langlois, C. Guillonneau, V.V. Tri, J.P. Meinngan and J. Maillard, *J. Heterocycl. Chem.*, 1982, **19**, 193.
108. J.R. Geigy SA, *Fr. Pat.*, 1 367 799, 1964; *Chem. Abs.*, 1964, **61**, p16055a.
109. C.L. Perrin and O. Nunez, *J. Am. Chem. Soc.*, 1986, **108**, 5997.
110. P. Cefelin, E. Sittler and O. Wichterle, *Collect. Czech. Chem. Commun.*, 1960, **25**, 2522; *Chem. Abs.*, 1961, **55**, 3608h.
111. T.V. Aleshnikova, A.F. Prokof'eva, V.V. Negrebetskii, A.F. Grapov and N.N. Mel'nikov, *Zh. Obshch. Khim.*, 1986, **56**, 2507; *Chem. Abs.*, 1988, **108**, 75487w.
112. Stamicarbon NV, *Belg. Pat.*, 609 822, 1962; *Chem. Abs.*, 1962, **57**, p16505g.
113. P. Javorski, Z. Vesela and S. Truchlik, *Chem. Zvest.*, 1978, **32**, 223; *Chem. Abs.*, 1978, **89**, 179837r.
114. J. Körösi, A. Lay and G. Szabo, *Ger. Pat.*, 2 119 163, 1971; *Chem. Abs.*, 1972, **76**, 25121.
115. S. Petersen and E. Tietze, *Liebigs Ann. Chem.*, 1959, **623**, 166; *Chem. Abs.*, 1960, **54**, 14258a.
116. R. Stollé, M. Merkle and F Hanusch, *J. Prakt Chem.*, 1934, **140**, 59; *Chem. Abs.*, 1934, **28**, 5445[6].
117. K. Nagarajan, V.P. Arya, S.J. Shenoy, R.K. Shah, A.N. Goud and G.A. Bhat, *Indian J. Chem., Sect. B*, 1977, **15**, 629; *Chem. Abs.*, 1978, **88**, 37693w.
118. S. Rajappa and R. Sreenivasan, *Indian J. Chem., Sect. B*, 1979, **174**, 334; *Chem. Abs.*, 1980, **92**, 215375m.
119. T. Sato and H. Wakatsuka, *Bull. Chem. Soc. Jpn.*, 1969, **42**, 1955.
120. F. Campagna, A. Carotti and G. Casini, *J. Heterocycl. Chem.*, 1990, **27**, 1973.
121. D.R. Shridhar, C.V.R. Sastry, P. Parihar, G.S. Thapar and A. Krishnamurthy, *Indian J. Chem., Sect. B*, 1985, **24**, 693; *Chem. Abs.*, 1986, **104**, 148718b.
122. H. Bredereck and K. Bredereck, *Chem. Ber.*, 1961, **94**, 2278; *Chem. Abs.*, 1961, **55**, 27372b.
123. G. Schwenker and R. Kolb, *Chem. Ber.*, 1975, **108**, 1142; *Chem. Abs.*, 1975, **83**, 10010f.
124. R. Richter, B. Tucker and H. Ulrich, *J. Org. Chem.*, 1983, **48**, 1694.
125. D. Bormann, *Chem. Ber.*, 1970, **103**, 1797; *Chem. Abs.*, 1970, **73**, 25206s.
126. G. Reinish, K. Dietrich and F. Dargazanli, *J. Prakt. Chem.*, 1969, **311**, 455; *Chem. Abs.*, 1969, **71**, 38784n.
127. U. Radics, J. Liebscher and M. Pätzel, *Synthesis*, 1992, 673.
128. J. Liebscher, M. Pätzel and Y.F. Kelboro, *Synthesis*, 1989, 672.
129. I. Hermecz, L. Vasvari-Debreczy, A. Horvath, M. Baloch, J. Kokosi, C. De Vos and L. Rodriguez, *J. Med. Chem.*, 1987, **30**, 1543.
130. R.G. Glushkov and O. Yu. Magidson, *Zh. Obshch. Khim.*, 1960, **30**, 649; *Chem. Abs.*, 1960, **54**, 24712a
131. L.V. Radvodovskaya, M.V. Erikova, N.A. Kiseleva, A.F. Grapov and N.N. Mel'nikov, *Zh. Vses. Khim. Ova.*, 1986, **31**, 105; *Chem. Abs.*, 1986, **105**, 226795m.
132. H. Reimlinger, J.J. Vanderwalle and W.R.F. Lingier, *Chem. Ber.*, 1970, **103**, 1934; *Chem. Abs.*, 1970, **73**, 25365t.
133. U. Radics, J. Liebscher, B. Ziemer and V. Rybakov, *Chem. Ber.*, 1992, **125**, 1389; *Chem. Abs.*, 1992, **117**, 48247x.
134. S. Petersen and E. Tietze, *Chem. Ber.*, 1957, **90**, 909; *Chem. Abs.*, 1958, **52**, 9158h.
135. T. Mukaiyama and S. Ono, *Tetrahedron Lett.*, 1968, 3569.
136. Sk.A. Ali and M.I.M. Wazeer, *J. Chem. Res. (S)*, 1992, 62; *(M)*, 1992, 614.

137. (a) G. Nawrath, *Angew. Chem.*, 1960, **72**, 1002; *Chem. Abs.*, 1961, **55**, 10306f; (b) Farbenfabriken Bayer AG, *Br. Pat.*, 922275, 1963; *Chem. Abs.*, 1963, **59**, p11261f; (c) Kanegafuchi Spinning Co. Ltd, *Belg. Pat.*, 652893, 1966; *Chem. Abs.*, 1966, **64**, p11084h.
138. A. Ciaperoni, L. Mariani and G.B. Gechele, *Chim. Ind. (Milan)*, 1968, **50**, 772; *Chem. Abs.*, 1968, **69**, 76583n.
139. S. Hünig, W. Lücke, V. Meuer and W. Grässmann, *Angew. Chem.*, 1963, **75**, 295; *Chem. Abs.*, 1963, **59**, 573g.
140. (a) R. Mazurkiewicz, *Acta Chim. Hung.*, 1984, **116**, 95; *Chem. Abs.*, 1984, **101**, 191663e; (b) R. Mazurkiewicz, *Acta Chim. Hung.*, 1990, **127**, 439; *Chem. Abs.*, 1991, **114**, 206693j.
141. A.V. Seleznev, D.A. Bravo-Zhivotovski, I.D. Kalikhman, V. Yu. Vitkovski, L.L. Ghaintseva and M.G. Voronkov, *Metalloorg. Khim.*, 1989, **2**, 472; *Chem. Abs.*, 1990, **112**, 77394v.
142. H. Möhrle and H. Dwuletzki, *Arch. Pharm. (Weinheim)*, 1986, **319**, 1049; *Chem. Abs.*, 1987, **106**, 32180g.
143. T.H. Koch, M.A. Geigel and C.-C. Tsai, *J. Org. Chem.*, 1973, **38**, 1090.
144. K. Gehrke, *Faserforsch. Textiltech.*, 1963, **14**, 468; *Chem. Abs.*, 1964, **61**, 4493f.
145. M. Seefelder, *Ger. Pat.* 1078568, 1960; *Chem. Abs.*, 1961, **55**, p16570i.
146. V.N. Sergeev, S.A. Artamkin, S.V. Pestunovich, A.I. Albanov, M.G. Voronkov and Yu.I. Baukov, *Zh. Obshch. Khim.*, 1992, **62**, 1819; *Chem. Abs.*, 1993, **118**, 234122b.
147. S.N. Kuz'min, L.A. Razvodovskaya and A.F. Grapov, *Zh. Obshch. Khim.*, 1992, **62**, 1785; *Chem. Abs.*, 1993, **118**, 213181b.
148. Y. Matsumura, J. Terauchi, Y. Yamamoto, T. Konno and T. Shono, *Tetrahedron*, 1993, **49**, 8503.
149. T. Shono, Y, Matsumura, O. Onomura, M. Ogaki and T. Kanazawa, *J. Org. Chem.*, 1987, **52**, 536.
150. C.J. Foti and D.L. Comins, *J. Org. Chem.*, 1995, **60**, 2656.
151. H. Bürger, T. Dittmar and G. Pawelke, *J. Fluorine Chem.*, 1995, **70**, 89.
152. R.W. Shaw and T. Gallagher, *J. Chem. Soc., Perkin Trans. 1.*, 1994, 3549.
153. A. Kinoshita and M. Mori, *Synlett*, 1994, 1020.
154. F. Campagna, A. Carotti, G. Casini and F. Palluotto, *Farmaco*, 1994, **49**, 653; *Chem. Abs.*, 1995, **122**, 81960.

CHAPTER 6

Tetrahydroazepinones

I. 1,3,4,5-TETRAHYDRO[2H]AZEPIN-2-ONES

The parent 1,3,4,5-tetrahydro[2H]azepin-2-one **1** (R = H) has been prepared by a Beckmann rearrangement of *anti* cyclohexen-2-one oxime (**2**, R = H), using oleum at 140 °C.[1] In a similar fashion, but using phosphorus pentachloride[2] or polyphosphoric acid,[3] the *anti*-oxime of isophorone (**2**, R = CH$_3$) gave the trimethyl analogue (**1**, R = CH$_3$).[2,3] The corresponding *syn*-oximes (**4**, R = H, CH$_3$) were shown to give the isomeric 1,5,6,7-tetrahydro[2H]azepin-2-ones (**3**, R = H, CH$_3$)[2,3] (see Section IV).

(1)   (2)   (3)   (4)

The Beckmann rearrangement has also been used under milder conditions with similar results. For example, heating the *anti*-oxime tosylate **5** in methanol under reflux gave the rearranged product **6**.[4] Under these mild conditions, however, the *anti*-oxime tosylates **7** (R = H, CH$_3$) failed to rearrange, although the analogous isomeric 1,5,6,7-tetrahydro products **9** (R = H, CH$_3$) were obtained from the *syn*-oxime tosylates **8** (R = H, CH$_3$).[4]

(5)   (6)   (7)

(8)   (9)

The Beckmann procedure has also been applied to a number of oximes (**10**, R = CH$_3$, C$_2$H$_5$, ^{i}Pr, ^{n}Bu, Ph, 4-nitrophenyl),[5] without prior separation of the *syn/anti* mixture. As might be expected, the products also consisted of mixtures of the 1,3,4,5-tetrahydro- and the 1,5,6,7-tetrahydro[2*H*]azepin-2-ones (**11** and **12**, R = CH$_3$, C$_2$H$_5$, ^{i}Pr, ^{n}Bu, Ph, 4-nitrophenyl),[5] respectively.

(10)   (11)   (12)

The Schmidt rearrangement has also been used to produce 1,3,4,5-tetrahydro[2*H*]azepin-2-ones; however, the products are mixtures of the 1,3,4,5- and the 1,5,6,7-products, as illustrated by the formation of both isomers **14** (R^1 = H, R^2 = CH$_3$ and R^1 = R^2 = H, CH$_3$) and **15** (R^1 = H, R^2 = CH$_3$ and R^1 = R^2 = H, CH$_3$) on treatment of the 3-chlorocyclohexenones **13** (R^1 = H, R^2 = CH$_3$ and R^1 = R^2 = H, CH$_3$) with sodium azide.[5,6]

In marked contrast to the above observations, ring expansion of the inseparable *syn/anti*-*N*-methylnitrones **16** and **17** gave exclusively the 1-methyl-1,3,4,5-tetrahydro[2*H*]azepin-2-ones **18** and **19**, using either *p*-toluenesulphonyl chloride[7] or chlorosulphonyl isocyanate.[8] In the latter case the authors[8] claimed

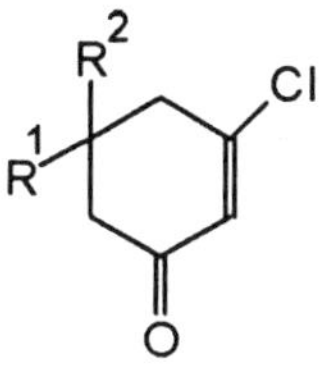

(13) 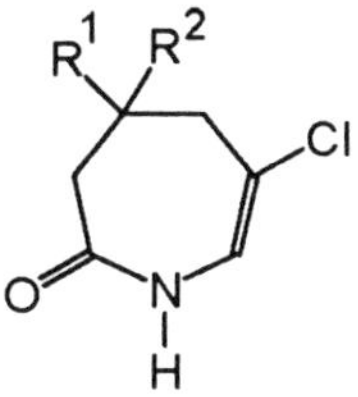   (14) 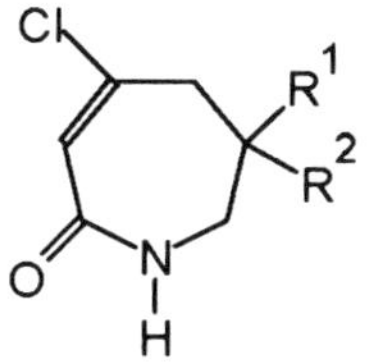   (15)

that the nature of the products was determined by the migratory aptitude of the substituents rather than the initial configuration of the N-methylnitrones.

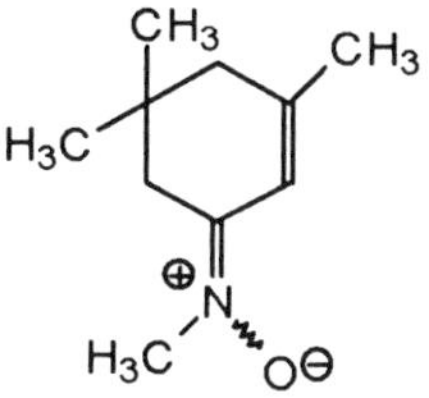

(16) 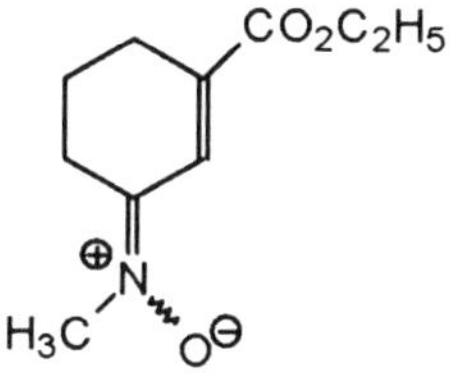   (17)

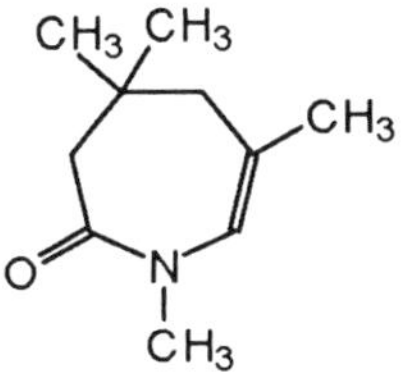

(18) 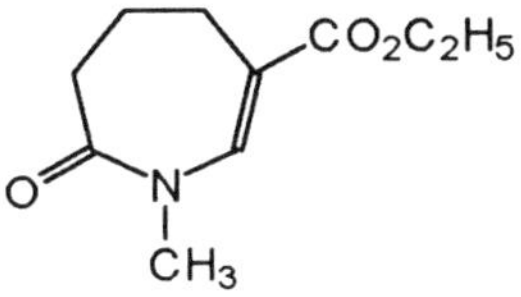   (19)

An interesting ring expansion procedure has been reported in which the 2-azidoethylthiocyclohexenones **20** (R = H, CH$_3$), on heating in xylene, rearranged to form the fused thiazolidines **21** (R = H, CH$_3$). Subsequently the trimethyl derivative **21** (R = CH$_3$) was desulphurated with nickel boride to produce the corresponding 1-ethyl-1,3,4,5-tetrahydro[2H]azepin-2-one **22**.[9]

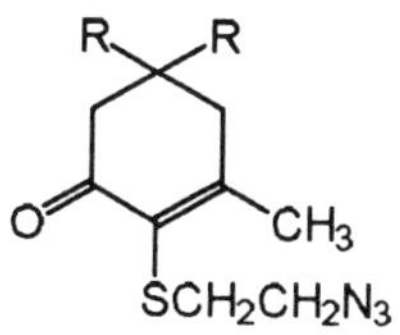

(20) 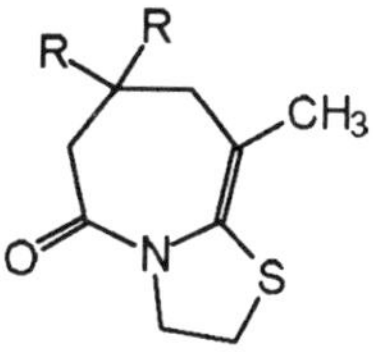   (21) 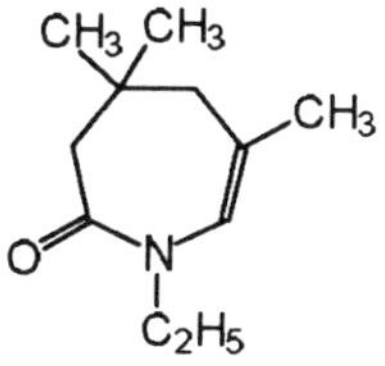   (22)

Two ring-closure procedures have been reported for the preparation of substituted 1,3,4,5-tetrahydro[2H]azepin-2-ones, both of which are potentially of wide application. In the first of these, cyclisation of substituted N-allyl-2-propenoic acid amides **23** (R^1 = R^2 = CH$_3$; R^3 = H, CH$_3$) with one equivalent of sodium hydride in xylene gave the tetrahydro[2H]azepinones **24** ($R^1 = R^2 = CH_3$; $R^3 = H, CH_3$) in high yields.[10] In the presence of methyl iodide and two equivalents of sodium hydride, N-alkylated products (**25**,

(23)   (24)   (25)

$R^1 = R^2 = CH_3$; $R^3 = H, CH_3$) were obtained and under these latter conditions the amides **23** ($R^1 = CH_3$, Ph, $R^2 = $ Ph, $R^3 = $ H) were converted directly into the N-alkylated azepinones **25** ($R^1 = CH_3$, Ph, $R^2 = $ Ph, $R^3 = $ H).[10] In these latter examples, no products corresponding to **24** were isolated.

The second ring closure involved the cyclisation of adipoyl chloride (**26**) with an appropriately substituted iminophosphorane (**27**, R = Ph, CH_2Ph, $CH_2CO_2C_2H_5$). In this case the product was a 1-substituted-7-chloro-1,3,4,5-tetrahydro[2H]azepin-2-one (**28**, R = Ph, CH_2Ph, $CH_2CO_2C_2H_5$).[11]

(26)   (27)   (28)

A list of substituted-1,3,4,5-tetrahydro[2H]azepin-2-ones is given in Table 1.

Table 1. 1,3,4,5-Tetrahydro[2H]azepin-2-ones

Structure	M.p. (°C)/b.p. (°C/mmHg)	Spectroscopic evidence	Ref.
(N-H)	133–135/2	IR, UV	1, 14
(N-CH$_2$Ph)	110/0.1	IR, MS, ^{1}H NMR	15
(7-Cl, N-H)	80–84	IR, UV	6
(7-C$_2$H$_5$, N-H)	Oil	IR, MS, ^{1}H NMR	4
(3-CO$_2$C$_2$H$_5$, N-CH$_3$)	90/0.05	IR, MS, ^{1}H NMR	7
(7-Cl, N-Ph)	80–82	IR, MS, ^{1}H NMR	11
(7-Cl, N-CH$_2$Ph)	35	IR, MS, ^{1}H NMR	11

Table 1 (continued)

Structure	M.p. (°C)/b.p. (°C/mmHg)	Spectroscopic evidence	Ref.
(7-chloro-1-(ethoxycarbonylmethyl)-2,3-dihydro-1H-azepin-2-one)	32	IR, MS, ^{1}H NMR	11
(3,7-dimethyl-1,3,4,5-tetrahydro-2H-azepin-2-one)		IR, ^{1}H NMR	13
(4,4-dimethyl-1,3,4,5-tetrahydro-2H-azepin-2-one)	78–80/0.01 64–65	IR, ^{1}H NMR IR, MS, ^{1}H, ^{13}C NMR	10 15
(7-chloro-4-methyl-1,3,4,5-tetrahydro-2H-azepin-2-one)	94–96	IR, ^{1}H NMR	5
(4,7-dimethyl-1,3,4,5-tetrahydro-2H-azepin-2-one)	71.5–72.5	IR, MS, ^{1}H NMR	5
(7-methyl-4-phenyl-1,3,4,5-tetrahydro-2H-azepin-2-one)	118.3–118.7	UV	2
(1,4,4-trimethyl-4,5-dihydro-1H-azepin-2(3H)-one)		IR, ^{1}H NMR IR, MS, ^{1}H, ^{13}C NMR	10 15

Table 1 (continued)

Structure	M.p. (°C)/b.p. (°C/mmHg)	Spectroscopic evidence	Ref.
4,4-dimethyl, N-COCF₃	Oil	IR, ¹H NMR	15
4-Ph, 4-CH₃, N-CH₃	Liquid	IR, ¹H NMR	10
4,4-diPh, N-CH₃	122	IR, ¹H NMR	10
4,4-diCH₃, 6-Cl	95–96	IR, UV	6
4,4-diCH₃, 6-CH₃	90.1–90.7 92–93 82	UV UV	2 3 10 5
4,4-diCH₃, 6-C₂H₅	72.5–73.5	IR, MS, ¹H NMR	5
4,4-diCH₃, 6-iC₃H₇	102–104	IR, MS, ¹H NMR	5

Table 1 (continued)

Structure	M.p. (°C)/b.p. (°C/mmHg)	Spectroscopic evidence	Ref.
(4,4-dimethyl-6-n-C$_4$H$_9$ azepinone)	45–46	IR, MS, ^{1}H NMR	5
(4,4-dimethyl-6-Ph azepinone)	149–151	IR, MS, ^{1}H NMR	5
(4,4-dimethyl-6-(4-NO$_2$-C$_6$H$_4$) azepinone)	209–220	IR, MS, ^{1}H NMR	5
(4,4-dimethyl-6-(4-OCH$_3$-C$_6$H$_4$) azepinone)	167–170	IR, MS, ^{1}H NMR	5
(1,4,4,6-tetramethyl azepinone)	50/0.05 49–50/0.05 56–57/0.02	IR, MS, ^{1}H NMR IR, ^{1}H NMR	7 8 10
(1-C$_2$H$_5$-4,4,6-trimethyl azepinone)		^{1}H NMR	9
(3-HO-4-CO$_2$C$_2$H$_5$-6-vinyl-7-Ph azepinone)	Amorphous solid	IR, UV, MS, ^{1}H NMR	12

Table 1 (continued)

Structure	M.p. (°C)/b.p. (°C/mmHg)	Spectroscopic evidence	Ref.
[structure: CH₃CO₂ and CO₂C₂H₅ substituted azepinone with Ph, N-H]	145–150	IR, UV, MS, ^{1}H NMR	12
[structure: CH₃CO₂ and CO₂C₂H₅ substituted azepinone with Ph, N-COCH₃]	123–128	IR, UV, MS, ^{1}H NMR	12

II. 1,3,4,7-TETRAHYDRO[2H]AZEPIN-2-ONES

There are relatively few examples in the literature of 1,3,4,7-tetrahydro[2H]azepin-2-ones, and although no methods of synthesis of a general nature have been published, a variety of specific ring-expansion and ring-closure reactions have been utilised, some of which potentially are capable of broader application.

The parent compound in the series (**29**, R = H) has been prepared by allowing a refluxing hydrochloric acid solution of the unsaturated aminonitrile **30** to pass through an alumina column at 300 °C. The reaction proceeds via hydrolysis of the nitrile followed by direct lactam formation from the intermediate unsaturated acid.[16]

(**29**) (**30**) (**31**)

(32)   **(33)**

The simplest substituted analogue, the 1-methyl derivative (**29**, R = CH$_3$), has been prepared by a Cope rearrangement of the cyclopropyl imidate **31**, followed by hydrolysis of the moisture-sensitive intermediate enol ether **32**.[17] This study was carried out to compare the kinetics of the reaction with those of the corresponding isoconjugate olefinic counterparts (e.g. **33**) and it was concluded that the imidates, as represented by **31**, are slower to react and require more forcing conditions.[17]

Ring expansion of the enamine lactam **34** with dimethyl acetylenecarboxylate has been used to give the enaminocaprolactam **35**, which on hydrolysis gave the corresponding 6-hydroxy-1,7-dihydro[2H]azepinone **36** (for full details see Chapter 9, Section III). Reduction of **36** with trimethylsilyl iodide gave the 1-benzyl-1,3,4,7-tetrahydro[2H]azepin-2-one **37**.[18]

(34)   **(35)**

(36)   **(37)**

Ring expansion of the vinylaziridine **38** with dichloroketene (**39**), generated *in situ* by dehydrochlorination of dichloroacetyl chloride, led to the formation of the 1,4-diphenyl-3,3-dichloro-1,3,4,7-tetrahydro[2H]azepin-2-one **40** in 40% yield, but the reaction does not appear to have been exploited further by using other substituted vinylaziridines.[19]

Claisen rearrangement of the vinyl-substituted oxazolidine **41** (R = *p*-CH$_3$C$_6$H$_4$SO$_2$, PhCH$_2$OCO), prepared *in situ* from the precursor phenyl selenides **42** (R = *p*-CH$_3$C$_6$H$_4$SO$_2$, PhCH$_2$OCO) by oxidative removal of selenium in the presence of DBU, gave the protected 7-substituted [2H]azepin-2-ones **43** (R = p-CH$_3$C$_6$H$_4$SO$_2$, PhCH$_2$OCO).[20] Removal of the protecting groups by appropriate means gave the 7-isobutyl-1,3,4,7-tetrahydro[2H]-azepin-2-one **43** (R = H) with retention of stereochemistry throughout the complete sequence of reactions (eight steps) from (S)-leucine (**44**), which was used as the starting material.[20]

Another ring-closure reaction reported for 1,3,4,7-tetrahydro[2H]azepin-2-ones is a modification of the previously quoted (see Chapter 5, Section II) molybdenum-catalysed procedure. In this case the dienamides **45** (R = H,

CH$_3$), on treatment with the catalyst system, gave the corresponding 1-benzyl-1,3,4,7-tetrahydro[2H]azepin-2-ones **46** (R = H, CH$_3$) with loss of ethylene.[21a]

The alternative use of a ruthenium catalyst has also been reported to effect the ring closure **45** (R = H) → **46** (R = H)[21b] and because of the greater tolerance

(45A)   (46A)

of this catalyst, the reaction has been extended to the cyclisation of the amino acid derivative **45A** which was cyclised to **46A** in 50% yield.[21c]

A list of the reported 1,3,4,7-tetrahydro[2H]azepine-2-ones is given in Table 2.

Table 2. 1,3,4,7-Tetrahydro[2H]azepin-2-ones

Structure	M.p. (°C)/b.p. (°C/mmHg)	Spectroscopic evidence	Ref.
			16
			17
		IR, MS, ^{1}H, ^{13}C NMR	21a, 21b
		IR, MS, ^{1}H, ^{13}C NMR	21a
		IR, MS, ^{1}H, ^{13}C NMR (supplementary material)	21c

Table 2 (continued)

Structure	M.p. (°C)/b.p. (°C/mmHg)	Spectroscopic evidence	Ref.
	88–90	α_D α_D, IR, MS, ^{1}H, ^{13}C NMR	20a 20b
	Oil	α_D α_D, IR, MS, ^{1}H, ^{13}C NMR	20a 20b
	115–117	α_D α_D, IR, MS, ^{1}H, ^{13}C NMR	20a 20b
	Oil	α_D α_D, IR, MS, ^{1}H, ^{13}C NMR	20a 20b
	173–177	IR, MS, ^{1}H, ^{13}C NMR	19
	Oil	IR, MS, ^{1}H NMR	18

III. 1,3,6,7-TETRAHYDRO[2H]AZEPIN-2-ONES

The parent, unsubstituted 1,3,6,7-tetrahydro[2H]azepin-2-one (47) has been prepared from 3-substituted caprolactams. Unexpectedly, 3-bromocaprolactam (48) in the presence of lutidine[22] or the 3-sulphur ylide 49, on treatment with sodium hydride followed by triethylborane and alkaline hydrogen peroxide,[23] both gave the β,γ-unsaturated ketone 47 as the major product along with a smaller amount of the expected α,β-unsaturated ketone 50. A mechanistic scheme involving the dienol 51 has been proposed[22] to account for these results.

(47)   (48)   (49)   (50)   (51)

Various ring-expansion procedures have been used to prepare substituted 1,3,6,7-tetrahydro[2H]azepin-2-ones. For example, the 1-benzyl-4-vinylazetidinones 52 (R = H, CH$_3$) were shown to undergo a base-catalysed [2,3] sigmatropic shift in the presence of lithium diisopropylamine to give the 1,3,6,7-tetrahydro[2H]azepin-2-ones 53 (R = H, CH$_3$) in high yield.[24] Relief of ring strain in going from a four-membered ring to a seven-membered ring was postulated as the reason for the ease of the reaction.

(52)   (53)

Vinylazetidines have also proved to be useful starting materials in studying the cobalt-catalysed carbonylation/ring-expansion reaction. At elevated temperature (90 °C) and 3.4 atm pressure of carbon monoxide, the vinylazetidines 54 (R = H, CH$_2$CH$_2$COCH$_3$, CH$_2$CH$_2$CO$_2$CH$_3$,, CH$_2$CH$_2$CN) rearranged with insertion of –CO– to yield the corresponding 1,3,6,7-tetrahydro[2H]azepin-2-ones **55** (R = H, CH$_2$CH$_2$COCH$_3$, CH$_2$CH$_2$CO$_2$CH$_3$, CH$_2$CH$_2$CN), in sharp contrast to azetidines containing a saturated substituent at position 4, which gave pyrrolidines as the main products.[25]

(54) (55) (56) (57)

As part of a study on the photochemistry of nitroalkanes, 4-phenyl-5-nitro-cyclohexene (**56**) was irradiated, undergoing ring expansion to give the N-hydroxyazepine **57** in 46% yield.[26]

Ring-closure techniques have also featured in the literature for the preparation of 1,3,6,7-tetrahydro[2H]azepin-2-ones. Cationic alkene cyclisation of the sulphoxide **58** with trifluoroacetic anhydride gave a mixture of saturated azepinones as well as the 1,3,6,7-tetrahydro product **60** (R = CH$_3$), which was isolated by chromatography.[27] Similarly, the chlorinated methylthio derivative **59** on treatment with SnCl$_4$ gave the lactam **60** (R = H).[27] A mechanism for the ring closure of the Pummerer intermediates **61** (R = H, CH$_3$) was discussed.

(58) (59) (60) (61)

The Cope rearrangement of cis-2,2-dimethyl-3-isobutenylcyclopentyl isocyanate (**62**) under forcing conditions (at 144 °C) gave the 3,6-dihydro-3,3,6,6-tetramethyl[2H]azepin-2-one **63**, which in turn was converted with water, methanol and thiophenol, respectively, to give the corresponding 7-substituted-1,3,6,7-tetrahydro-3,3,6,6-tetramethyl[2H]azepin-2-ones **64** (R = OH,

(62) (63) (64)

OCH$_3$, SPh).[28,29] Catalytic hydrogenation of **63** with Adam's catalyst reduced the 1,7-double bond, yielding 1,3,6,7-tetrahydro-3,3,6,6-tetramethyl[2*H*]azepin-2-one **64** (R = H).[28,29]

More recently, as part of a study into inhibitors of HIV-1 protease, the 1,3,6,7-tetrahydro[2*H*]azepin-2-one **64A** (R = SiPh$_2$^{t}Bu) was prepared by hydrolysis of the ester **64B** followed by ring closure using DPPA. Further elaboration of the deprotected azepine **64A** (R = H) led to the peptidomimetic azepine **64C**.[30a] In a similar manner, ring closure of the ester **64D** led to the 1,3,6,7-tetrahydro-tetramethyl[2*H*]azepin-2-one **64E**, which was also elaborated to several peptidomimetic azepines for testing as potential inhibitors of HIV-1 protease.[30a]

(64A)   **(64B)**   **(64C)**

(64D)   **(64E)**

A full listing of the published 1,3,6,7-tetrahydro-tetramethyl[2*H*]azepin-2-ones is given in Table 3.

Tetrahydroazepinones

Table 3. 1,3,6,7-Tetrahydro[2H]azepin-2-ones

Structure	M.p. (°C)/b.p. (°C/mmHg)	Spectroscopic evidence	Ref.
	79	IR, ^{1}H NMR ^{1}H NMR	22 23
		IR, ^{1}H NMR	24
	Oil	IR, MS, ^{1}H NMR	27
		IR, MS, ^{1}H NMR	26
	109–111	IR, MS, ^{1}H, ^{13}C NMR	25
		IR, ^{1}H NMR	24
		IR, ^{1}H NMR	27

Table 3 (continued)

Structure	M.p. (°C)/b.p. (°C/mmHg)	Spectroscopic evidence	Ref.
(azepinone with PhCH₂CO₂CH₂– and –CH₂CONH₂ substituents, N-Ph)			30b
(azepinone with PhCH₂CO₂CH₂– and –CH₂CO₂H substituents, N-Ph)			30b
(azepinone with PhCH₂CO₂CH₂– and –CH₂CH₂OH substituents, N-Ph)			30b
(azepinone with –CH₂CH₂OH substituent, N-CH(tBuO₂C)CH(CH₃)₂)		^{1}H NMR	30a
(azepinone with –CH₂CO₂CH₃ substituent, N-CH(tBuO₂C)CH(CH₃)₂)		^{1}H NMR	30a
(azepinone with –CH(CH₂Ph)CO₂CH₃ substituent, N-CH(tBuO₂C)CH(CH₃)₂)		^{1}H NMR	30a

Table 3 (continued)

Structure	M.p. (°C)/b.p. (°C/mmHg)	Spectroscopic evidence	Ref.
[structure: 7-membered lactam with CH₂CH₂OSi(tC₄H₉)Ph₂ substituent and N-CH(CO₂tBu)iPr]		^{1}H NMR	30a
[structure: 7-membered lactam with CH(NPhth)CH₃ substituent and N-CH(CH₃)CH₂OH]		^{1}H NMR	30a
[structure: 4,5-dimethyl tetrahydroazepinone, N-CH₂CH₂COCH₃]	14–115	IR, MS, ^{1}H, ^{13}C NMR	25
[structure: 4,5-dimethyl tetrahydroazepinone, N-CH₂CH₂CO₂CH₃]		IR, MS, ^{1}H, ^{13}C NMR	25
[structure: 4,5-dimethyl tetrahydroazepinone, N-CH₂CH₂CN]		IR, MS, ^{1}H, ^{13}C NMR	25

Table 3 (continued)

Structure	M.p. (°C)/b.p. (°C/mmHg)	Spectroscopic evidence	Ref.
	239–242	IR, MS, ^{1}H NMR	32
	147–148	IR, MS, ^{1}H NMR	31
	113–114	IR, UV, ^{1}H NMR	28, 29
	144	IR, UV, ^{1}H NMR	28, 29
	Oil	IR, ^{1}H NMR	28, 29
	102–103	IR, ^{1}H NMR	28, 29

Table 3 (continued)

Structure	M.p. (°C)/b.p. (°C/mmHg)	Spectroscopic evidence	Ref.
[structure: 7-Br, 3-CH₃, 6,6-di-CH₃ azepinone]		¹H NMR	31
[structure: 7-Br, 3-CH₂Br, 6,6-di-CH₃ azepinone]		¹H NMR	31

IV. 1,5,6,7-TETRAHYDRO[2H]AZEPIN-2-ONES

As early as 1952 it was recognised that Beckmann rearrangement of *syn*-oximes, such as **65** ($R^1 = H$, $R^2 = CH_3$)[33] and **65** ($R^1 = H$, $R^2 = Ph$ and $R^1 = R^2 = CH_3$)[2] led to 1,5,6,7-tetrahydro[2H]azepin-2-ones (**66**, $R^1 = H$, $R^2 = CH_3$, Ph and $R^1 = R^2 = CH_3$), whereas the corresponding *anti*-oximes gave the 1,3,4,5-tetrahydro[2H]azepin-2-ones **67** ($R^1 = H$, $R^2 = Ph$ and $R^1 = R^2 = CH_3$)[2] (see Section I for details). Later many other workers exploited this reaction to give the monomethyl derivative **66** ($R^1 = R^2 = H$),[4] the trimethyl derivative **66** ($R^1 = R^2 = CH_3$)[3,5,34–36] and the unsubstituted parent compound **68**.[1,37]

(65)   (66)   (67)   (68)

As discussed earlier (Section I) the mixed *syn-* and *anti-*oximes **10** (R = CH$_3$, C$_2$H$_5$, ^{i}Pr, ^{n}Bu, Ph, 4-nitrophenyl) gave both 1,5,6,7-tetrahydro and 1,3,4,5-tetrahydro isomers (**12** and **11**, respectively, where R = CH$_3$, C$_2$H$_5$, ^{i}Pr, ^{n}Bu, Ph, 4-nitrophenyl).[5]

Other derivatives prepared by this procedure included the 4-chloro derivatives **15** (R^1 = R^2 = H, CH$_3$ and R^1 = H, R^2 = CH$_3$)[38,5] and the acid **69**.[39]

(10)   (11)   (12)

(15)   (69)

Ring expansion of appropriately substituted cyclohexenones using the Schmidt rearrangement has been used to give the 1,5,6,7-tetrahydro[2*H*]azepin-2-ones **15** (R^1 = R^2 = H, CH$_3$) and **70** (R^1 = H, CH$_3$, R^2 = H)[6,40] and the 6-aryl derivatives **71** (R = H, OCH$_3$, NO$_2$).[41] In the preparation of the 4-chloro derivatives **15** (R^1 = R^2 = H, CH$_3$), the isomeric 1,3,4,5-tetrahydro compounds **14** (R^1 = R^2 = H, CH$_3$) are also formed[6] (see also Section I).

(70)   (71)   (14)

Treatment of the dichlorocarbene adduct **72** from piperidone with LiAlH$_4$ and base gave the expected ring expansion product **73**; however, the yield was poor (9%) and the method produced a number of substituted piperidone by-products, making isolation difficult.[42]

(72) (73)

The 5-methylthioazabicycloheptanes **74** (R = H, NO$_2$) underwent a rapid exothermic rearrangement on treatment with trifluoroacetic acid to give the corresponding 1,5,6,7-tetrahydro[2H]azepin-2-ones **77** (R = H, NO$_2$) in yields of over 90%,[43] illustrating how successful this approach can be. The proposed mechanism is protonation of the β-lactam carbonyl group to give the intermediate **75** (R = H, NO$_2$), followed by rearrangement to a carbocation **76** (R = H, NO$_2$) which loses a proton.[43]

(74) (75)

(76) (77)

The thiolactone **78**, on treatment with trimethylsilyl iodide, underwent an acyl transfer process from the S atom to the N atom, generating a new C–N bond as it lost the *N*-Boc protecting group and giving the tetrahydroazepinone **79**. The driving force was claimed to be the electrophilic character α to the sulphur atom.[44]

(78)   (79)

Palladium-catalysed carbonylation, using $Pd(OCOCH_3)_2$ and PPh_3 as catalyst, has been used as a general means of cyclising iodoalkylamines (**80**, $n = 1$–3; X = I, Br).[45] In the presence of $^nBu_3N$ and carbon monoxide, the products are five-, six- or seven-membered lactams, depending on the length of the alkyl chain. In the specific case of the alkylamine **80** ($n = 3$; X = I) the product consisted of a mixture of the two lactams **81** and **82**, from which the 1,5,6,7-tetrahydro[2H]azepin-2-one **82** was separated by chromatography in 23% yield.[45]

(80)   (81)   (82)

Elimination of halogen substituents from 3,3-dibromo- or 3,3-dichlorocaprolactams has been shown to be a useful method for the preparation of 3-substituted-1,5,6,7-tetrahydro[2H]azepin-2-ones. For example, treatment of 3,3-dichlorocaprolactam (**83**, $R^1 = Cl$, $R^2 = H$) with piperidine or morpholine gave the 3-piperidino and 3-morpholino derivatives **84** (R = H; X = CH_2, O).[46] Similar results were obtained on treatment of the N-methyl analogue **83** ($R^1 = Cl$, $R^2 = CH_3$) and the dibromo derivative **83** ($R^1 = Br$, $R^2 = H$) with piperidine, giving the corresponding products (**84**, R = CH_3, H; X = CH_2).[47,48]

(83)   (84)

A variety of elimination reactions has also been used to produce 3-substituted-1,5,6,7-tetrahydro[2H]azepin-2-ones;[22,49–51] full details of the products from these reactions, along with a number of other derivatives described in the literature,[52–57] are given in Table 4.

Table 4. 1,5,6,7-Tetrahydro[2H]azepin-2-ones

Structure	M.p. (°C)/b.p. (°C/mmHg)	Spectroscopic evidence	Ref.
	58; 138–140/2 Oil 60–65/0.5	IR, ^{1}H NMR	1 22 37 6, 23
	Oil	^{1}H NMR	42
	96–98; 133–140/0.1		49
	67–69; 112–113/0.04		49
	83	IR, MS, ^{1}H NMR	50
	156–158 158–160	IR, ^{13}C NMR	46 48

Table 4 (continued)

Structure	M.p. (°C)/b.p. (°C/mmHg)	Spectroscopic evidence	Ref.
(3-morpholino-azepan-2-one)	140–141		46
(3-phenyl-dihydroazepinone)	138–139	IR, MS, ^{1}H, ^{13}C NMR	51
(4-methyl-dihydroazepinone)	81–82 Tosylate 142–143	IR, UV Tosylate IR, ^{1}H NMR	40 4
(4-methoxy-dihydroazepinone)	70–72; 119–125/0.012	IR, UV, ^{1}H NMR	38
(4-chloro-azepan-2-one)	128–129.5 128.5–129	IR, ^{1}H NMR	6 38
(3-chloro-N-methyl-dihydroazepinone)	98/0.03	IR, ^{1}H NMR	50
(3-piperidino-N-methyl-dihydroazepinone)	137–140/1–2	IR, UV	47

Table 4 (continued)

Structure	M.p. (°C)/b.p. (°C/mmHg)	Spectroscopic evidence	Ref.
	Oil	IR, MS, ^{1}H NMR	45
	128–131	^{1}H NMR	53
	106	^{1}H NMR	53
	120–121		56
	200–203	IR, UV, ^{1}H NMR	54
	240–243	IR, UV, ^{1}H NMR	54
	173–177	IR, UV, ^{1}H NMR	54

Table 4 (continued)

Structure	M.p. (°C)/b.p. (°C/mmHg)	Spectroscopic evidence	Ref.
4-chlorophenyl-NH, CN, azepinone structure	189–194	IR, UV, ^{1}H NMR	54
$C_2H_5O_2CCH_2NH$, CN, azepinone structure	102–104	IR, UV, ^{1}H NMR	57
4,6-dimethyl azepinone	71–72.5 72.5–74.5	UV, IR	33 5
methyl-phenyl azepinone	146–146.6 142–146	UV IR, MS, ^{1}H NMR	2 41
methyl-(4-methoxyphenyl) azepinone	189–192	IR, MS, ^{1}H NMR	41
methyl-(4-nitrophenyl) azepinone	200–202	IR, MS, ^{1}H NMR	41

Table 4 (continued)

Structure	M.p. (°C)/b.p. (°C/mmHg)	Spectroscopic evidence	Ref.
(4-methyl-3-mercapto tetrahydroazepinone)	Oil	IR, ^{1}H NMR	44
(4-chloro-6-methyl tetrahydroazepinone)	91–92	IR, ^{1}H NMR	5
(4-(4-bromophenyl)-6-carboxy tetrahydroazepinone)	232	IR	39
(6,6-dimethyl-1-n-propyl tetrahydroazepinone)			52
(3-phenoxy-4-methylthio-5-benzylidene tetrahydroazepinone)	193–195	IR, MS, ^{1}H NMR	43
(3-phenoxy-4-methylthio-5-(3-nitrobenzylidene) tetrahydroazepinone)	225–230	IR, ^{1}H NMR	43

Table 4 (continued)

Structure	M.p. (°C)/b.p. (°C/mmHg)	Spectroscopic evidence	Ref.
4,6,6-trimethyl	108.8–109.1 112–113 112–113	UV UV IR, ^{1}H NMR	2 3 34 5, 35, 36
4-CH$_2$Br-6,6-diMe		MS, ^{1}H NMR	31
4-C$_2$H$_5$-6,6-diMe	66.5	IR, MS, ^{1}H NMR	5
4-iC$_3$H$_7$-6,6-diMe	108–110	IR, MS, ^{1}H NMR	5
4-nC$_4$H$_9$-6,6-diMe	110/0.5	IR, MS, ^{1}H NMR	5
4-Ph-6,6-diMe	163.5–164.5	IR, MS, ^{1}H NMR	5

Table 4 (continued)

Structure	M.p. (°C)/b.p. (°C/mmHg)	Spectroscopic evidence	Ref.
4-(4-methoxyphenyl)-6,6-dimethyl tetrahydroazepinone	153–154	IR, MS, ^{1}H NMR	5
4-(4-nitrophenyl)-6,6-dimethyl tetrahydroazepinone	191–202	IR, MS, ^{1}H NMR	5
4-methoxy-6,6-dimethyl tetrahydroazepinone	148–149	IR, ^{1}H NMR	41
4-ethoxy-6,6-dimethyl tetrahydroazepinone	112–113 112–113	IR, ^{1}H NMR	41 23
4-(propargyloxy)-6,6-dimethyl tetrahydroazepinone	133.5–134	MS, IR, ^{1}H NMR	55
4-(allyloxy)-6,6-dimethyl tetrahydroazepinone	83–84.5	IR, ^{1}H NMR	55

Table 4 (continued)

Structure	M.p. (°C)/b.p. (°C/mmHg)	Spectroscopic evidence	Ref.
CH₃CH=CHCH₂O-[azepinone with 2×CH₃]	128–130	IR, ¹H NMR	55
CH₂=C(CH₃)CH₂O-[azepinone with 2×CH₃]	78–84	IR, ¹H NMR	55
(CH₃)₂C=CHCH₂O-[azepinone with 2×CH₃]	88–91	IR, ¹H NMR	55
PhCH=CHCH₂O-[azepinone with 2×CH₃]	163–165	IR, ¹H NMR	55
PhCH₂CH(n-C₄H₉)CH₂O-[azepinone with 2×CH₃]	Oil	IR, MS, ¹H NMR	55
Cl-[azepinone with 2×CH₃]	84.5–85.5 84.5–85.5	IR, ¹H NMR	6 38 41, 55

Table 4 (continued)

Structure	M.p. (°C)/b.p. (°C/mmHg)	Spectroscopic evidence	Ref.
4-(phenylamino)-6,6-dimethyl tetrahydroazepinone	232		41
4-(2-methylphenylamino)-6,6-dimethyl tetrahydroazepinone	227–228		41
4-(3-nitrophenylamino)-6,6-dimethyl tetrahydroazepinone	228–230		41
4-(4-nitrophenylamino)-6,6-dimethyl tetrahydroazepinone	242–243		41
4-(benzylamino)-6,6-dimethyl tetrahydroazepinone	296–297		41

Table 4 (continued)

Structure	M.p. (°C)/b.p. (°C/mmHg)	Spectroscopic evidence	Ref.
(3,4-dimethoxyphenyl)-(CH₂)₂NH-substituted 5,5-dimethyl-azepan-2-one	174–175		41
4,7,7-trimethyl-1-bromo-1H-azepin-2(3H,7H)-one	Liquid	IR, ^{1}H NMR	31
5-bromo-4,7,7-trimethyl-1-bromo-1H-azepin-2(3H,7H)-one	113	MS, ^{1}H NMR	31
4-bromomethyl-5-bromo-7,7-dimethyl-1-bromo-1H-azepin-2(3H,7H)-one	171–173	MS, ^{1}H NMR	31

VIII. 1,2,6,7-TETRAHYDRO[3H]AZEPIN-3-ONES

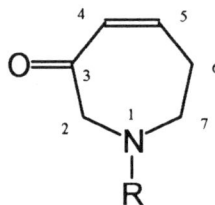

The single representative of this series to appear in the literature (**85**) was isolated as a by-product from the intramolecular [2 + 2] cycloaddition of the ketiminium salt **87** generated *in situ* from the α-N-tosylaminoamide **86**.[58] In general, this reaction leads to azabicyclic ketones, but in this specific case treatment of **87** with trifluoroacetic anhydride gave the azabicyclic ketone **88** in 71% yield, along with 7% of 1-tosyl-5-methyl-1,2,6,7-tetrahydro[3*H*]azepin-3-one (**85**). Characterisation of the by-product **85** was claimed by IR, ^{1}H, ^{13}C NMR, MS, and elemental analysis, but unfortunately no physical data were quoted in the publication.[58]

(85) (86)

(87) (88)

XII. 1,2,3,7-TETRAHYDRO[4*H*]AZEPIN-4-ONES

The preparation of the 1,2,3,7-tetrahydro[4*H*]azepin-4-one **90** (R^1 = CN; R^2 = H) was claimed in 1961 from the reaction of the β,β'-iminodipropionic acid dinitrile **89** (R^1 = CN, R^2 = H) with ethyl oxalate in the presence of sodium ethoxide,[59] but the structural assignment of the product was erroneous and was subsequently shown to be the corresponding pyrroline (**91**).[60] The other tetrahydro[4*H*]azepin-4-one derivatives **90** (R^1 = CO$_2$CH$_3$, R^2 = H; R^1 = CN, R^2 = COCH$_3$) claimed in the earlier paper,[59] therefore, by analogy are all wrongly assigned and consequently will not be discussed further.

The only genuine preparation of a 1,2,3,7-tetrahydro[4*H*]azepin-4-one was achieved by an intramolecular Friedel–Crafts cyclisation, at room temperature, of the readily accessible 3-[*N*-(2-chloroprop-2-enyl)propanoic acid **92**

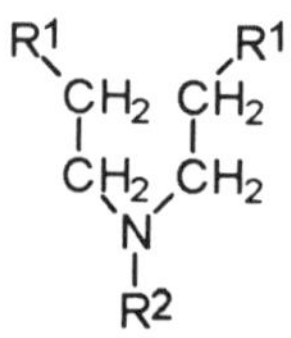

(89) 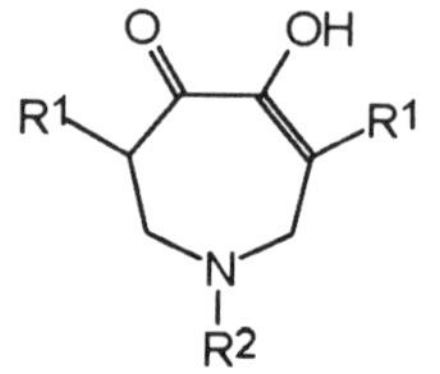 (90) 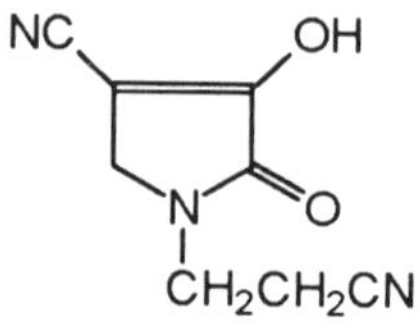 (91)

(R = OH) via the acid chloride **92** (R = Cl) to give 6,6-dichloro-N-tosyl-azepan-4-one **93**.[61] Compound **93** was converted into the 6-chloro- and 6-methoxy-1,2,3,7-tetrahydro[4H]azepin-4-ones **94** (R = Cl, OCH$_3$) on treatment with sodium carbonate in aqueous tetrahydrofuran and in methanol, respectively.[61]

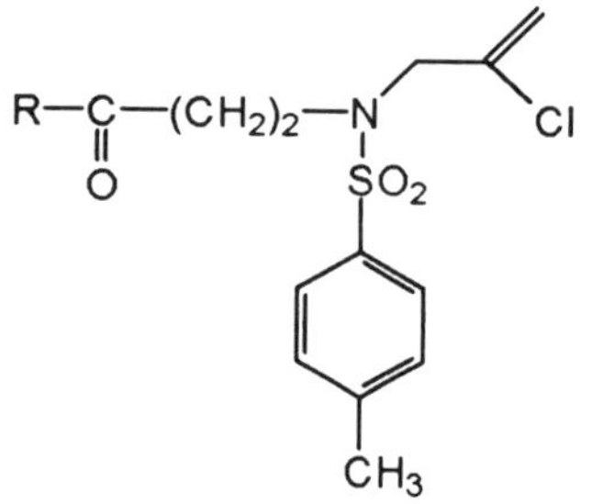

(92) 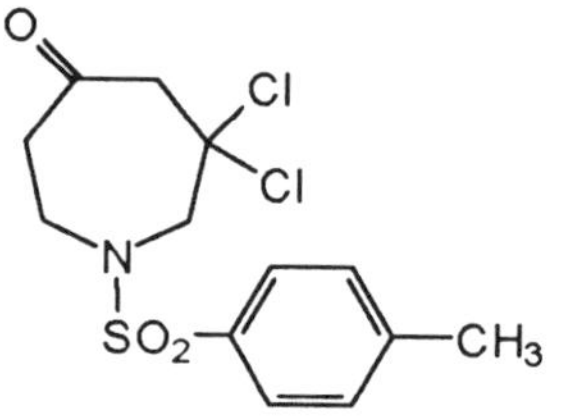 (93) 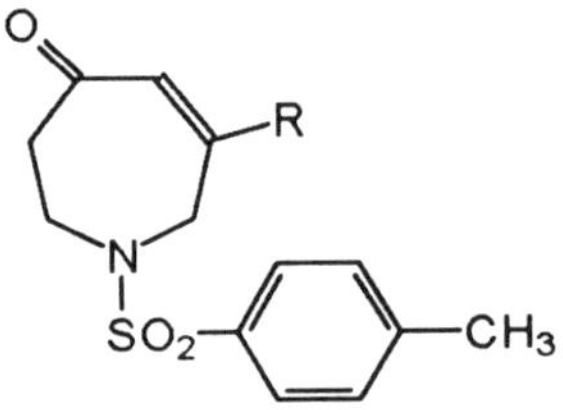 (94)

The two derivatives characterised[61] are given in Table 5.

Table 5. 1,2,3,7-Tetrahydro[4H]azepin-4-ones

Structure	M.p. (°C)	Spectroscopic evidence	Ref.
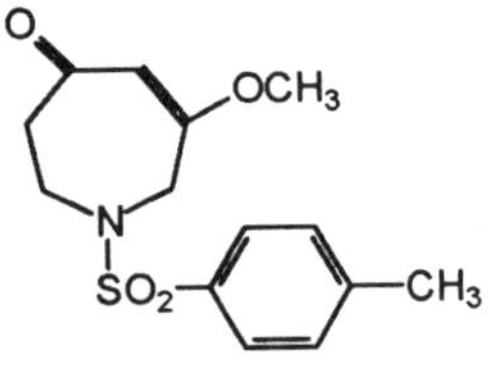 (OCH$_3$ derivative)	122–123	MS, IR, ^{1}H NMR	61
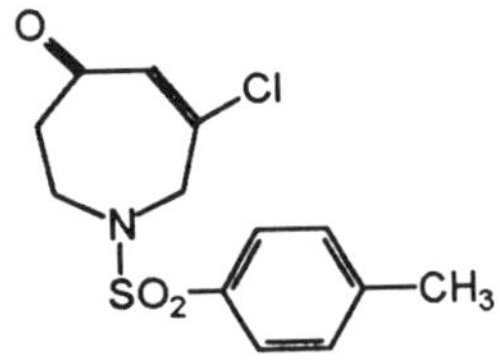 (Cl derivative)	98	IR, ^{1}H NMR	61

XIII. 1,5,6,7-TETRAHYDRO[4H]AZEPIN-4-ONES

Methods for the preparation of 1,5,6,7-tetrahydro[4H]azepin-4-ones are based on ring-expansion procedures. For example, Curtius rearrangement of the azides **95** (R = H, CH$_3$) in the presence of methanol[62] or ethanol[62,63] led to the corresponding 2-substituted-1,5,6,7-tetrahydro[4H]azepin-4-ones **96** (R^1 = H, CH$_3$; R^2 = OCH$_3$, OC$_2$H$_5$), whereas with benzene as the solvent, rearrangement resulted in the formation of the diones **97** (R = H, CH$_3$). In the presence of aniline, the Curtius reaction on the azide **95** (R = CH$_3$) led to the amine **96** (R^1 = CH$_3$; R^2 = NHPh).[63] The ketones **96** (R^1 = H, CH$_3$; R^2 = OCH$_3$, OC$_2$H$_5$) existed in solution in equilibrium with the tautomeric 3,5,6,7-tetrahydro[4H]azepin-4-ones **98** (R^1 = H, CH$_3$; R^2 = OCH$_3$, OC$_2$H$_5$)[62,63] (see Section XV).

(95) (96) (97) (98)

A possible mechanism for the conversion of vinyl azides such as **95** (R = H) into the corresponding tetrahydroazepines **96** (R^1 = H, R^2 = OCH$_3$) and **98** (R^1 = H, R^2 = OCH$_3$) has been proposed.[64]

In an unusual ring-expansion/rearrangement reaction, the 6-methyltetrahydropyridazin-3-ones **99** (R = Ph, cyclohexyl), on treatment with dimethyl

(99) (100)

acetylenedicarboxylate, yielded the corresponding 1-anilino- and 1-cyclohexyl-amino-1,5,6,7-tetrahydro[4H]azepin-4-ones **100** (R = Ph, cyclohexyl).[65]

It was postulated that the reaction might have occurred through a Michael addition of the secondary amine group of the pyridazinone to the dimethyl acetylenedicarboxylate, followed by attack of a carbanion on the carbonyl group to give a bicyclic intermediate which in turn underwent ring opening to give the product. No simple Michael addition products were observed, however.[65]

A two-carbon homologation of N-methylpyrrolidone (**101**) using lithium triphenylsilylacetylide (**102**) led to 1-methyl-1,5,6,7-tetrahydro[4H]azepin-4-one (**103**) in a yield of 70%.[66] The method is of broad application and has been used for the preparation of eight- and nine-membered rings and for the azepine ring system.[66]

(101)   LiC≡CSiPh₃   (103)
         (102)

Flash vacuum thermolysis of 4,5-dihydroisoxazole-5-spirocyclobutanes (**104**, R^1 = CH₃, Ph; R^2 = H) induced rearrangement to produce 2-substituted-1,5,6,7-tetrahydro[4H]azepin-4-ones (**105**, R = CH₃, Ph) as the main products.[67,68] The reaction has the disadvantage that it also produces 1-alkenyl-pyrrolidin-2-ones (**106**, R = CH₃, Ph) as by-products, and requires severe conditions (700 °C) to obtain the best results. Under the same conditions the benzyl-substituted isoxazole **104** (R^1 = CH₃, R^2 = CH₂Ph) gave a mixture of the 7- and 5-benzyl-1-methyl-1,5,6,7-tetrahydro[4H]azepin-4-ones **107** (R^1 = CH₂Ph, R^2 = H and R^1 = H, R^2 = CH₂Ph) in poor yields.[68]

(104)   (105)   (106)

On prolonged standing at room temperature, the 2-methyl derivatives **105** (R = CH₃) and **107** (R^1 = CH₂Ph, R^2 = H and R^1 = H, R^2 = CH₂Ph) rearranged to the corresponding azadienols **108** (R^1 = R^2 = H; R^1 = CH₂Ph, R^2 = H; R^1 = H, R^2 = CH₂Ph).[68]

A list of 1,5,6,7-tetrahydro[4H]azepin-4-ones reported in the literature is given in Table 6.

Table 6. 1,5,6,7-Tetrahydro[4H]azepin-4-ones

Structure	M.p. (°C)	Spectroscopic evidence	Ref.
	81.5–82	IR, MS, ^{1}H, ^{13}C NMR	66
	90–91	IR, UV, ^{1}H NMR	63 62, 64
	78–78.5	IR, UV, ^{1}H NMR	63
	87–88	^{1}H, ^{13}C NMR IR, MS, ^{1}H, ^{13}C NMR	67 68
	144–146	^{1}H, ^{13}C NMR IR, MS, ^{1}H, ^{13}C NMR	67 68

Table 6 (continued)

Structure	M.p. (°C)	Spectroscopic evidence	Ref.
		IR, MS, ^{1}H, ^{13}C NMR	68
		IR, MS, ^{1}H, ^{13}C NMR	68
	146–146.5	IR, UV, MS, ^{1}H NMR	63 62
	127.5–128.5	IR, UV, ^{1}H NMR	63
	214–214.5	IR, UV, ^{1}H NMR	63
	177–178	IR, MS, ^{1}H NMR	65

Tetrahydroazepinones 399

Table 6 (continued)

Structure	M.p. (°C)	Spectroscopic evidence	Ref.
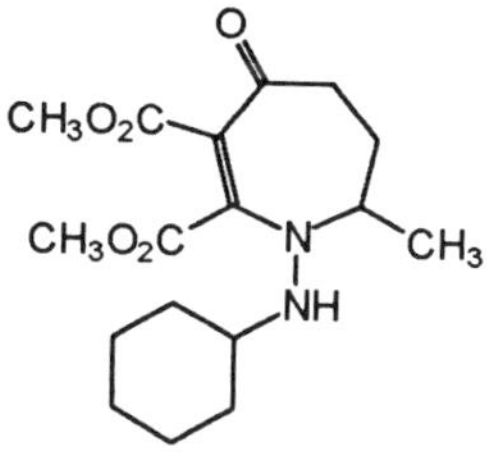 CH₃O₂C-, CH₃O₂C- azepinone with NH-cyclohexyl, CH₃	126–127	IR, MS, ¹H NMR	65
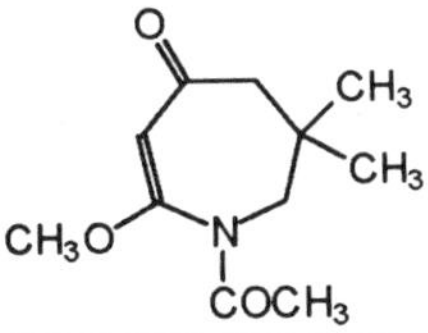 azepinone with two CH₃, CH₃O, COCH₃	82.5–83	IR, UV, ¹H NMR	63

XIV. 2,3,5,6-TETRAHYDRO[4H]AZEPIN-4-ONES

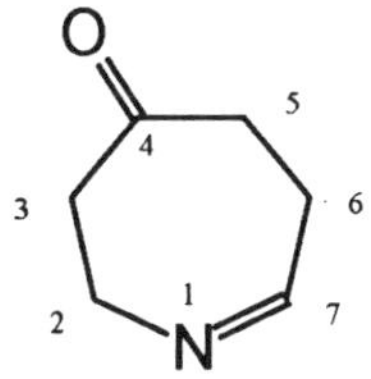

There is only one reference[69] in the literature to 2,3,5,6-tetrahydro[4H]azepin-4-ones and only the 2,5-dimethyl-7-ethoxy derivative (**109**) is described. The compound was prepared as an epimeric mixture at C-2 and C-5, either by hydrogenation of the azatropone **110** or by alkylation of the azepane-2,5-dione **111** with triethyloxonium fluoborate. Compound **109** is an oil (b.p. 42/0.2 mmHg), the structure of which was confirmed by IR and ¹H NMR spectra and elemental analysis.[69]

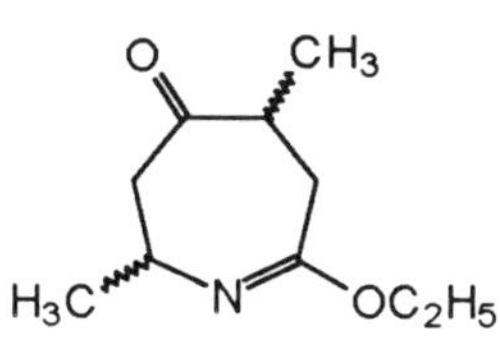

(**109**) 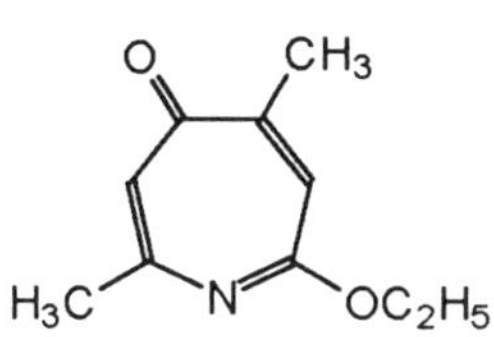   (**110**) 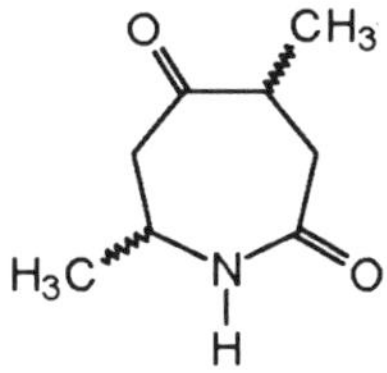   (**111**)

XV. 3,5,6,7-TETRAHYDRO[4H]AZEPIN-4-ONES

The only reference to the preparation of derivatives of 3,5,6,7-tetrahydro[4H]azepin-4-one is to the enol ethers **98** (R^1 = H, CH_3; R^2 = OCH_3, OC_2H_5),[62] which were formed by a Curtius reaction in methanol or ethanol on the azides **95** (R = H, CH_3). The 3,5,6,7-tetrahydro[4H]azepin-4-ones **98** (R^1 = H, CH_3; R^2 = OCH_3, OC_2H_5) exist as the minor tautomers in solution along with the major tautomeric 1,5,6,7-tetrahydro[4H]azepin-4-ones **96** (R^1 = H, CH_3; R^2 = OCH_3, OC_2H_5) (see Section XIII). The existence of the compounds **98** (R^1 = H, CH_3; R^2 = OCH_3, OC_2H_5) was demonstrated by

(95)   (96)   (98)

examination of their ^{1}H NMR spectra, but they have not been isolated in a pure state.[62]

REFERENCES

1. O.D. Strizmakov, E.N. Zil'berman and S.V. Svetozarskii, *Zh. Obshch. Khim.*, 1965 **35**, 628; *Chem. Abs.*, 1965, **63**, 5535a.
2. R.S. Montgomery and G. Dougherty, *J. Org. Chem.*, 1952, **17**, 823.
3. R.H. Mazur, *J. Org. Chem.*, 1961, **26**, 1289.
4. T. Sato, H. Wakatsuka and K. Amano, *Tetrahedron*, 1971, **27**, 5381.
5. G.I. Hutchison, R.H. Prager and A.D. Ward, *Aust. J. Chem.*, 1980, **33**, 2477.
6. Y. Tamura and Y. Kita, *Chem. Pharm. Bull.*, 1971, **19**, 1735.
7. R.H. Prager, K.D. Raner and A.D. Ward, *Aust. J. Chem.*, 1984, **37**, 381.
8. S.P. Joseph and D.N. Dhar, *Tetrahedron*, 1988, **44**, 5209.
9. A.G. Schultz and R. Ravichandran, *J. Org. Chem.*, 1980, **45**, 5008.
10. M. Bortolussi, R. Bloch and J.M. Conia, *Tetrahedron Lett.*, 1977, 2289.
11. T. Aubert, M.I. Farnier and R. Guilard, *Synthesis*, 1990, **2**, 149.
12. T. Sano, Y. Horiguchi, K. Tanaka and Y. Tsuda, *Chem. Pharm. Bull.*, 1985, **33**, 5197.

13. W. Theilacker, K. Ebke, L. Seidl and S. Schwerin, *Angew. Chem.*, 1963, **75**, 208.
14. A.R. Doumaux Jr. and D.J. Trecker, *J. Org. Chem.*, 1970, **35**, 2121.
15. M.R. Hatswell, R.H. Prager and A.D. Ward, *Aust. J. Chem.*, 1993, **46**, 135.
16. Yu.D. Smirnov and A.P. Tomilov, *Zh. Org. Khim.*, 1969, **5**, 864; *Chem. Abs.*, 1969, **71**, 38299h.
17. L.A. Paquette and G.D. Ewing, *J. Am. Chem. Soc.*, 1978, **100**, 2908.
18. G.W. Heinicke, A.M. Morella, J. Orban, R.H. Prager and A.D. Ward, *Aust. J. Chem.*, 1985, **38**, 1847.
19. M. Ishida, H. Muramaru and S. Kato, *Synthesis*, 1989, **7**, 562.
20. (a) P.A. Evans, A.B. Holmes and K. Russell, *Tetrahedron: Asymmetry*, 1990, **1**, 593; (b) A.B. Holmes, P.A. Evans and K. Russell, *J. Chem. Soc.*, *Perkin Trans. 1*, 1994, 3397; (c) P.A. Evans, A.B. Holmes and K. Russell, *Tetrahedron Lett.*, 1992, **33**, 6857.
21. (a) G.C. Fu and R.M. Grubbs. *J. Am. Chem. Soc.*, 1992, **114**, 7324; (b) G.C. Fu, S.T. Nguyen and R.M. Grubbs. *J. Am. Chem. Soc.*, 1993, **115**, 9856; (c) S.J. Miller and R.H. Grubbs, *J. Am. Chem. Soc.*, 1995, **117**, 5855.
22. H.K. Reimschuessel, J.P. Sibilia and J.V. Pascale, *J. Org. Chem.*, 1969, **34**, 959.
23. T. Duong, R.H. Prager, A.D. Ward and D.I.B. Kerr, *Aust. J. Chem.*, 1976, **29**, 2651.
24. T. Durst, R. Van den Elzen and M.J. LeBelle, *J. Am. Chem. Soc.*, 1972, **94**, 9261.
25. D. Roberto and H. Alper, *J. Am. Chem. Soc.*, 1989, **111**, 7539.
26. K. Yamada, S. Tanaka, K. Naruchi and M. Yamamoto, *J. Org. Chem.*, 1982, **47**, 5283.
27. H. Ishibashi, M. Ikeda, H. Maeda, K. Ishiyama, M. Yoshida, S. Akai and Y. Tamura, *J. Chem. Soc.*, *Perkin Trans. 1*, 1987, 1099.
28. T. Sasaki, S. Eguchi and M. Ohno, *J. Org. Chem.*, 1972, **37**, 466.
29. T. Sasaki, S. Eguchi and M. Ohno. *J. Am. Chem. Soc.*, 1970, **92**, 3192.
30. (a) K.A. Newlander, J.F. Callahan, M.L. Moore, T.A. Tomaszek, Jr, and W.F. Huffman, *J. Med. Chem.*, 1993, **36**, 2321; (b) J.F. Callahan, J.W. Bean, J.L. Burgess, D.S. Eggleston, S.M. Hwang, K.D. Kopple, P.F. Koster, A. Nichols, C.E. Peishoff, J.M. Samanen, J.A. Vasko, A. Wong and W.F. Huffman, *J. Med. Chem.*, 1992, **35**, 3970.
31. D. Gravel, J. Hebert, J. Bilodeau, E. Cavalieri and J.P. Daris, *Can. J. Chem.*, 1974, **52**, 645.
32. J.B. Taylor, D.R. Harrison and F. Fried, *J. Heterocycl. Chem.*, 1972, **9**, 1227.
33. E.C. Horning, V.I. Stromberg and H.A. Lloyd, *J. Am. Chem. Soc.*, 1952, **74**, 5153.
34. T.H. Koch, M.A. Geigel and C.-C. Tsai, *J. Org. Chem.*, 1973, **38**, 1090.
35. S.J. Neeson and P.J. Stevenson, *Tetrahedron Lett.*, 1988, **29**, 3993.
36. R. Grigg, V. Santhakumar, V. Sridharan, P. Stevenson, A. Teasdale, M. Thornton-Pett and T. Worakun, *Tetrahedron*, 1992, **47**, 9703.
37. F.J. Donat and A.L. Nelson, *J. Org. Chem.*, **22**, 1957, 1107.
38. Y. Tamura, Y. Kita and M. Terashima, *Chem. Pharm. Bull.*, 1971, **19**, 529.
39. A. Essawy, M. Abdalla and A. Deeb, *Rev. Roum. Chim.*, 1981, **26**, 601; *Chem. Abs.*, 1981, **95**, 97152y.
40. K. Mitsuhashi and K. Nomura, *Chem. Pharm. Bull.*, 1965, **13**, 951; *Chem Abs.*, 1965, **63**, 13096e.
41. T. Duong, R.H. Prager, J.M. Tippet, A.D. Ward and D.I.B. Kerr, *Aust. J. Chem.*, 1976, **29**, 2667.
42. T. Oishi, M. Fukui, R. Kenkyusho, Y. Ban and M. Honda, *Heterocycles*, 1976, **5**, 281.
43. J.L. Fahey, B.C. Lange, J.M. van der Veen, G.R. Young and A.K. Bose. *J. Chem. Soc.*, *Perkin Trans. 1*, 1977, 1117.
44. E. Vedejs and J.S. Stults, *J. Org. Chem.*, 1988, **53**, 2226.

45. M. Mori, Y. Washioka, T. Urayama, K. Yoshiura, K. Chiba and Y. Ban, *J. Org. Chem.*, 1983, **48**, 4058.
46. R.G. Glushkov, V.A. Volskova, V.G. Smirnova and O. Yu. Magidson, *Dokl. Akad. Nauk SSSR*, 1969, **187**, 327; *Chem. Abs.*, 1969, **71**, 112753v
47. R.G.Glushkov, V.G. Smirnova, K.A. Zaitseva, N.A. Novitskaya, M.D. Mashkovskii and G.N. Pershin, *Khim.-Farm. Zh.*, 1974, **8**(3), 14; *Chem. Abs.*, 1974, **81**, 13368y.
48. W.J. Brouillette and H.M. Einspahr, *J. Org. Chem.*, 1984, **49**, 5113.
49. J.R. Geigy AG, *Br. Pat.*, 949 955, 1964; *Chem. Abs.*, 1964, **61**, p1762c
50. C. Lambert, B. Caillaux and H.G. Viehe, *Tetrahedron*, 1985, **41**, 3331.
51. M.S. Akhtar, W.J. Brouillette and D.V. Waterhous, *J. Org. Chem.*, 1990, **55**, 5222.
52. G.B. Gill, G. Pattenden and S.J. Reynolds, *Tetrahedron Lett.*, 1989, **30**, 3229.
53. T. Beisswenger and F. Effenberger, *Chem. Ber.*, 1984, **117**, 1513.
54. R.G. Glushkov and T.V. Stezhko, *Khim. Geterotsikl. Soedin.*, 1978, 1252; *Chem. Abs.*, 1979, **90**, 22784u.
55. B.A. Mooney, R.H. Prager and A.D. Ward, *Aust. J. Chem.*, 1980, **33**, 2717.
56. R.G. Glushkov, V.G. Smirnova, I.M. Zasosova, T.V. Stezhko, I.M. Ovcharova and T.F. Vlasova, *Khim. Geterotsikl. Soedin.*, 1978, 374; *Chem. Abs.*, 1978, **89**, 43306j.
57. R.G. Glushkov and T.V. Stezhko, *Khim. Geterotsikl. Soedin.*, 1980, 1097; *Chem. Abs.*, 1981, **94**, 47260r.
58. B. Gobeaux and L. Ghosez, *Heterocycles*, 1989, **28**, 29.
59. Von W. Treibs and A. Lange, *J. Prakt. Chem.*, 1961, **14**, 208.
60. A.H. Rees, *J. Chem. Soc.*, 1965, 1749.
61. F.A. Fraser, G.R. Proctor and J. Redpath, *J. Chem. Soc., Perkin Trans. 1*, 1992, 445.
62. Y. Tamura, Y. Yoshimura and Y. Kita, *Chem. Pharm. Bull.*, 1971, **19**, 1069.
63. Y. Tamura, Y. Yoshimura and Y. Kita, *Chem. Pharm. Bull.*, 1972, **20**, 871.
64. A. Hassner, N.H. Wiegand and H.E. Gottlieb, *J. Org. Chem.*, 1986, **51**, 3176.
65. S.N. Ege, M.L.C. Carter, D.F. Ortwine, S.S.S.P. Chou and J.F. Richman, *J. Chem. Soc., Trans. Perkin 1*, 1977, 1252.
66. K. Suzuki, T. Ohkuma and G. Tsuchihasi, *J. Org. Chem.*, 1987, **52**, 2929.
67. A. Goti, A. Brandi, F. De Sarlo and A. Guarna, *Tetrahedron Lett.*, 1986, **27**, 5271.
68. A. Goti, A. Brandi, F. De Sarlo and A. Guarna, *Tetrahedron*, 1992, **48**, 5283.
69. E.J. Moriconi and I.A. Maniscalco, *J. Org. Chem.*, 1972, **37**, 208.

CHAPTER 7

Tetrahydroazepinediones

I. 2,3,4,5-TETRAHYDRO[1H]AZEPINE-2,3-DIONES

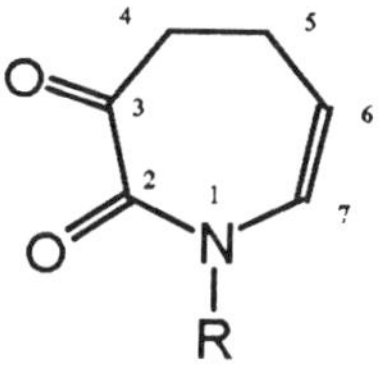

The preparation of the 7-phenyl derivatives **1** (R = Ph, OC_2H_5, $OCOCH_3$, H), the sole representatives of the 2,3,4,5-tetrahydro[1H]azepine-2,3-diones, was achieved by de-ethoxycarbonylation of the dihydroazatropolones **2** (R = Ph, OC_2H_5, $OCOCH_3$, H), which in turn were prepared by base-catalysed ring expansion of the photoadducts **3** (R = β-Ph, α-OC_2H_5, α-$OCOCH_3$, H) formed from the olefins **4** (R = Ph, OC_2H_5, $OCOCH_3$, H) and the dioxopyrroline **5**.[1] The products were obtained in yields of 63–65% from **2**. Direct de-ethoxycarbonylation of the diphenyl adduct **3** (R = β-Ph) also produced the

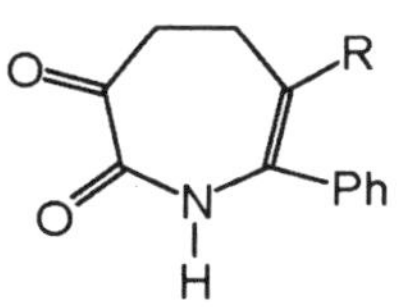

(1) 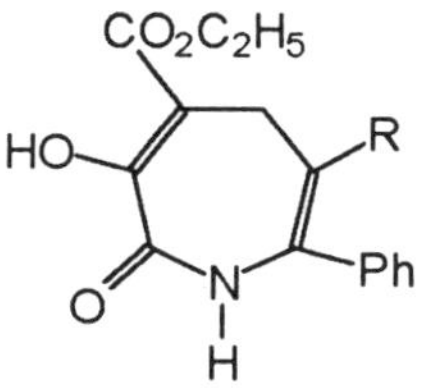   (2) 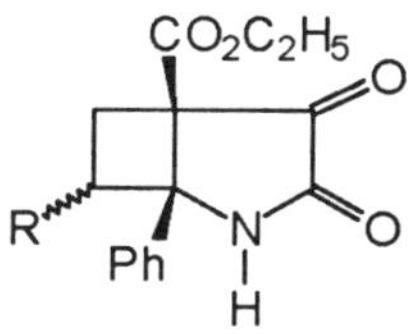   (3)

6,7-diphenyl-2,3,4,5-tetrahydro[1H]azepine-dione **1** (R = Ph) in a yield of 31% from **3**. Details of the compounds are listed in Table 1.

(4) **(5)**

Table 1. 2,3,4,5-Tetrahydro[1H]azepine-2,3-diones

Structure	M.p. (°C)	Spectroscopic evidence	Ref.
	162–164	IR, UV, MS, ^{1}H NMR	1
	101–103	IR, UV, MS, ^{1}H NMR	1
	121–126	IR, UV, MS, ^{1}H NMR	1
	204–208	IR, UV, MS, ^{1}H NMR	1

V. 2,3,4,5-TETRAHYDRO[1H]AZEPINE-2,4-DIONES

Irradiation of the azabicycloheptane-3,5-diones **6** (R^1 = CH_3, CH_2Ph; R^2 = H) using a high-pressure mercury lamp gave the corresponding 2,3,4,5-tetrahydro[1H]azepine-2,4-diones **7** (R = CH_3, CH_2Ph) by a cyclopropane ring-opening mechanism in yields of 47% and 34%, respectively.[2] The reaction, however, is limited in that the corresponding 7-phenylazabicycloheptane-3,5-diones **6** (R^1 = CH_3, CH_2Ph; R^2 = Ph), on irradiation, led to complex mixtures from which no ring expansion products could be isolated.[2]

The physical data for the two products (**7**, R = CH_3, CH_2Ph), the only representatives of the 2,3,4,5-tetrahydro[1H]azepine-2,4-diones in the literature, are listed in Table 2.

Table 2. 2,3,4,5-Tetrahydro[1H]azepine-2,4-diones

Structure	B.p. (°C/mmHg)	Spectroscopic evidence	Ref.
(R = CH_3)	140/13	IR, MS, ^{1}H NMR	2
(R = CH_2Ph)	170/0.3	IR, MS, ^{1}H NMR	2

VIII. 2,3,4,5-TETRAHYDRO[1H]AZEPINE-2,5-DIONES

The unsubstituted parent compound (8) in this series was isolated in the course of a study on the photolysis of N-alkylthioalkyl- and N-alkoxyalkylsuccinimides (9, $n = 1-3$, $X = S$, $R = CH_3$; $n = 1$, $X = O$, R = alkyl).[3] Whereas the ethers 9 ($n = 1$, $X = O$, R = alkyl) gave bicyclic products on irradiation, the thioether 9 ($n = 1$, $X = S$, $R = CH_3$) failed to react and the thioethers 9 ($n = 2-3$, $X = S$, $R = CH_3$) gave ring-expanded products. The 2,3,4,5-tetrahydro[1H]azepine-2,5-dione 8 was formed from the thioether 9 ($n = 2$, $X = S$, $R = CH_3$) but was only isolated in a 3% yield. A similar yield of the saturated azepanedione 10 was also isolated from the same reaction.[3]

(8)   (9)   (10)

The only other representative of this series, 4,6,7-trimethyl-2,3,4,5-tetrahydro[1H]azepine-2,5-dione (15), has been prepared from the benzoquinone 11 by two alternative pathways, A (11 → 12 → 13 → 14 → 15) and B (11 → 12 → 15), as illustrated.[4]

The ring expansion of the benzoquinone **11** was achieved by a Schmidt reaction, and the alkylation reaction was performed using triethyloxonium fluoborate. Pathway B represents the most efficient route to **15**, but pathway A was investigated in order to examine spectroscopically the intermediates (**13** and **14**) which differed from their corresponding 4,7-dimethyl analogues in that the 2-ketone in the 4,7-dimethyl analogue of **12** formed an enol ether rather than the 5-ketone as in **13**.[4] Table 3 gives the data on the two known 2,3,4,5-tetrahydro[1H]azepine-2,5-diones.

Table 3. 2,3,4,5-Tetrahydro[1H]azepine-2,5-diones

Structure	M.p. (°C)	Spectroscopic evidence	Ref.
	155–157	IR, UV, MS, ^{1}H NMR	3
	101–102	IR, UV, ^{1}H NMR	4

IX. 2,5,6,7-Tetrahydro[1H]azepine-2,5-diones

Misiti et al.[5] claimed that treatment of the 2,5-dihydro[1H]azepine-2,5-diones **16** (R = H, CH$_3$) with bromine gave the bromo-2,5,6,7-tetrahydro[1H]azepine-2,5-dione derivatives **17** (R = Br, CH$_3$, CH$_2$Br). A subsequent publication by Rickards and Smith,[6] however, demonstrated that Misiti et al.'s 2,5-dihydro[1H]azepine-2,5-diones **16** (R = H, CH$_3$) had been assigned the wrong structures, and were the 2,5-dihydro[1H]azepine-2,5-diones **18** (R = H, CH$_3$) (see also Chapter 10, Section V). This correction invalidates the originally proposed structures for the two bromo compounds (**17**, R = Br, CH$_3$) and, in the absence of further evidence, the existence of 2,5,6,7-tetrahydro[1*H*]azepine-2,5-diones must be considered doubtful since no other analogues have been reported.

(16)   (17)   (18)

XIV. 2,3,4,7-TETRAHYDRO[1*H*]AZEPINE-2,7-DIONES

The 3,3-diphenyl-2,3,4,7-tetrahydro[1*H*]azepine-2,7-dione **19** was prepared by cyclisation of the nitrile ester **20**, which in turn was synthesised from diphenylacetonitrile (**21**) and the methyl ester of γ-bromocrotonic acid (**22**) in the presence of sodamide. The ring-closure reaction to **19** was carried out by hydrolysing the nitrile ester **20** with aqueous methanol and completing the cyclisation of the nitrile acid with sulphuric acid. The product (**19**) was prepared for comparison with the corresponding tetrahydropyridine and no other azepine derivatives were synthesised by this procedure.[7]

(19)   (20)   (21)   (22)

The *trans*-3,4-dibromo-2,3,4,7-tetrahydro[1*H*]azepine-2,7-dione **23** was prepared by the action of bromine on 1-methylazepine-2,7-dione (**24**). ^{1}H NMR and IR spectroscopy demonstrated that the dibromo compound was the 1,2-addition product (**23**) rather than the 1,4-addition product (**25**).[8] The physical data are presented in Table 4.

Table 4. 2,3,4,7-Tetrahydro[1*H*]azepine-2,7-diones

Structure	M.p. (°C)	Spectroscopic evidence	Ref.
	173		7
	78–79	IR, UV, ^{1}H NMR	8

XV. 2,3,6,7-TETRAHYDRO[1*H*]AZEPINE-2,7-DIONES

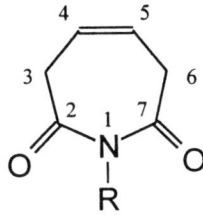

The only compound reported in this series is 4-phenyl-2,3,6,7-tetrahydro[1*H*]azepine-2,7-dione (**26**), which was prepared by cyclisation of the dinitrile **27** to 2-amino-7-bromo-5-phenyl[3*H*]azepine hydrobromide (**28**) using anhydrous hydrogen bromide, followed by hydrolysis using aqueous dimethylformamide.[9] The product (**26**) melted at 157–158 °C and had IR and ^{1}H NMR spectra consistent with the structure proposed.

(**26**)   (**27**)   (**28**)

XXIII. 2,3,4,5-TETRAHYDRO[1*H*]AZEPINE-4,5-DIONES

There are only two representatives of the 2,3,4,5-tetrahydro[1*H*]azepine-4,5-diones in the literature, the 6,7-diphenyl-3,3-dimethyl-substituted derivative **29** (R = H) and the 2,6,7-triphenyl-3,3-dimethyl analogue **29** (R = Ph).[10]

(**29**)   (**30**)   (**31**)

The diphenyl compound **29** (R = H) was prepared along with its enol tautomer **31** (R = H) by hydrolysis of the azeto[1,2a]pyrrole **30** (R = H). A yield of 30% was achieved using 2M HCl in ethanol at room temperature. By contrast, the 2,6,7-triphenyl-*endo*-azeto[1,2a]pyrrole **30** (R = Ph) produced only the enol tautomer **31** (R = Ph) under the same conditions, but gave a 10% yield of the 2,5-dione tautomer **29** (R = Ph) under milder hydrolysis conditions using silica.[10] The structure of the 2,6,7-triphenyl compound **29** (R = Ph), m.p. 214 °C, was confirmed by its IR, ^{1}H, and ^{13}C NMR spectra, but no data were quoted in the publication for the 6,7-diphenyl analogue **29** (R = H).[10]

XXIV. 3,4,5,6-TETRAHYDRO[2H]AZEPINE-4,5-DIONES

The 2,7-diphenyl-6-isopropyl-3,3-dimethyl-substituted derivative **32** is the only representative of this series. It was prepared from the azeto[1,2a]pyrrole **33** in the same manner as the related derivatives **29** described above.[10] In this case a yield of 95% was achieved and no other tautomers were formed. The product (**32**) melted at 165–166 °C and was characterised by its IR, ^{1}H and ^{13}C NMR spectra.

(32)   (33)

REFERENCES

1. Y. Horiguchi, T. Sano and Y. Tsuda, *Heterocycles*, 1985, **26**, 1509.
2. E. Sato, Y. Ikeda and Y. Kanaoka, *Chem. Pharm. Bull.*, 1983, **31**, 1362.
3. H. Nakai, Y. Sato, T. Mizoguchi, M. Yamazaki and Y. Kanaoke, *Heterocycles*, 1977, **8**, 345.
4. E.J. Moriconi and I.A. Maniscalco, *J. Org. Chem.*, 1972, **37**, 208.
5. D. Misiti, H.W. Moore and K. Folkers, *Tetrahedron*, 1966, **22**, 1201.
6. R.W. Rickards and R.M. Smith, *Tetrahedron Lett.*, 1966, 2361.
7. E. Urech, E. Tagmann, E. Sury and K. Hoffmann, *Helv. Chim. Acta*, 1953, **36**, 1809.
8. R. Shapiro and S. Nesnow, *J. Org. Chem.*, 1969, **34**, 1695.
9. W.A. Nasutavicus, S.W. Tobey and F. Johnson, *J. Org. Chem.*, 1967, **32**, 3325.
10. F. Stierli, R. Prewo, J.H. Bieri and H. Heimgartner, *Helv. Chim. Acta*, 1983, **66**, 1366.

CHAPTER 8

Dihydroazepines

I. 2,3-DIHYDRO[1H]AZEPINES

PREPARATION

This group of compounds is fairly well represented in the literature, their apparent stability presumably being attributable to the conjugated dienamine structure therein. Many of the syntheses reported for 2,3-dihydro[1H]azepines involve ring expansions of either five- or six-membered ring compounds.

One of the earliest and most ingenious approaches involved the double cycloaddition of dimethyl acetylenedicarboxylate (DMAD) with 2-alkylbenzothiazoles. Thus the initial adduct **1** (R = H) was desulphurised by Raney nickel

Dihydroazepines

yielding the *N*-phenyldihydroazepine 2. Although desulphurisation of the adduct 1 (R = CH$_3$) was not reported, there seems little doubt that this could be successfully achieved.[1]

Ring expansion of various pyrrole derivatives has been a popular strategy. 2-Methylthio-1-methyl-2-pyrroline (3), but not the corresponding dimethylacetal, reacted easily with DMAD yielding the dihydroazepine 4 (R = SMe), along with other products.[2] This work has been repeated and extended[3] by desulphurisation to give 4 (R = H), and it was inferred that if *N*-acylpyrrolines were employed, reduced enamine activity caused the reaction to stop at the [2 + 2] cycloaddition stage; such intermediates were isolated from the higher homologues of 3.[3]

In similar fashion, *N*-acylpyrrolines (5) carrying an amino group underwent cycloaddition/ring expansion with DMAD and, less efficiently, with methyl propiolate.[4] Better yields of products 6 (R = H, CO$_2$Me) were obtained after relatively short reaction times in DMSO rather than xylene. The iminophosphoranes 7 also underwent acetylenic ester ring expansions in good yields, giving 8 (R = H, CO$_2$Me, CO$_2$Et).[5] The latter were shown to react with DMAD yielding ultimately indole derivatives by ring expansion to azonine

compounds, followed by ring collapse.[6] Acid treatment of 8 gives the free amino compounds 9; detailed spectroscopic details for all dihydroazepines were provided.[5,6] It has been shown that the iminophosphoranes 8 react with isocyanates to produce pyrimido[4,5-*b*] azepines;[7] this usefully complements the already available methods.[8]

It was demonstrated in 1973[9] that 2,3-dihydro[1*H*]azepines arise from thermal rearrangements of biscyclopropapyrrolidines. Thus 10 was obtained from both *syn*- (11) and *anti*- (12) isomers. The large difference in activation energies for these processes were convincingly explained by the more favourable interaction of the two cyclopropyl groups in the *syn* isomer (**11**) compared with the *anti* isomer (**12**) in which the relevant p orbitals lie at a mutual angle of 60°.

(11)   (10)   (12)   (13)

The electron-withdrawing tendency of the CO_2CH_3 group must also contribute to the energy barrier since the *N*-methyl *anti* isomer gave **13** at a substantially lower temperature (280 °C).

Ring expansion of cyclic vinylaziridines (**14**) is a thermal process leading to 2,3-dihydroazepines (**15**);[10] the nitrogen atom could be substituted by phenyl or benzyl but analogues lacking ester substituents behaved differently. These materials undergo solvent-dependent reactions with DMAD leading either to substitution or to ring expansion.

(14)   (15)   (16)   (17)

During studies on allenic amines, it was discovered that the initial [2 + 2] cycloaddition products (**16**) from acetylenic esters and ketones were converted easily into **17** (X = $COCH_3$, CO^nPr, CO_2Me; R = H or CH_3).[11] Full details have not been published.

A very interesting, novel approach[12] involves a one-step conversion by reaction of 1-aminoacryl derivatives with certain 2-pyranone compounds. Thus **18** and **19** in methanol at 25 °C gave dihydroazepines (**20**), presumably via Michael addition, decarboxylation, electrocyclisation and proton transfer as shown. This seems to be the only example of its kind.

Other approaches to 2,3-dihydro[1*H*]azepines involve manipulations of dihydroazepinones, their derivatives or [1*H*]azepines. For example, among the products obtained when palladium acetate in acetic acid reacted with *N*-ethoxycarbonyl[1*H*]azepine (**21**) was the 2,3-diacetoxydihydroazepine **22**, of undetermined stereochemistry.[13] The unstable [3*H*]azepine **23** yielded the dihydroazepine **24** on catalytic hydrogenation; this served to establish the constitution of **23**, itself the product of dihydropyridine ring expansion[14] (see Chapter 11).

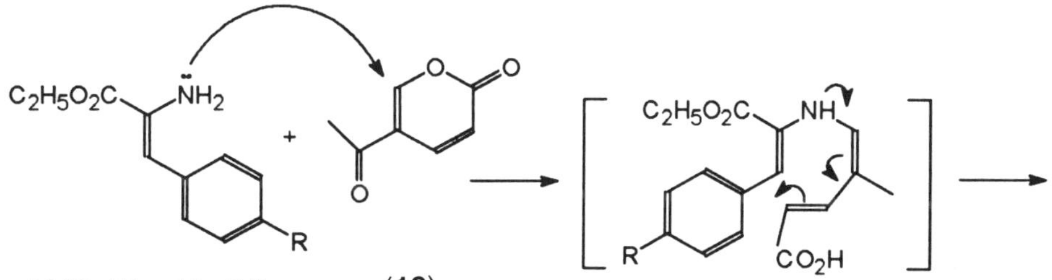

Dihydroazepines

(18) (R = H, Cl) **(19)** → [intermediate with $C_2H_5O_2C$, NH, CO_2H, R] →

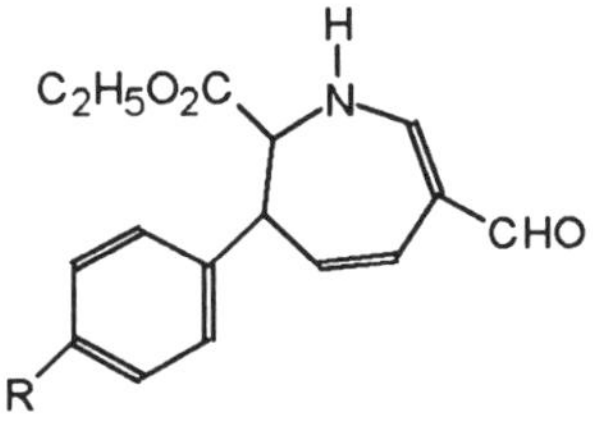

(20) (R = H, Cl)

There are several examples of LiAlH$_4$ reduction of 1,3-dihydro[2H]azepin-2-ones (e.g. **25**, R,R = O) yielding the corresponding dihydroazepines (**25**, R = R = H)[15] (see Chapter 9).

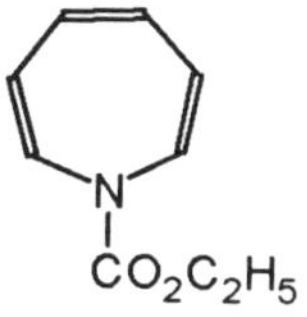

(21) 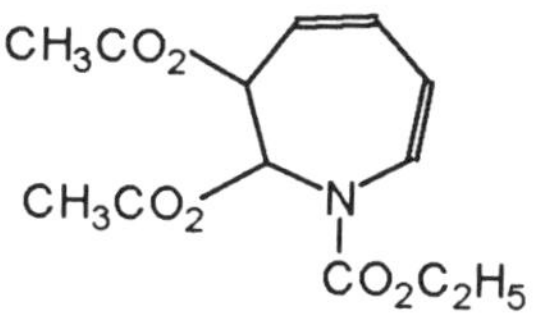   **(22)**

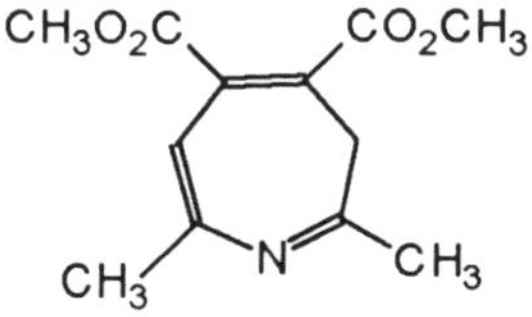

(23) 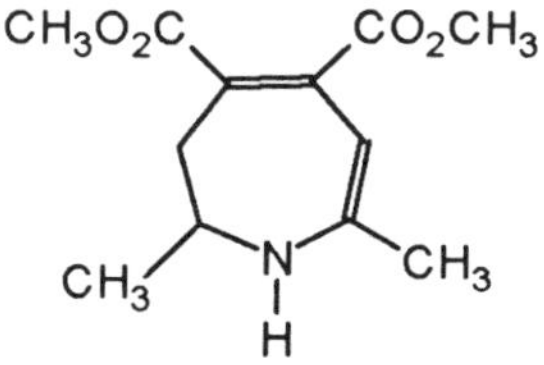   **(24)**

The naturally occurring yellow pigment muscaflavin (**26**) from fly agaric has been synthesised from a pyridylalanine; the key step in this biomimetic procedure involves closure of N to C-7.[18]

For completeness, we mention here the fact that aminoazepines (e.g. **27**) (Chapter 11) condense with certain reactive methylene compounds such as dimedone to give products (e.g. **28**) which are effectively dihydroazepines.[16]

PROPERTIES

Most 2,3-dihydro[1H]azepines are oils or low-melting solids. Although relatively stable, particularly below 0 °C, witness the production of **10** at 350 °C,[9] nevertheless they do tend to decompose on standing and from early work they have been shown to form stable crystalline salts (see Table 1). The site of protonation was shown to be C-4.[15b] Irradiation (UV) converts the bases by valence-bond isomerisation into 2-azabicyclo[3.2.0]hept-6-enes. For example, the N-methyl derivative of **25** gave **29** as the major isomer.[15] Diels–Alder reactions take place readily[10b,15] even with DMAD, which ordinarily reacts with enamines by [2 + 2] cycloadditions followed by ring expansion.[17] However, **8** reacts with dienophiles in [2 + 2] fashion at the C-4–C-5 double bond.[6] Dehydrogenation attempts (DDQ) on compounds **6** were unsuccessful; only in one case did a ring contraction take place which could be explained by assuming that dehydrogenation to the corresponding azepine was followed by an azanorcaradiene rearrangement.[4] Not surprisingly, catalytic hydrogenation of the 2,3-dihydro[1H]azepine ring can lead to an azepane.[9,19]

Table 1 lists those 2,3-dihydro[1H]azepines whose identities are fully supported by chemical analyses and, in most cases, by spectroscopic measurements.

Dihydroazepines 417

Table 1. 2,3-Dihydro[1H]azepines

Structure	M.p. (°C)/b.p. (°C/mmHg)	Derivatives and m.p. (°C)	Spectroscopic evidence	Ref.
(NH)	62/11		IR, UV	19
(N-CH₃)			IR, ^{1}H NMR	9, 19
(N-C₂H₅, with CH₃ groups)	61–64/1	HClO₄ 99–101 N-Phenylmaleimide 149–150	^{1}H NMR	15b
(N-CH₃, with CH₃ groups)	50–54/1	HClO₄ 106–106.5 N-Phenylmaleimide 98–98.5	^{1}H NMR	15b
(N-CH₃, CO₂CH₃)			IR, ^{1}H, ^{13}C NMR	11
(N-CO₂CH₃)		N-Phenylmaleimide 177.5–178	UV, IR, ^{1}H NMR	9
(N-Ph, CO₂CH₃, CO₂CH₃)	123		UV, IR, ^{1}H NMR	10a

Table 1 (continued)

Structure	M.p. (°C)/b.p. (°C/mmHg)	Spectroscopic evidence		Ref.
(azepine with CO_2CH_3, CO_2CH_3, N-CH$_2$Ph)	110	N-Phenyl-maleimide 248	^{1}H, ^{13}C NMR	10a, 10b
(azepine with two CO_2CH_3, N-CH$_3$)			IR, ^{1}H NMR	3
(azepine with CHO, HO_2C, CO_2H, NH)	120		MS, ^{1}H NMR	18
(azepine with CHO, CH_2O_2C, CO_2CH_3, NH)	104–105.5	Semicarb. 189–191	UV, MS, IR	18
(azepine with CH_3CO_2, CH_3CO_2, N-$CO_2C_2H_5$)	Oil		MS, ^{1}H NMR	13
(azepine with CO_2CH_3, CO_2CH_3, CH_3, CH_3, NH)	98–99		IR, UV	14

Dihydroazepines

Table 1 (continued)

Structure	M.p. (°C)/b.p. (°C/mmHg)	Spectroscopic evidence	Ref.
4,5-bis(CO₂CH₃), 3,2-bis(CO₂CH₃), N-Ph dihydroazepine	127	IR	1
4-CO₂CH₃, 3-CO₂CH₃, 2-SCH₃, 7-CH₃, N-H dihydroazepine	163–164	MS, ^{1}H NMR, UV	2, 3
5-CO₂CH₃, 3-CO₂CH₃, 2-(N=PPh₃), N-Tos dihydroazepine	209	UV, IR, ^{1}H, ^{13}C NMR	5, 6
5-CO₂CH₃, 3-CO₂CH₃, 2-NH₂, N-Tos dihydroazepine	135	UV, IR, ^{1}H, ^{13}C NMR	5, 6
5-CO₂C₂H₅, 3-CO₂CH₃, 2-(N=PPh₃), N-Tos dihydroazepine	178–179	UV, IR, ^{1}H, ^{13}C NMR	5
5-CO₂C₂H₅, 3-CO₂CH₃, 2-NH₂, N-Tos dihydroazepine	121	UV, IR, ^{1}H, ^{13}C NMR	5

Table 1 (continued)

Structure	M.p. (°C)/b.p. (°C/mmHg)	Spectroscopic evidence	Ref.
(7-membered ring with N-Tos, C=N-PPh$_3$, CO_2CH_3, $CO_2C_2H_5$, CO_2CH_3 substituents)	254	UV, IR, ^{1}H, ^{13}C NMR	5
(7-membered ring with N-Tos, C=N-PPh$_3$, CO_2CH_3, $CO_2C_2H_5$, $CO_2C_2H_5$ substituents)	254	UV, IR, ^{1}H, ^{13}C NMR	5
(7-membered ring with N-Tos, C-NH$_2$, CO_2CH_3, $CO_2C_2H_5$, $CO_2C_2H_5$ substituents)	234	UV, IR, ^{1}H, ^{13}C NMR	5
(dihydroazepine with phenyl, CHO, $C_2H_5O_2C$ substituents, NH)	118–120	^{1}H, ^{13}C NMR	12
(dihydroazepine with 4-chlorophenyl, CHO, $C_2H_5O_2C$ substituents, NH)	155–157	^{1}H, ^{13}C NMR	12

Dihydroazepines

Table 1 (continued)

Structure	M.p. (°C)/b.p. (°C/mmHg)	Spectroscopic evidence	Ref.
(2-hydroxyphenyl-substituted dihydroazepine with CO_2CH_3 and CH_3O_2C groups, NH)	'Schaum'	^{1}H, ^{13}C NMR	12
(dihydroazepine with CN, CO_2CH_3, CO_2CH_3, NH_2, $CO_2C_2H_5$)	150–151	^{1}H, ^{13}C NMR, MS, IR	4
(dihydroazepine with CN, CO_2CH_3, CO_2CH_3, NH_2, CH_3, $CO_2C_2H_5$)	166–168	^{1}H, ^{13}C NMR, MS, IR	4
(dihydroazepine with CN, CO_2CH_3, CO_2CH_3, NH_2, Ph, $CO_2C_2H_5$)	198–199	^{1}H, ^{13}C NMR, MS, IR	4
(dihydroazepine with CN, CO_2CH_3, NH_2, $CO_2C_2H_5$)	123	^{1}H, ^{13}C NMR, MS, IR	4
(dihydroazepine with CN, CO_2CH_3, NH_2, CH_3, $CO_2C_2H_5$)	137	^{1}H, ^{13}C NMR, MS, IR	4

Table 1 (continued)

Structure	M.p. (°C)/b.p. (°C/mmHg)	Spectroscopic evidence	Ref.
(Ph, CN, CO₂CH₃, NH₂, N-CO₂C₂H₅ azepine)	159	^{1}H, ^{13}C NMR, MS, IR	4
(dimedone-azepine)	113–116	IR, ^{1}H, ^{13}C NMR, UV	16
(phenyl pyrazolone-azepine)	140–141 HBF₄ 128–129	IR, ^{1}H, ^{13}C NMR, UV	16
(C₂H₅O₂C, CN azepine)	74–75	IR, ^{1}H, ^{13}C NMR, UV	16
(CH₂OH, CH₃O₂C, CO₂CH₃ azepine)	Oil		18

II. 2,5-DIHYDRO[1H]AZEPINES

Several groups of workers have made contributions to this area within the last 20 years. Cope rearrangements of *cis*-iminovinylcyclopropanes (e.g. **30**) would be expected to yield 2,5-dihydro[1H]azepines. In this case **30** did give **31** which was air- and moisture-sensitive,[20] but a more favourable equilibrium in the rearrangement involves having a π-stabilising group attached to the cyclopropane.[21] Thus the cyclopropyl compound **32**, prepared *in situ*, rearranged in 95% yield to **33**. Vinylaziridines have been employed in two ways undergoing cycloadditions with alkynes. First, vinylaziridines (e.g. **34**, $R^1 = R^3 = CH_3$, $R^2 = Ph$) react with alkynylphosphonium salts (e.g. **35**, $R^4 = CH_3$ or Ph) to give dihydroazepines (**36**) (see Table 2).[22] This is to be seen as a hetero-Cope rearrangement. Second, vinylaziridines **37** react even at −20 °C with DMAD giving **38**.[23,24] It is thought that initially **37** adds to DMAD in Michael fashion, but thereafter it is not possible to distinguish between the options.[24] Recently the cyclopropylazirine **37A** was thermolysed to the tricyclic amine **37B**, which reacted with phenyl isocyanate to yield the dihydroazepine **38A**.[77]

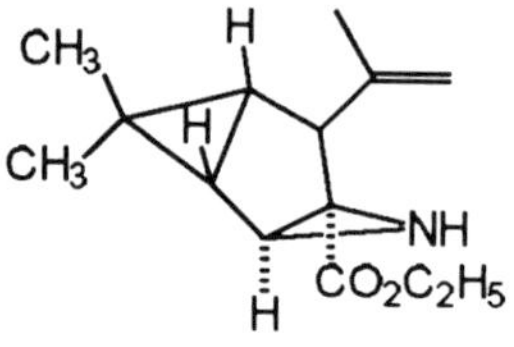

(37B) 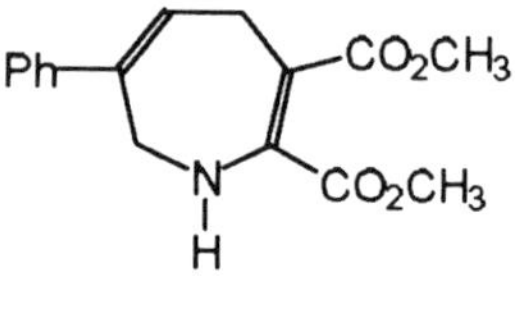   (38) 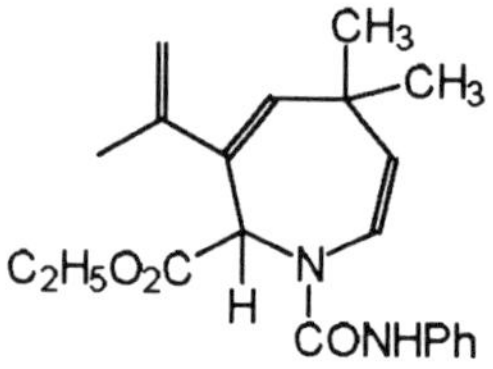   (38A)

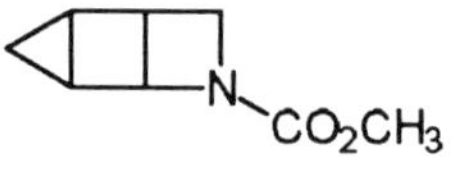

(39) 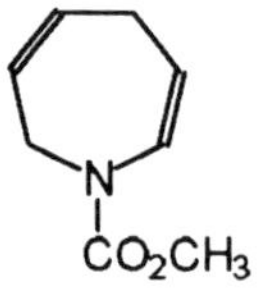   (40)

Finally in this section, one of the simplest examples of a 2,5-dihydro[1H]-azepine (**40**) was reported[25,26] to arise by thermolysis of the tricyclic system **39** in refluxing xylene (see Table 2).

Table 2. 2,5-Dihydro[1H]azepines

Structure	M.p. (°C)/b.p. (°C/mmHg)	Spectroscopic evidence	Ref.
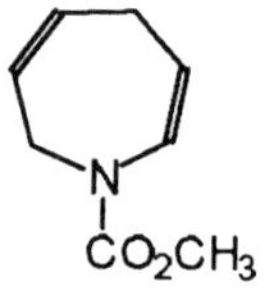	Oil	IR, MS, ^{1}H NMR	26
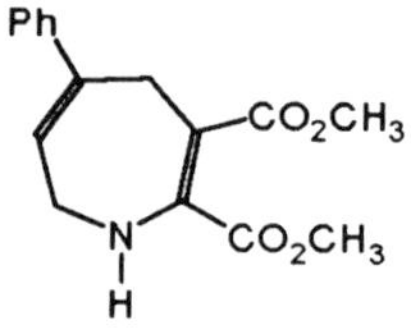	122–123	IR, ^{1}H NMR	24

Table 2 (continued)

Structure	M.p. (°C)/b.p. (°C/mmHg)	Spectroscopic evidence	Ref.
(azepine with CHO, CH₃, CH₂Ph)			21
(azepine with Ph, CH₃, PPh₃Br, CH₃, NH)	242–246	IR, ^{1}H, ^{13}C, ^{31}P NMR, MS	22
(azepine with CH₃, Ph, PPh₃Br, CH₃, Ph, NH)	214–222	IR, ^{1}H, ^{13}C, ^{31}P NMR, MS	22
(azepine with Ph, PPh₃Br, CH₃, Ph, NH)	130–135	IR, ^{1}H, ^{13}C, ^{31}P NMR, MS	22
(azepine with Ph, PPh₃Br, CH₃, CH₃, N-CH₂-C₆H₄Cl)	140–160	IR, ^{1}H, ^{13}C, ^{31}P NMR	22
(azepine with isopropenyl, CH₃, CH₃, C₂H₅O₂C, CONHPh)	209–211	^{1}H NMR	77

III. 2,7-DIHYDRO[1H]AZEPINES

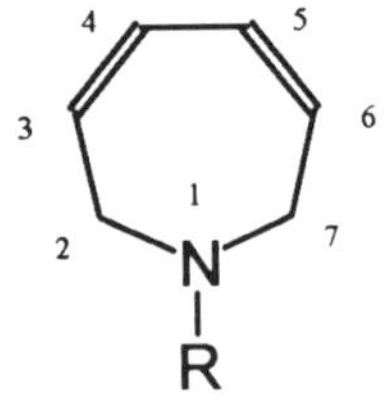

There are, surprisingly, only two examples.[27] The anion derived from **41** reacted with DMAD to provide the enolic product **42**. Such ring expansions generally

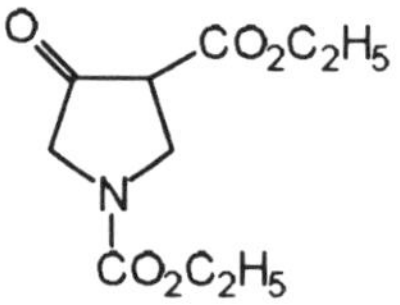

(41) 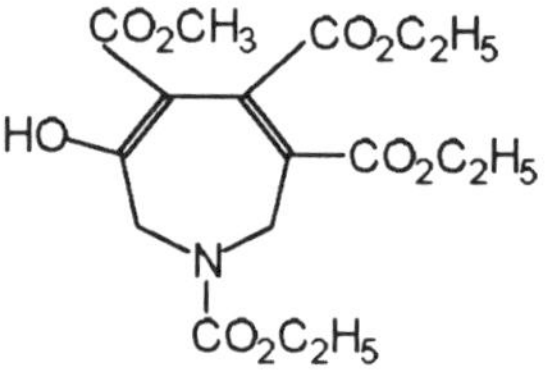   (42)

yield enolic products,[28-30] although decarboxylation of the ester group at C-4 generally reveals a ketone carbonyl group at C-3; in this case, hydrolysis gave unstable products.[31]

Recently it has been shown[79] that the *cis,cis* dienes **42A** (X = CH_2Br) react with various nitrogen nucleophiles to give 2,7-dihydro[1H]azepines **42B** (Y = CH_2Ph, ^nBu, tosyl; $R^1 = R^3 = {}^tBu$; $R^2 = R^4 = H$). This synthesis has the advantage of versatility since the key dienes were obtained by catechol

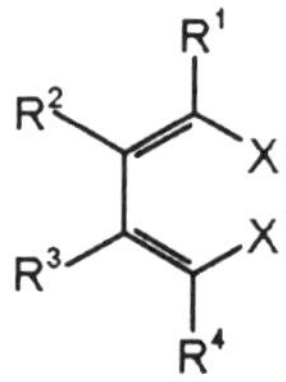

(42A) 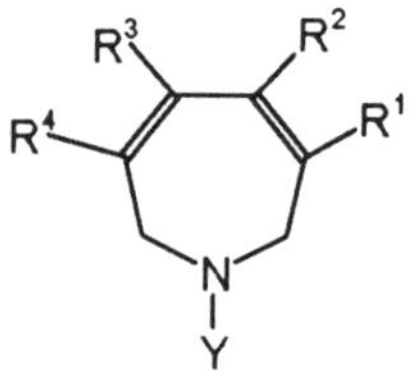   (42B)

oxidation [$Pb(OAc)_4$] and subsequent mild manipulations (**42A**; X = $CO_2CH_3 \rightarrow CH_2OH \rightarrow CH_2Br$).

IV. 4,5-DIHYDRO[1H]AZEPINES

PREPARATION

This is a thoroughly explored area. Treatment of 4-chloromethyl-1,4-dihydropyridines with potassium cyanide gave cyanodihydroazepines (e.g. **43** → **44**) as far back as 1920,[32] but it was only in 1962 that it was shown that the product was **44** and that the reaction was best conducted at room temperature[33] or below.[34] A study of the kinetics[35] of this ring expansion, taken with the isolation of the [4H]azepine **45**,[36] implicated **45** in the process since cyanide ion rapidly converted **45** into **44**. Sodium ethoxide also solvolyses **43** giving the ethoxydihydroazepine **46**.[37] The diacetyl analogue **47** was also shown to react with

(47) (48) (49)

aqueous alcoholic KCN to give the expected dihydroazepine **48**.[38] The dimethyl ester corresponding to **43** reacted with NaBH$_4$ in acetonitrile giving **49**,[39] whereas in liquid ammonia over 7 days it yielded **50** and methyl 2-methylpyrrole-3-carboxylate.[40] Later it was demonstrated that the dihydropyridine methyl and ethyl esters **51** reacted with sodium cyanide in DMSO or with

(50) (51) (52)

(53) (54)

methanol containing triethylamine to give the dihydroazepines **52** (X = CN and OMe, respectively).[41] Furthermore, the carbon nucleophile from acetonylacetone could also[42] participate in the ring expansion reaction, producing for example **52** [X = CH(COCH$_3$)$_2$].

To summarise, dihydropyridines such as **43** and **51**, under controlled conditions, react with a variety of basic nucleophiles yielding 4,5-dihydro[1H]azepines; SH$^-$ is an exception where a sulphur-bridged product (**53**) has been established.[39] It is accepted that the ring expansion proceeds initially by abstraction of the proton from the >NH group, followed by its conversion into the [4H]azepine structure (e.g. **54**).[43,44] The latter was shown to react with cyanide ion yielding **44** (methyl esters), as pointed out earlier.[35]

Aryl-substituted 4,5-dihydro[1H]azepines were obtained from the appropriate dihydropyridines (**43**, **51**) by reaction with aryl Grignard reagents; the products (**55**) contain the *cis* isomers as major products but alkyl Grignards gave predominantly the *trans* isomers,[45] as seen above.[41] Solvolysis of the tosylate **56** gave the exomethylene derivative **57**.[46]

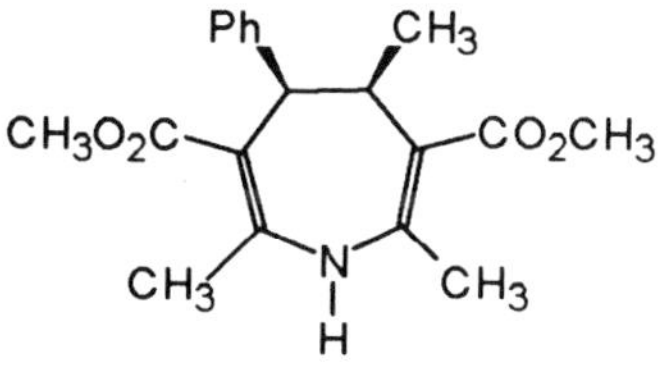

(55) 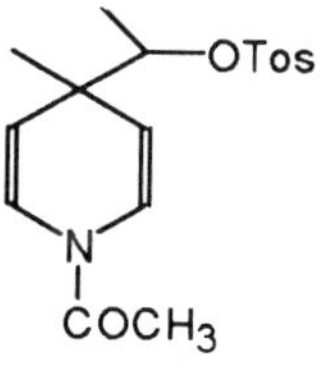 (56) 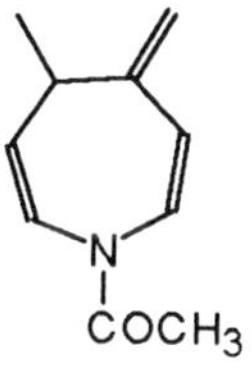 (57)

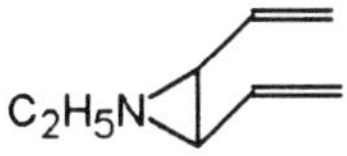

(58) 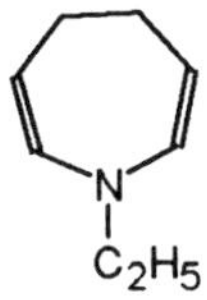 (59)

Apart from routes involving dihydropyridines, there have been several other approaches to the 4,5-dihydro[1H]azepine ring system. One of the earliest of these[47] involved Cope rearrangement of *cis*-divinylaziridine (**58**) to **59**. A later similar study on the *N*-methyl analogue of **58** and the dipropenylaziridines also states that 4,5-dihydro[1H]azepines were identified on scanty evidence.[48,49] Flash vacuum pyrolysis of the aziridine **60** gave in poor yield the dihydroazepine

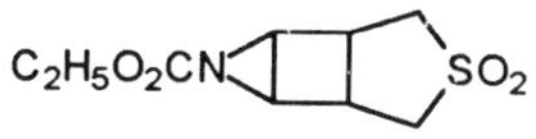

(60) 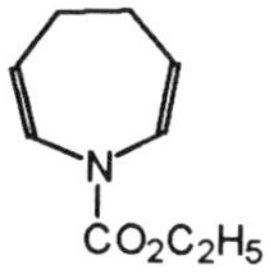 (61) 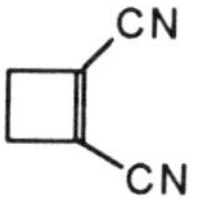 (62)

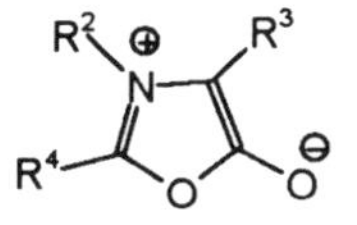

(63) 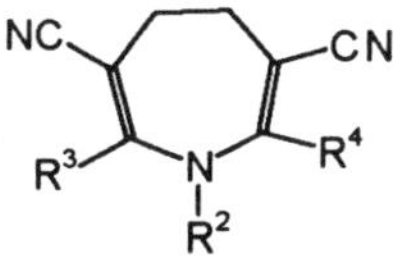 (64)

61, which was stable enough to be isolated.[50] Dicyanocyclobutene (**62**) undergoes 1,3-dipolar cycloadditions with mesoionic oxazolones (e.g. **63**, $R^2 = CH_3$ or H, $R^3, R^4 = CH_3$, Ph or $PhCH_2$)[51,52] giving dihydroazepines (**64**). The nitrile ylid **65** has also been reacted with **62** to give **64** ($R^3 = R^4 = Ph$, $R^2 = H$).[53] In one case, an intermediate bicyclic carboxylic acid (**66**) was isolated from an oxazolone reaction; its structure was confirmed by X-ray analysis and it was then converted thermally into the expected dihydroazepine.[54a] This reaction has been rationalised using MO calculations.[54b]

(65) (66)

(67) (68)

There is only one synthetic method for 4,5-dihydro[1H]azepines which depends on direct cyclisation of acyclic precursors; the (2Z,4E)-α-allyldienenitriles **67** in refluxing xylene were converted into the dihydroazepines (**68**).[55] It was considered that the first step is a [1, 6] hydrogen shift which leads to a helical 8π transition state suitable for cyclisation.

Conversions of already formed azepine rings are known. Thus, the tetrahydroazepinones **69** reacted with Vilsmeier reagent ($POCl_3$/DMF) to give **70**.[56] On the other hand, the [1H]azepine **71** adds iodine azide (IN_3) yielding

(69) (70) (71)

(72) (73)

72,[57] whilst **71** undergoes reductive silylation with chlorosilanes and magnesium in HMPA producing **73**.[58-60] Both regio- and stereochemistry were deduced from ^{1}H NMR spectra. The chlorosilanes used were CH_3SiCl, Ph_2SiCl_2 and $(CH_3)_2SiCl_2$ and yields were moderate.[59] Temperature-dependent ^{1}H and

Dihydroazepines

^{13}C NMR spectra were used[60] to infer restricted rotation between the nitrogen atom and the carbonyl group.

PROPERTIES

The least substituted N-alkyl members of this series appear to be rather reactive,[47,49] not entirely surprising given that they are doubly enaminic. On the other hand, N-acyl compounds (e.g. **61**) are more easily handled, are fully characterised and have been utilised by several groups.[50,57,59] Attachment of substituents, particularly electron-withdrawing groups, at C-2, C-3, C-6 and C-7, have a beneficial effect on stability, as one might expect.

4,5-Dihydro[1H]azepines are sensitive to many reagents. Thus the early studies[32] on conversions **43** → **44** failed at temperatures above ambient since it was later found that the dihydroazepine could be converted into a pyrrole (**74**)[34] and ethyl acrylate. Similarly, the tricyano compound **75** gave the dicyanopyrrole **76**.[34] Heating[41] 4-substituted 4,5-dihydro[1H]azepines (**44**, **46**) gave

(**74**)   (**75**)

(**76**)   (**77**)

[4H]azepines (e.g. **77**) at around 265 °C; these then isomerised at 320 °C to [3H]azepines, which are thermodynamically more stable (see Chapter 11).[36] A 4-unsubstituted compound (**49**) reacted differently with KOH in ethanol or water; it gave first an amino diester (**78**) and then 1-acetyl-2-methylcyclopentene

(**78**)   (**79**)

(80)

(81)

(79),[37] whilst in aqueous acid, **80** was isolated. A lactone (**81**) was obtained by treatment of **49** with NBS.[43]

Amongst the original experiments[63] aimed at dehydrogenation of 4,5-dihydro[1*H*]azepines, sodium nitrite in acetic acid was employed. The final product was eventually shown to be the furo[2,3-*b*]pyridine **82**,[62] hot aqueous ethanolic silver nitrite proving to be a superior medium for production of the furopyridine; it was accompanied by a minor product formulated as a pyrrolo[2,3-*b*]pyridine (**83**). This problem was addressed later[64] and by treating

(82)

(83)

(84)

(85)

(86)

44 with an acidic ion-exchange resin, the previously suspected[62] cyano octanedione diester (**84**) intermediate was isolated. There are cases where ring contraction back to 1,4-dihydropyridine derivatives has been seen,[41] particularly with hydrogen halides. An interesting transannular cyclisation, **85** → **86**,[42] gives useful access to 2-azabicyclo[3.2.1]oct-3-enes. Not surprisingly, catalytic hydrogenation of 4,5-dihydro[1*H*]azepines yielded azepanes; for syntheses of the latter carrying particular substituents, this has to be an attractive method.[36,37] Table 3 displays most of the 4,5-dihydro[1*H*]azepines reported in the literature to date.

Table 3. 4,5-Dihydro[1H]azepines

Structure	M.p. (°C)/b.p. (°C/mmHg)	Spectroscopic evidence	Ref.
N-CH₃ derivative		^{1}H NMR	49
N-C₂H₅ derivative	69–70/20	IR	47
N-tBu derivative		^{1}H NMR	49
4,5-diMe, N-CH₃ derivative		^{1}H NMR	49
4,5-diMe, N-tBu derivative		^{1}H NMR	49
N-CO₂C₂H₅ derivative	Oil	^{1}H NMR	50
4-C₂H₅, 2-CN, N-Ph derivative	Oil	IR, ^{1}H, ^{13}C NMR, MS	55

Table 3 (continued)

Structure	M.p. (°C)/b.p. (°C/mmHg)	Spectroscopic evidence	Ref.
(azepine with C_3H_7, CN, N-Ph)	Oil	IR, ^{1}H, ^{13}C NMR, MS	55
(azepine with CH$_2$Ph, CN, N-Ph)	Oil	IR, ^{1}H, ^{13}C NMR, MS	55
(azepine with CH$_2$Ph, CN, N-nBu)	Oil	IR, ^{1}H, ^{13}C NMR	55
(azepine with CH$_3$, C$_2$H$_5$, CN, N-Ph)	Oil	IR, ^{1}H, ^{13}C NMR, MS	55
(azepine with Ph$_2$Si-OH, Ph$_2$Si-OH, N-CO$_2$C$_2$H$_5$)	Oil	UV, IR, ^{1}H NMR, MS	59
(azepine with Ph$_2$Si-O-SiPh$_2$, N-CO$_2$C$_2$H$_5$)	174–175	UV, IR, ^{1}H NMR, MS	59

Table 3 (continued)

Structure	M.p. (°C)/b.p. (°C/mmHg)	Spectroscopic evidence	Ref.
	99–100	UV, IR, ^{1}H NMR, MS	59
	Oil	UV, IR, ^{1}H NMR, MS	59
	Oil	UV, IR, ^{1}H NMR, MS	59
	149–151	IR, ^{1}H NMR	57
	102–103.5	IR, ^{1}H NMR, MS	54
	216	UV	53

Table 3 (continued)

Structure	M.p. (°C)/b.p. (°C/mmHg)	Spectroscopic evidence	Ref.
OHC, CHO, Cl, Cl, N-CH$_2$Ph	147–148	IR, ^{1}H NMR, MS	56
OHC, CHO, Cl, Cl, N-CH$_2$CO$_2$C$_2$H$_5$	Unstable oil	IR, ^{1}H NMR	56
OHC, CHO, Cl, Cl, N-Ph	111	IR, ^{1}H NMR, MS	56
CN, C$_2$H$_5$O$_2$C, CO$_2$C$_2$H$_5$, CH$_3$, CH$_3$, N-H	106–107	UV, IR	34
CN, NC, CN, CH$_3$, CH$_3$, N-H	153–153.5	UV, IR	34
OC$_2$H$_5$, C$_2$H$_5$O$_2$C, CO$_2$C$_2$H$_5$, CH$_3$, CH$_3$, N-H		UV	36
CH$_3$O$_2$C, CO$_2$CH$_3$, CH$_3$, CH$_3$, N-H	91–92	UV, IR, ^{1}H NMR	36, 37

Table 3 (continued)

Structure	M.p. (°C)/b.p. (°C/mmHg)	Spectroscopic evidence	Ref.
(CH$_3$O$_2$C, CH$_3$, fused lactone, NH dihydroazepine)	222–223	UV, IR, ^{1}H NMR	43
(CH$_3$O$_2$C, CH$_3$, CN, CO$_2$CH$_3$, CH$_3$, NH dihydroazepine)	187–188	UV, IR, ^{1}H NMR	37
(CH$_3$O$_2$C, CH$_3$, COCH$_3$/CHCO$_2$CH$_3$ substituent, CO$_2$CH$_3$, CH$_3$, NH dihydroazepine)	138–139	UV, IR, MS, ^{1}H NMR	40
(CH$_3$CO, CH$_3$, CN, COCH$_3$, CH$_3$, NH dihydroazepine)	151	UV, IR, ^{1}H NMR	61
(CH$_3$O$_2$C, CH$_3$, CH$_3$, CN, CO$_2$CH$_3$, CH$_3$, NH dihydroazepine)	147–148.5	IR, MS, ^{1}H NMR	41
(CH$_3$O$_2$C, CH$_3$, CH$_3$, OCH$_3$, CO$_2$CH$_3$, CH$_3$, NH dihydroazepine)	157–159	IR, MS, ^{1}H NMR	41

Table 3 (continued)

Structure	M.p. (°C)/b.p. (°C/mmHg)	Spectroscopic evidence	Ref.
	180–181	IR, MS, ¹H NMR	41
	192.5–193.5	UV, IR, MS, ¹H NMR	41
	Oil	¹H NMR	45
	135–137	¹H NMR	45
	86–88	UV	37
	82–84	¹H NMR	45

Table 3 (continued)

Structure	M.p. (°C)/b.p. (°C/mmHg)	Spectroscopic evidence	Ref.
(4-tBu substituted dihydroazepine diester)	125–126	¹H NMR	45
(4-cyclohexyl substituted dihydroazepine diester)	112.5–113.5	¹H NMR	45
(4-nBu substituted dihydroazepine diester)	Oil	¹H NMR	45
(4-(2-CF₃-phenyl) substituted dihydroazepine diester)	157–158.5	¹H NMR	45
(4-(3-CF₃-phenyl) substituted dihydroazepine diester)	93–94	¹H NMR	5
(4-(4-CF₃-phenyl) substituted dihydroazepine diester)	105.5–106.5	¹H NMR	45

Table 3 (continued)

Structure	M.p. (°C)/b.p. (°C/mmHg)	Spectroscopic evidence	Ref.
	117–118	^{1}H NMR	45
	122–122.5	^{1}H NMR	45
	107–108	^{1}H NMR	45
	58–60	^{1}H NMR	45
	104–106	UV, IR, ^{1}H NMR, MS	42
	147–148.5	UV, IR, ^{1}H NMR, MS	42

Table 3 (continued)

Structure	M.p. (°C)/b.p. (°C/mmHg)	Spectroscopic evidence	Ref.
	99–100.5	UV, IR, ^{1}H NMR, MS	42
	125.5–127	UV, IR, ^{1}H NMR, MS	42
		UV, ^{1}H NMR	51
	243–245	IR, ^{1}H NMR	52
	93–94	IR, ^{1}H NMR	52
		IR, MS, ^{1}H NMR	52
	115–116	IR, MS, ^{1}H NMR	52

V. 3,4-DIHYDRO[2H]AZEPINES

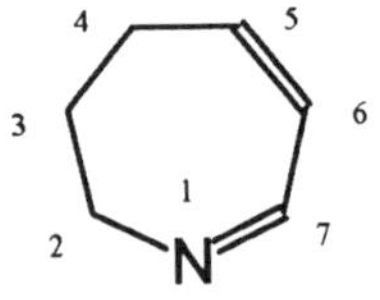

The only bona fide synthesis of the 3,4-dihydro[2H]azepine ring system involved isophorone (**87**), either by Schmidt rearrangement in methanol or by Beckmann rearrangement of the *syn*-oxime (**88** → **89**) followed by treatment with trimethyloxonium fluoroborate.[65]

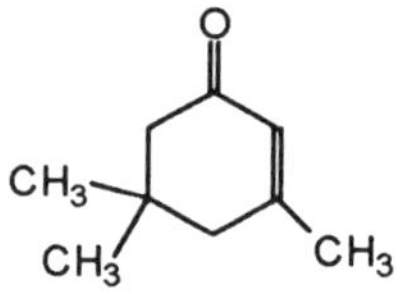

(**87**) 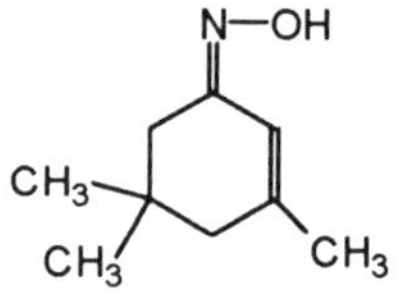　　(**88**) 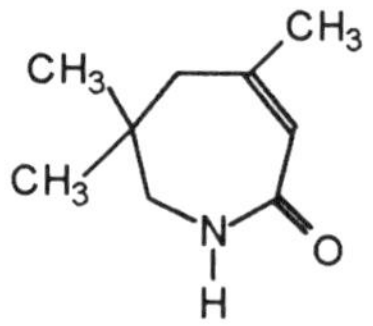　　(**89**)

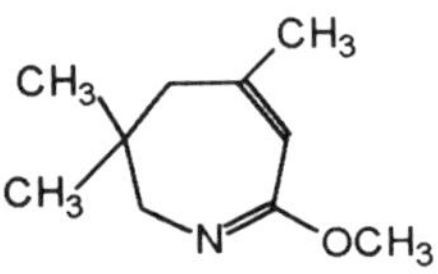

(**90**) 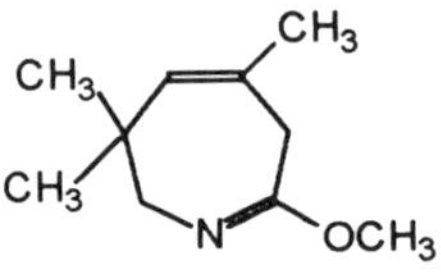　　(**91**)

The product (**90**) was accompanied by a small percentage of the isomer **91**, which was laboriously separated (see Section VI). Pyrolysis of **90** was only fruitful using sensitisers; with acetophenone as sensitiser, a dimeric structure

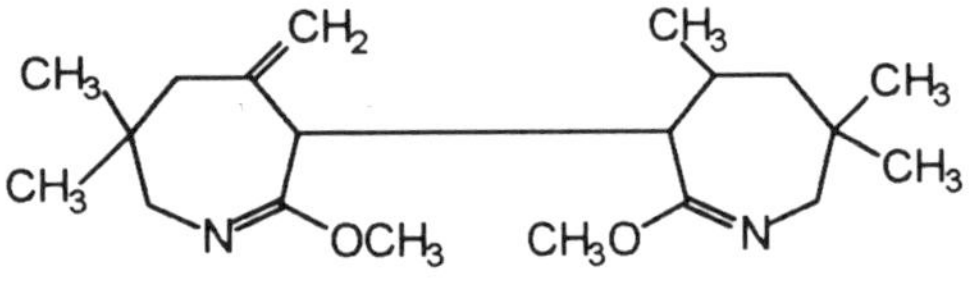

(**92**)

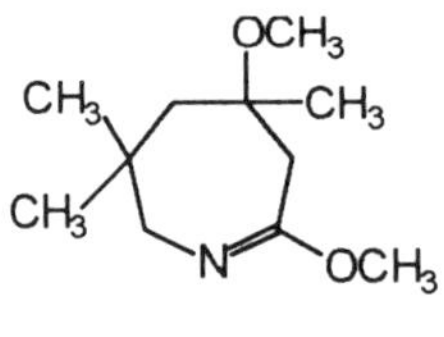

(**93**) 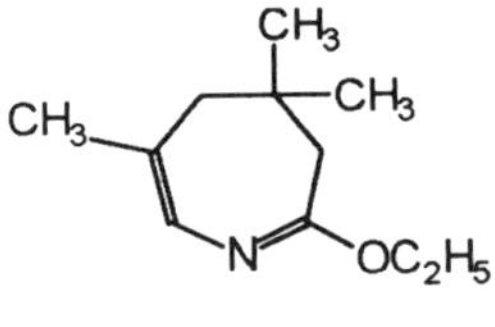　　(**94**)

Dihydroazepines

(92) was seen, but in the presence of methanol, the latter added to the 3,4-bond giving 93. There was no evidence for valence isomerisation such as was seen in the case of 94;[66] see Section VIII.

Slow isomerism (95 → 96; R = H or vinyl) was reported,[67] in the case of 1,2-divinylaziridines facile thermal valence isomerisation to 5,6-dihydro[2H]-azepines (95), but 3,4-dihydro[2H]azepines (96) were not isolated; see Section VII. Similarly, certain tetrahydroazepinones (97) are said to tautomerise to 98

(95) (96) (97)

(98) (99) (100)

on standing neat or, more rapidly, in solution. However, although the ^{1}H NMR evidence is fairly convincing, structures such as 98 were not isolated.[68]

There is one example (99) of a 3,4-dihydro[2H]azepine which arose from acid-catalysed hydration of a [3H]azepine.[43] Hot acidic methanol caused ring contraction to a 1,4-dihydropyridine (47; OCH$_3$ for Cl); it was presumed that dehydration of 99 preceded the ring contraction which is typical of [3H]- and [4H]azepines in this series (Chapter 11). It was shown[15] that when some 2,3-dihydro[1H]azepines were protonated, this took place at C-4 leading to perchlorates of cations such as 100. Many other examples are mentioned in the patent literature.[74-76]

There is one report[78] of a direct cyclisation leading to a 3,4-dihydro[2H]-azepine. Thus the one-pot reduction (dibal) of 100A to the corresponding

(100A) (100B)

amino aldehyde followed by hydrolysis (pH 5.5–6.5) gave the dihydroazepine **100B**.

VI. 3,6-DIHYDRO[2H]AZEPINES

This is a little known subject. As stated above, **91**, which arose as a by-product during the preparation of **90**,[65] was not characterised adequately, being estimated as only 80% pure. On the other hand, a cyclic amidine (**101**) does seem to

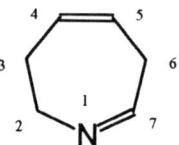

(**101**)

have been isolated by reaction of the appropriate lactam[70] with N,N-dimethylcarbamoyl chloride.[69] Such amidines were shown to react with diphenylcyclopropenone giving cyclopentyl annulations across the N–C-2 bond.[69]

VII. 5,6-DIHYDRO[2H]AZEPINES

As mentioned in Section V, a rapid thermal valency isomerisation (**102** → **95**) has been reported.[67] The products (**95**, R = H and vinyl) were adequately

(**102**)   (**103**)

identified by elemental analysis (one case), IR, ^{19}F NMR and ^{1}H NMR spectroscopy. However, they did isomerise fairly rapidly to **96**, which are obviously thermodynamically more stable. The work cited in the previous section on the cyclic amidines[69] also led to formation of a 5,6-dihydro[2H]azepine (**103**) by treatment of the appropriate lactam[71] with N,N-dimethylcarbamoyl chloride.

VIII. 4,5-DIHYDRO[3H]AZEPINES

Several distinct methods have been employed to approach this ring system. First the *anti*-oxime of isophorone (**87**) underwent Beckmann rearrangement to **104**, which gave **105** when reacted with triethyloxonium tetrafluoroborate.[66] Second,

(**104**)   (**105**)   (**106**)   (**107**)

selective catalytic hydrogenation over 5% palladium on charcoal of the 2-dimethylamino[3H]azepine **106** gave **107**. Another example of the second approach is to be found in the work carried out on the 8π anion obtained from another [3H]azepine (**108**) by treatment with various bases.[72,73] Whilst

(**108**)   (**109**)

in several cases, the anion from **108** reacted usefully with several electrophiles giving C-3 substitution, low-temperature treatment of **108** with n-butyllithium in the presence of potassium *tert*-butoxide led to addition of the reagent across the C-4–C-5 bond giving **109**.

Photochemical irradiation of the available 3,6-dihydro[2H]azepines revealed a difference in behaviour. Whereas **105** on sensitised irradiation was converted through isolable 7-azabicyclo[3.2.0]hept-6-ene derivatives (e.g. **110**),[66] **107** gave

(110)   (111)   (112)

111 and **112** but no azabicycloheptene, although one might argue that it could have been an intermediate.

Photochemical reaction of benzonitrile with 1-methoxycyclopentene yielded the 4,5-dihydro[3H]azepine **113** in around 40% yield.[80]

(113)   (114)   (115)   (116)

Lastly, chromium complexes (**114**; R^1 = Ph, 2-furyl) have recently been shown to react with azadienes (**115**; R^2 = C_2H_5, Ph, p-tolyl, cyclopropyl, p-ClC$_6$H$_4$) providing dihydroazepines (**116**) in acceptable yields (52–90%).[81] Several examples were quoted.

REFERENCES

1. R.M. Acheson, M.W. Foxton and G.R. Miller, *J. Chem. Soc.*, 1965, 3200.
2. T. Oishi, S. Murakami and Y. Ban, *Chem. Pharm. Bull.*, 1972, **20**, 1740.
3. H. Takahata, A. Tomiguchi, A. Hagiwara and T. Yamazaki, *Chem. Pharm. Bull.*, 1982, **30**, 3959.
4. H. Matsunaga, M. Sonoda, Y. Tomioka and M. Yamazaki, *Chem. Pharm. Bull.*, 1984, **32**, 2596.
5. H. Wamhoff and G. Hendrikx, *Chem. Ber.*, 1985, **118**, 863.
6. H. Wamhoff, F.-J. Fassbender, G. Hendrikx, H. Puff and P. Woller, *Chem. Ber.*, 1986, **119**, 2114.
7. H. Wamhoff, H. Wintersohl, S. Stolben, J. Paasch, Z. Nai-Jue and G. Fang, *Liebigs Ann. Chem.*, 1990, 901.
8. G.R. Proctor, in A. Rosowski, (Ed.), '*Azepine Ring Systems Containing Two Rings*', Wiley, New York, 1984, pp. 741–742.
9. S.R. Tanny and F.W. Fowler, *J. Am. Chem. Soc.*, 1973, **95**, 7320.

10. (a) W. Eberbach and J.C. Carré, *Chem. Ber.*, 1983, **116**, 563; (b) W. Eberbach, J.C. Carré and H. Fritz, *Tetrahedron Lett.*, 1977, **50**, 4385; (c) W. Eberbach and J.C. Carré, *Tetrahedron Lett.*, 1980, **21**, 1145.
11. W. Klop and L. Brandsma, *J. Chem. Soc., Chem. Commun.*, 1983, 988.
12. V. Kvita, H. Sauter and G. Rihs, *Helv. Chim. Acta.*, 1989, **72**, 457.
13. K. Saito, M. Kozaki and K. Takahashi, *Heterocycles*, 1990, **31**, 1491.
14. T.J. Van Bergen and R.M. Kellogg, *J. Org. Chem.*, 1971, **36**, 978.
15. (a) L.A. Paquette, *Tetrahedron Lett.*, 1963, **29**, 2027; (b) L.A. Paquette, *J. Am. Chem. Soc.*, 1964, **86**, 4092.
16. S. Bátori, R. Gomppcr, J. Meier and H.-U. Wagner, *Tetrahedron*, 1988, **44**, 3309.
17. G.A. Berchtold and G.F. Uhlig, *J. Org. Chem.*, 1963, **28**, 1459.
18. H. Barth, G. Burger, H. Dopp, M. Kobayashi and H. Musso, *Liebigs Ann. Chem.*, 1981, 2164.
19. E. Vogel, R. Erb, G. Lenz and A. Bothner-By, *Liebigs Ann. Chem.*, 1965, **682**, 1.
20. L.A. Paquette and G.D. Ewing, *J. Am. Chem. Soc.*, 1978, **100**, 2908.
21. R.K. Boeckman, Jr, M.D. Shair, J.R. Vargas and L.A. Stolz, *J. Org. Chem.*, 1993, **58**, 1295.
22. M.A. Calcagno and E.E. Schweizer, *J. Org. Chem.*, 1978, **43**, 4207.
23. A. Hassner, R. D'Costa, A.T. McPhail and W. Butler, *Tetrahedron Lett.*, 1981, **22**, 3691.
24. A. Hassner, W. Chau and R. D'Costa, *Isr. J. Chem.*, 1982 **72**, 76.
25. J. Kurita, K. Iwata, M. Hasebe and T. Tsuchiya, *J. Chem. Soc., Chem. Commun.*, 1983, 941.
26. J. Kurita, K. Iwata, H. Sakai and T. Tsuchiya, *Chem. Pharm. Bull.*, 1985, **33**, 4572.
27. M. McHugh and G.R. Proctor, *J. Chem. Res. (S)*, 1985, 246; *(M)*, 2230.
28. A.J. Frew and G.R. Proctor, *J. Chem. Soc., Perkin Trans. 1*, 1980, 1245.
29. A.J. Frew, G.R. Proctor and J.V. Silverton, *J. Chem. Soc., Perkin Trans. 1*, 1980, 1251.
30. A. Chenna, J. Donnelly, K.J. McCullough, G.R. Proctor and J. Redpath, *J. Chem. Soc., Perkin Trans. 1*, 1990, 261.
31. F.A. Fraser, *PhD Thesis*, University of Strathclyde, Glasgow, 1990.
32. E. Benary, *Chem. Ber.*, 1920, **53**, 2218.
33. E. Bullock, B. Gregory, A.W. Johnson, P.J. Brignell, U. Eisner and H. Williams, *Proc. Chem. Soc.*, 1962, 122.
34. P.J. Brignell, E. Bullock, U. Eisner, B. Gregory, A.W. Johnson and H. Williams, *J. Chem. Soc.*, 1963, 4819.
35. P.J. Brignell, U. Eisner and H. Williams, *J. Chem. Soc.*, 1965, 4226.
36. M. Anderson and A.W. Johnson, *Proc. Chem. Soc.*, 1964, 263.
37. M. Anderson and A.W. Johnson, *J. Chem. Soc.*, 1965, 2411.
38. R.C. Allgrove and U. Eisner, *Tetrahedron Lett.*, 1967, 499.
39. J. Ashby and U. Eisner, *J. Chem. Soc. C*, 1967, 1706.
40. J. Ashby, L.A. Cort, J. Elvidge and U. Eisner, *J. Chem. Soc. C*, 1968, 2311.
41. B. Gregory, E. Bullock and T.-S. Chen, *Can. J. Chem.*, 1979, **57**, 44.
42. B. Gregory, E. Bullock and T.-S. Chen, *Can. J. Chem.*, 1985, 843.
43. M. Anderson and A.W. Johnson, *J. Chem. Soc. C*, 1966, 1075.
44. E. Bullock, B. Gregory and M.T. Thomas, *Can. J. Chem.*, 1977, **55**, 693.
45. D.A. Claremon, D.E. McClure, J.P. Springer and J.J. Baldwin, *J. Org. Chem.*, 1984, **49**, 3871.
46. P.M. Atlani, J.F. Biellman and J. Moron, *Tetrahedron*, 1973, **29**, 391.
47. E.L. Stogryn and S.J. Brois, *J. Org. Chem.*, 1965, **30**, 88.
48. J.C. Pommelet and J. Chuche, *Tetrahedron Lett.*, 1974, 3897.

49. (a) J.C. Pommelet and J. Chuche, *Can. J. Chem.*, 1976, **54**, 1571; (b) N. Manisse and J. Chuche, *J. Am. Chem. Soc.*, 1977, **99**, 1272.
50. R.A. Aitken, J.I.G. Cadogan, I. Gosney, B.J. Hamill and L.M. McLaughlin, *J. Chem. Soc., Chem. Commun.*, 1982, 1164.
51. H.-D. Martin and M. Hekman, *Angew. Chem., Int. Ed. Engl.*, 1972, **11**, 926.
52. I.J. Turchi, C.A. Maryanoff and A.R. Mastrocola, *J. Heterocycl. Chem.*, 1980, **17**, 1593.
53. H.-D. Martin, F.-J. Mais, B. Mayer, H.-J. Hecht, M. Hekman and A. Steigel, *Monatsh. Chem.*, 1983, **114**, 1145.
54. (a) C.A. Maryanoff, C.B. Karash, I.J. Turchi, E.R. Corey and B.E. Maryanoff, *J. Org. Chem.*, 1989, **54**, 3790; (b) C.A. Maryanoff and I.J. Turchi, *Heterocycles*, 1993, **35**, 649.
55. J.-M. Fang, C.-C. Yang and Y.-W. Wang, *J. Org. Chem.*, 1989, **54**, 481.
56. T. Aubert, M. Farnier, I. Meunier and R. Guilard, *J. Chem. Soc., Perkin Trans. 1*, 1989, 2095.
57. T. Sasaki, K. Kanematsu and Y. Yukimoto, *J. Org. Chem.*, 1972, **37**, 890.
58. K. Saito and K. Takahashi, *Heterocycles*, 1979, **12**, 263.
59. K. Saito, H. Kojima, T. Okudaira and K. Takahashi, *Bull. Chem. Soc. Jpn.*, 1983, **56**, 175.
60. S. Itoh, Y. Yokoyama, T. Okumoto, K. Saito, K. Takahashi, K. Satake, H. Takamuko, M. Kimura and S. Morosawa, *Bull. Chem. Soc. Jpn.*, 1990, **63**, 3162.
61. R.C. Allgrove, L.A. Cort, J.A. Elvidge and U. Eisner, *J. Chem. Soc. C*, 1971, 434.
62. E. Bullock, B. Gregory and A.W. Johnson, *J. Chem. Soc.*, 1964, 1632.
63. E. Bullock, B. Gregory and A.W. Johnson, *J. Am. Chem. Soc.*, 1962, **84**, 2260.
64. B. Gregory, E. Bullock and T.-S. Chen, *Can. J. Chem.*, 1977, **55**, 4061.
65. T.H. Koch, M.A. Geigel and C.-C. Tsai, *J. Org. Chem.*, 1973, **38**, 1090.
66. T.H. Koch,and D.A. Brown, *J. Org. Chem.*, 1971, **36**, 1934.
67. E.L. Stogryn and S.J. Brois, *J. Am. Chem. Soc.*, 1967, **89**, 605.
68. A. Goti, A. Brandi, F. De Sarlo and A. Guarna, *Tetrahedron*, 1992, **48**, 5283.
69. T. Eicher and R. Rohde, *Synthesis*, 1985, 619.
70. H.K. Reimschussel, J.P. Sibilia and J.V. Pascale, *J. Org. Chem.*, 1969, **34**, 959.
71. Y.D. Smirnov and A.P. Tomilov, *Zh. Org. Khim.*, 1969, 864.
72. J.W. Streef and H.C. Van der Plas, *Tetrahedron Lett.*, 1979, **24**, 2287.
73. J.W. Streef, H.C. Van der Plas, A. Van Veldhuizen and K. Goubits, *Recl. Trav. Chim. Pays-Bas*, 1984, **103**, 225.
74. L.A. Paquette, *US Pat.*, 3 239 505; *Chem. Abs.*, 1966, **64**, 19575f.
75. L.A. Paquette, *US Pat.*, 3 267 092; *Chem. Abs.*, 1966, **65**, 16952b.
76. L.A. Paquette, *US Pat.*, 3 267 093; *Chem. Abs.*, 1967, **67**, 3004y.
77. K. Isomura, H. Kawasaki, K. Takehara and H. Taniguchi, *Heterocycles*, 1995, **40**, 511.
78. M.E.M. Cromwell, R. Gebhard, X. Li, E.S. Batenburg, J.C.P. Hopman, J. Lugtenburg and R.A. Mathies, *J. Am. Chem. Soc.*, 1992, **114**, 10860.
79. J.G. Walsh, P.J. Furlong and D.G. Gilheany, *J. Chem. Soc., Chem. Commun.*, 1994, 67.
80. J. Mattay, J. Runsink, R. Heckendorn and T. Winkler, *Tetrahedron*, 1987, **43**, 5781.
81. J. Barluenga, M. Tomás, A. Ballesteros, J. Santamaria and F. López-Ortiz, *J. Chem. Soc., Chem. Commun.*, 1994, 321.

CHAPTER 9

Dihydroazepinones

I. 1,3-DIHYDRO[2H]AZEPIN-2-ONES

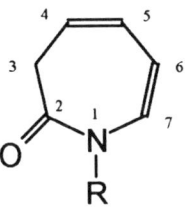

A great deal is known on this subject. This is due partly to the ease of preparation of 1,3-dihydro[2H]azepin-2-ones, and partly because of their stability.

PREPARATION

At about the same time (1962–63) in Germany and the USA, a novel entry to 1,3-dihydro[2H]azepin-2-ones was disclosed.[1–3] In this unusual procedure, a cold ethereal solution of chloramine is added to vigorously stirred sodium salts of 2,6-dialkylphenols held at 125–150 °C.[1] Full experimental details have been provided;[4] for example, 2,4,6-trimethylphenol gives 3,5,7-trimethyl-1,3-dihydro[2H]azepin-2-one (**1**).

Many other 3,7-dialkyl-1,3-dihydro[2H]azepin-2-ones were claimed,[5–10] but phenols lacking 2,6-disubstitution did not participate. Details of the mechanism have been addressed[11] and it was shown that in the case of unsymmetrically disubstituted 2-methylphenols, increasing the size of the 6-substituent increased the degree to which amination proceeded at the methyl-bearing carbon atom C-2. It is generally accepted that chloramine provides an electrophilic nitrogen-containing species to react at the nucleophilic carbon centres (C-2,

C-6) of the phenol (**2, 2'**). Along the reaction pathway one can visualise an azanorcaradiene structure (**3**) plausibly leading to the observed products (e.g. **1**). 7-Aryl-1,3-dihydro[2H]azepin-2-ones were also obtained from phenolic precursors.[12,13] Thus the spiroquinol ethers **4** obtained by oxidative coupling of phenols (**5**) reacted with some primary amines to give good yields of the 7-aryl derivatives (**6**). Formation of the latter were rationalised[13] as proceeding via an azanorcaradiene intermediate (**7**), so there is a similarity between this procedure and the phenoxide and chloramine approach, even though the amino components are differently charged in the two cases.

Phenylnitrene has been a fruitful source of azepines (Chapter 11). In certain circumstances, generation of phenylnitrenes can lead to the formation of 1,3-dihydro[2H]azepine-2-ones. Thus reductive treatment of pentafluoronitrosobenzene with triethyl phosphite in anisole gave **8**[14] in poor yield along with other products.

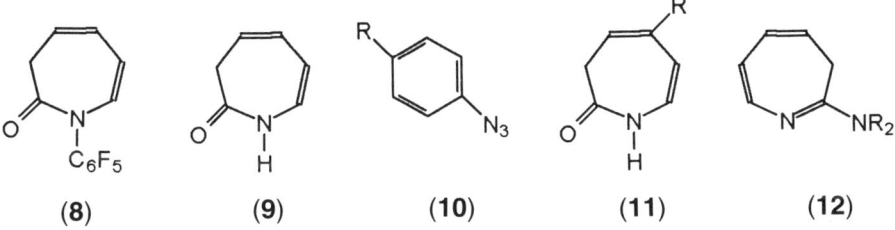

(8) (9) (10) (11) (12)

Photolysis and thermolysis of phenyl azide in acetic acid was also shown to yield the dihydroazepinone ring product **9** amongst several other materials.[15] The Salford group have improved on this approach; based on earlier work,[16] it has been shown that photolysis of phenyl azides carrying electron-withdrawing groups in aqueous THF is a very useful method for making 1,3-dihydro[2H]azepin-2-ones.[17,18] Thus **10** gave **11** (R = electron-withdrawing groups; see Table 1) in generally acceptable yields. Intramolecular nitrene insertion was invoked to explain the formation of acetyl-substituted 1,3-dihydro[2H]azepinones when 2-alkylindazoles were irradiated in dilute acid solution.[36]

It is known that hydrolysis of 2-amino[3H]azepines **12** gives 1,3-dihydro[2H]azepin-2-ones. Thus **12** (R = H) gave **11** (R = H);[19] treatment of **12** (R = H) with H_2S gave the thia analogue (**15**). Diethylaminoazepines (**12**, R = C_2H_5, various substituents; see Table 1) were quaternised on the ring nitrogen and hydrolysed in good yield to 1,3-dihydro[2H]azepin-2-ones[20] whilst 2-acetoxy[3H]azepines were also shown[21] to undergo hydrolysis to give **11** (R = H) in excellent yields.

Finally on the preparative side, Curtius rearrangement, performed on cis-2-vinylcyclopropyl carbonylazide (**13**), led to **11** (R = H), presumably via

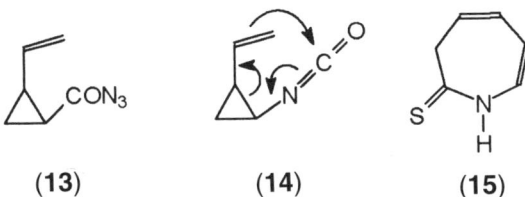

(13) (14) (15)

the cis-vinyl isocyanate (**14**) as shown. This reaction proceeds at 80 °C to completion and can be thought of as an aza-Cope rearrangement (**14** → **11**). The trans-vinylcyclopropyl isocyanate was isolated (b.p. 85–86 °C/165 mmHg) and proved to be stable up to 200 °C.[22,23]

PROPERTIES

Reduction of 1,3-dihydro[2H]azepin-2-ones is fairly predictable. Catalytic hydrogenation gave the appropriate caprolactam,[1,19,22] whilst $LiAlH_4$ reduction gave 2,3-dihydro[1H]azepines[24,25] (Chapter 8). Where the substitution pattern allows, both of these represent an efficient two-step synthesis from a phenol. Treatment of **1** with hydrochloric acid yielded a lactone (**16**);[2] N-alkylation was

(16)   (17)   (18)

(19)   (20)

brought about using sodium hydride in DMF to obtain **17**, for example,[2] which could also be reacted by direct reaction of 2,4,6-trimethylphenol with chloromethylamine.[2] Although **17** appeared to deprotonate uniquely using the methylsulphinyl carbanion (Na^+ $^-CH_2SOCH_3$) in DMSO producing a highly coloured red solution,[18] electrophilic addition by methyl iodide failed.[26] However, when deprotonation was achieved using $CD_3SOCD_2^-$ Na^+/CD_3SOCD_3 followed by quenching with D_2O, convincing ^1H NMR evidence was obtained for the exchange (almost complete) of four hydrogen atoms as shown (**19**). This striking result illuminating the acidity of the 7-CH_3 hydrogen atoms was rationalised as being due to the proximity of the nitrogen atom. The other methyl hydrogens at C-3 and C-5 did not exchange at all. Eventually, O-alkylation of **1** was achieved using triethyloxonium tetrafluoroborate.[23,27] The 2-ethoxy product **20** hydrolysed in dilute acid giving the lactone **16** and prolonged treatment of **20** with piperidine led to the 2-amino[3H]azepine **12** [$R_2 = -(CH_2)_5-$]. P_2S_5 was used to convert **1** into the thia analogue **21**, which also reacted with triethyloxonium tetrafluoroborate giving the thia analogue of **20**.[27]

Although **1** failed to undergo Diels–Alder reaction with either maleic anhydride or N-phenylmaleimide, it gave a poor yield (19.6%) of **22** with DMAD and a quantitative yield of the adduct **23** (R = H) with TCE. The N-methyl compound **17** also reacted satisfactorily with TCE giving **23** (R = CH_3); the corresponding N-ethyl analogue was obtained in 91% yield.[28] On the other hand, hexafluoro-2-butyne and **1** reacted differently; a low yield of a

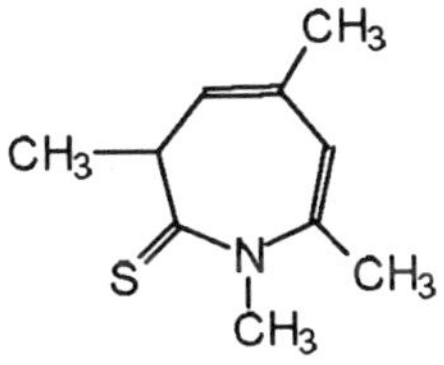

(21) 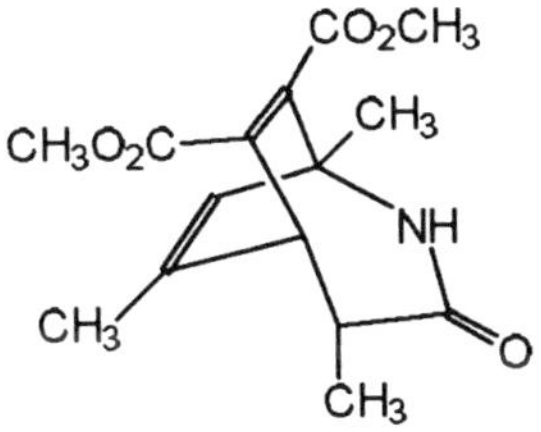 (22)

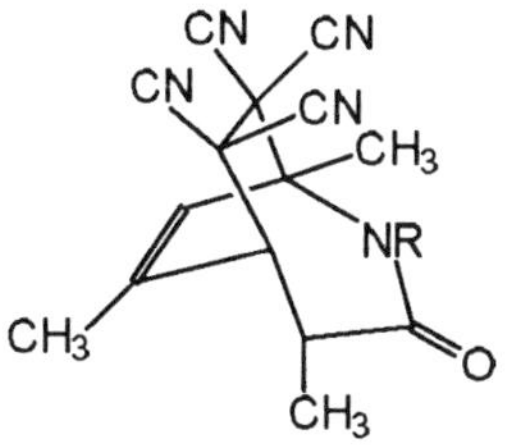

(23) 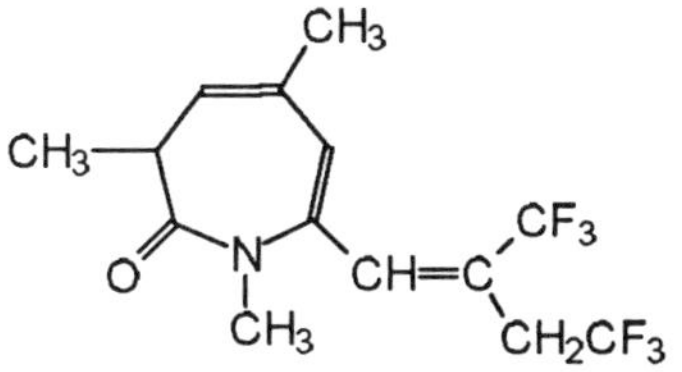 (24)

7-methyl-substituted compound (**24**) was isolated, a further testimony to the acidity of the hydrogen atoms in the 7-methyl group of compounds such as **1** and **17**.

A study of chlorinations involving 1,3-dihydro[3*H*]azepin-2-ones using *N*-chlorosuccinimide in refluxing methylene chloride[29] revealed that in molecules lacking a 5-substituent, the monochlorinated product proved to be the 6-chloro isomer (e.g. **25**), whereas in the structure including a 5-substituent, the major product was the 4-chloro compound **26**. This was explained in terms of the principle of least motion.[29]

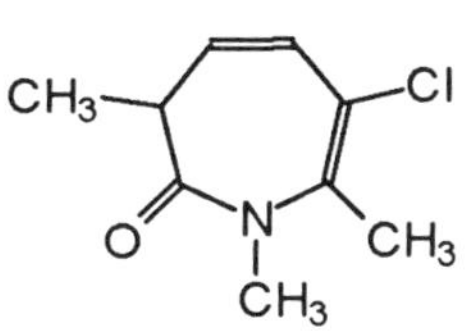

(25) 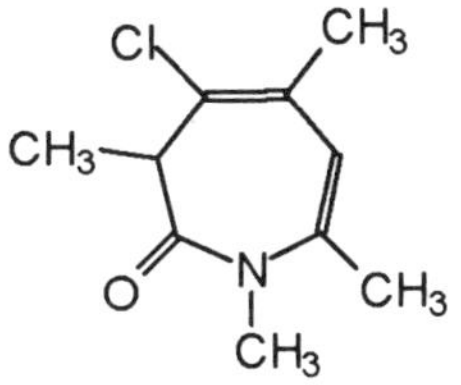 (26) 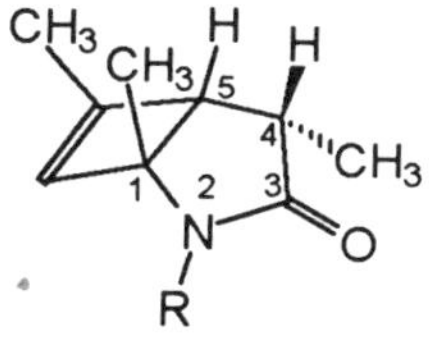 (27) 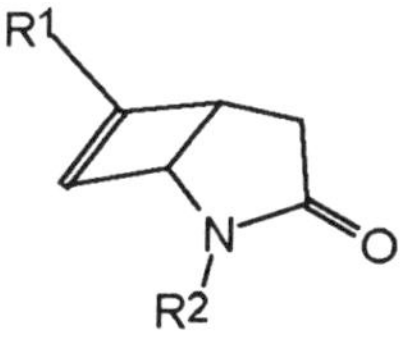 (28)

Photochemistry of 1,3-dihydro[2*H*]azepin-2-ones has been a popular area of study. Similarities in structure compared with cycloheptadienes and cycloheptatrienes encouraged the expectation that the azepinones might yield 2-azabicyclo[3.2.0]heptenones on irradiation, and this proved to be correct.

Compound **1**, for example, gave **27** (R = H) in THF or methanol and the *N*-methyl variant (**17**) gave **27** (R = CH$_3$).[24,30] Assignment of the stereochemistry as in **27** arises through the observation that the coupling between the 5H and neighbouring 4H in the ^{1}H NMR spectrum was 10 Hz. The dihedral angle between the relevant C–H bonds must therefore be low (Karplus equation). Overall this is a very suitable two-step synthesis of 2-azabicyclo[3.2.0] heptenes from phenols.

It was confirmed[31] that irradiation of **1** and of **17** in diethyl ether gave the same products (**27**, R = H and CH$_3$), but a later re-examination detected minor unstable by-products thought to arise from decarbonylation.[32] In this work small amounts of isomers epimeric at C-4[27] were isolated by GLC; as expected, the coupling constant between H-5 and H-4 was below 3 Hz, consonant with these protons being mutually *trans*. Thus the methyl group at C-4 is on the convex (*exo*) face of the bicyclic structure. Photolysis of **9**[23] and **11** (R = Ph)[20] provided the typical products **28** (R^1 = R^2 = H and R^1 = Ph, R^2 = CH$_3$, respectively).

7-Aryl-1,3-dihydro[2*H*]azepin-2-ones (**6**) on irradiation gave different products[33] depending upon the solvent. Thus the expected product (**29**) was

(29) (30)

obtained in benzene, but in chloroform, imines (**30**, R = ^{t}Bu, R^2 = CH$_3$ or H) were isolated. In methanol mixtures of two enol forms of **30** were seen.

Dihydroazepinones

Table 1. 1,3-Dihydro[2H]azepin-2-ones

Structure	M.p. (°C)/b.p. (°C/mmHg)	Spectroscopic evidence	Ref.
(azepin-2-one)	47–49	^{1}H NMR, IR	18a, 19, 21, 23
(2-thione)	106–107.5		19
(N-CH$_3$)	50–54/0.6 94–95/12	IR, UV	20, 23
(N-C$_6$F$_5$)	128–130/0.4	IR	14
(3-OCH$_3$)	135–136	IR, ^{1}H NMR, MS	36
(3-CN)	148–149	IR, ^{1}H NMR, MS	18b
(4-CO$_2$H)	144–145	IR, ^{1}H NMR	18a

Table 1 (continued)

Structure	M.p. (°C)/b.p. (°C/mmHg)	Spectroscopic evidence	Ref.
5-Ph azepinone	141–142	UV	20
5-(4-pyridyl) azepinone	163–165	IR, UV	20
5-CN azepinone	115	^{1}H NMR	18a
5-CF_3 azepinone	50	^{1}H NMR	18a
5-CO_2CH_3 azepinone	112	^{1}H NMR	18a
5-$CO_2C_2H_5$ azepinone	82	^{1}H NMR	18a

Table 1 (continued)

Structure	M.p. (°C)/b.p. (°C/mmHg)	Spectroscopic evidence	Ref.
4-SO$_2$NH$_2$ dihydroazepinone	171	^{1}H NMR, X-ray, MS	18a, 35
4-CO$_2$CHPh$_2$ dihydroazepinone	154	^{1}H NMR	18a
4-CO$_2$H dihydroazepinone	180	^{1}H NMR	18a
4-SO$_2$NHC(=NH)NH$_2$ dihydroazepinone	190–192	IR, ^{1}H NMR, MS	35
7-COCH$_3$ dihydroazepinone	100–105/0.001	R, ^{1}H NMR, MS	36
7-CN dihydroazepinone	143	IR, ^{1}H NMR, MS, X-ray	18b

Table 1 (continued)

Structure	M.p. (°C)/b.p. (°C/mmHg)	Spectroscopic evidence	Ref.
4-COCH₃, 1-CH₃ azepinone	54–56	UV, ^{1}H NMR	20
4-CON(C₂H₅)₂, 1-CH₃ azepinone	71–73	UV, ^{1}H NMR	20
4-Ph, 1-CH₃ azepinone	42–44	UV, ^{1}H NMR	20
4-CH₃O, 1-CH₃ azepinone	82.3/0.3	UV, ^{1}H NMR	20
5-Ph, 1-CH₃ azepinone	92–93	UV, ^{1}H NMR	20
5-Ph, 1-CH₃ azepinethione	115	UV, ^{1}H NMR	20

Table 1 (continued)

Structure	M.p. (°C)/b.p. (°C/mmHg)	Spectroscopic evidence	Ref.
(1-methyl, 5-Cl dihydroazepinone)	68–72/0.2	UV, ^{1}H NMR	20
(1-methyl, 5-CON(C_2H_5)$_2$ dihydroazepinone)	87–89	UV, ^{1}H NMR	20
(1-methyl, 5-Ph dihydroazepinone)	75–76	UV, ^{1}H NMR	20
(3-CO$_2$CH$_3$, 5-Cl dihydroazepinone)	153	^{1}H NMR	18a
(3-CN, 5-CH$_3$ dihydroazepinone)	105	IR, ^{1}H NMR, MS,	18b
(3-CO$_2$CH$_3$, 6-CO$_2$CH$_3$ dihydroazepinone)	138	^{1}H NMR	18a
(3-CN, 6-Cl dihydroazepinone)	156	IR, ^{1}H NMR, MS,	18b

Table 1 (continued)

Structure	M.p. (°C)/b.p. (°C/mmHg)	Spectroscopic evidence	Ref.
3-isopropyl-7-methyl-1,3-dihydro-2H-azepin-2-one	90–91	IR, UV, ^{1}H NMR	11
3-methyl-7-isopropyl-1,3-dihydro-2H-azepin-2-one	113.5–114	IR, UV, ^{1}H NMR	11
3-tert-butyl-7-methyl-1,3-dihydro-2H-azepin-2-one	131–131.5	IR, UV, ^{1}H NMR	11
3,7-dimethyl-1,3-dihydro-2H-azepin-2-one	121–122	IR, UV	1, 2
3,7-diethyl-1,3-dihydro-2H-azepin-2-one	88.5–89	IR, UV	2
4,7-dicarboxy-1,3-dihydro-2H-azepin-2-one	225	^{1}H NMR	18a
5-chloro-7-cyano-1,3-dihydro-2H-azepin-2-one	189–190	IR, ^{1}H NMR, MS, X-ray	18b

Table 1 (continued)

Structure	M.p. (°C)/b.p. (°C/mmHg)	Spectroscopic evidence	Ref.
(7-methyl-2-oxo-2,3-dihydro-1H-azepine-2-carbonitrile)	136–137	IR, ^{1}H NMR, MS	18b
(5-chloro-2-oxo-2,3-dihydro-1H-azepine-2-carbonitrile)	170	IR, ^{1}H NMR, MS	18b
(dicyano dihydroazepinone)	155–157	IR, ^{1}H NMR, MS	18b
(trimethyl dihydroazepinone)	69–71/1.1	UV, ^{1}H NMR	20
(trimethyl dihydroazepinone)	132	IR, UV	2, 4
(trimethyl dihydroazepinone)	109–115/13		2

Table 1 (continued)

Structure	M.p. (°C)/b.p. (°C/mmHg)	Spectroscopic evidence	Ref.
3,5,7-trimethyl-1-methyl azepinone	121–122/12		2
3,5,7-trimethyl-1-ethyl azepinone	116–120/12	IR, UV	28
3-methyl-6-chloro-7-methyl azepinone (NH)	158–158.5	IR, ^{1}H NMR	29
3-t-butyl-6-chloro-7-methyl azepinone (NH)	183.5–184	IR, ^{1}H NMR	29
3-methyl-6-chloro-7-methyl-1-methyl azepinone	–	UV, IR, ^{1}H NMR	29
3,5-dimethyl-7-(CH(CF$_3$)CH$_2$CF$_3$)-1-ethyl azepinone	132–133	IR, UV	28
3-methyl-4-chloro-5-methyl-7-methyl azepinone (NH)	174.5–175	IR,UV, ^{1}H NMR	29

Table 1 (continued)

Structure	M.p. (°C)/b.p. (°C/mmHg)	Spectroscopic evidence	Ref.
3,5,7-trimethyl-6-chloro-1,3-dihydro-2H-azepin-2-one	141–143	IR, ^{1}H NMR	29
3,5,7-trimethyl-1,3-dihydro-2H-azepine-2-thione	136.5–137	UV, ^{1}H NMR	27
1-(2-acetoxyethyl)-3,5,7-trimethyl-1,3-dihydro-2H-azepin-2-one	116–120/0.4	UV, IR	34
1-(2-hydroxyethyl)-3,5,7-trimethyl-1,3-dihydro-2H-azepin-2-one	134/0.35	IR	34
1-[2-(diethylamino)ethyl]-3,5,7-trimethyl-1,3-dihydro-2H-azepin-2-one	123/0.3 HCl 159–160		34
1-[3-(dimethylamino)propyl]-3,5,7-trimethyl-1,3-dihydro-2H-azepin-2-one	121–127/0.3 HCl 169–169.5		34

Table 1 (continued)

Structure	M.p. (°C)/b.p. (°C/mmHg)	Spectroscopic evidence	Ref.
	126–138/0.13 HCl 197–198		34
	150–158/0.15 HCl 222–224 MeI 218–219		34
	136–147/0.2 MeI 197–198		34
	129–150/0.3 HCl 247–249 MeI 180–182		34
	146–151/0.15 HCl 171–173		34

Dihydroazepinones

Table 1 (continued)

Structure	M.p. (°C)/b.p. (°C/mmHg)	Spectroscopic evidence	Ref.
	136–164/0.2 HCl 172–173		34
	153–155	IR, ^{1}H, ^{13}C NMR, MS, UV	13
	162–164	IR, UV, ^{1}H NMR	13
	209–211	IR, UV, ^{1}H NMR, MS	13
	144–146	IR, UV, ^{1}H NMR	13

Table 1 (continued)

Structure	M.p. (°C)/b.p. (°C/mmHg)	Spectroscopic evidence	Ref.
(azepinone with $^tC_5H_{11}$ substituents, N-$^nC_3H_7$, 2,4,6-tri-$^tC_5H_{11}$-phenol)	180–182	IR, UV, ^{1}H NMR	13
(azepinone with CPh$_3$ and $^tC_4H_9$ substituents, N-$^nC_3H_7$, di-CPh$_3$-$^tC_4H_9$-phenol)	211–214	IR, UV, ^{1}H NMR	13
(azepinone with CPh$_3$ and $^tC_4H_9$ substituents, N-cyclohexyl, di-CPh$_3$-$^tC_4H_9$-phenol)	188–191	IR, UV, ^{1}H NMR	13
(azepinone with $^tC_5H_{11}$ substituents, N-cyclohexyl, tri-$^tC_5H_{11}$-phenol)	153–156	IR, UV, ^{1}H NMR	13
(azepinone with CPh$_3$ and $^tC_4H_9$ substituents, N-CH$_3$, di-CPh$_3$-$^tC_4H_9$-phenol)	143–145	IR, UV, ^{1}H NMR	13

Table 1 (continued)

Structure	M.p. (°C)/b.p. (°C/mmHg)	Spectroscopic evidence	Ref.
(structure: 7-membered lactam with CH3, OC2H5, CH3, CH3, CH3 substituents)	143–145	IR, UV, ^{1}H NMR	37

II. 1,5-DIHYDRO[2H]AZEPIN-2-ONES

All of the work in this section is due to Sano and his co-workers. Basically the photochemically induced [2 + 2]cycloaddition of alkenes with 2-aryl pyrroline-4,5-diones (31) gives rise to the 2-azabicyclo[3.2.0]heptane-3,4-dione ring system (32).[38] The regiochemistry and, where appropriate, stereochemistry of this reaction were supported by X-ray analysis of the analogue 33.[39] Pyridones

(31)   (32)   (33)

were also formed amongst the products[40] in variable yields. All of the 1,5-dihydro[2H]azepin-2-ones reported were obtained from these 2-azabicyclo[3.2.0]heptane-3,4-diones either via imidic esters (34)[41] when treated with SnCl$_4$, then hydrolysed (5% HCl), or more commonly from the bicycles using DBU.[42–45] Imidic esters (34) were obtained from diones (32) by treatment with triethyloxonium fluoroborate [(C$_2$H$_5$)$_3$O$^+$ BF$_4^-$] (Meerwein's reagent) and the initial SnCl$_4$ induced products were [4H]azepines (35)[41] (see Chapter 11).

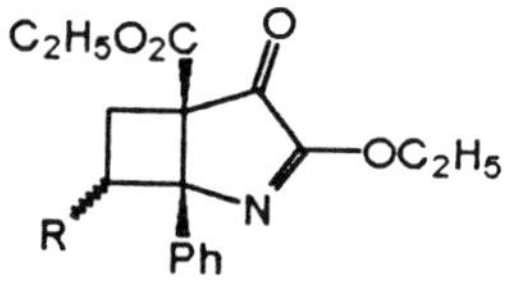

(34) 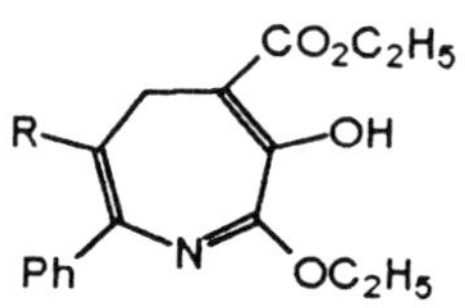 **(35)**

1,5-Dihydro[2H]azepin-2-ones (**36**) obtained thus carry a hydroxy group at C-3 and, on dehydrogenation with DDQ yield 'azatropolones' (Chapter 12):

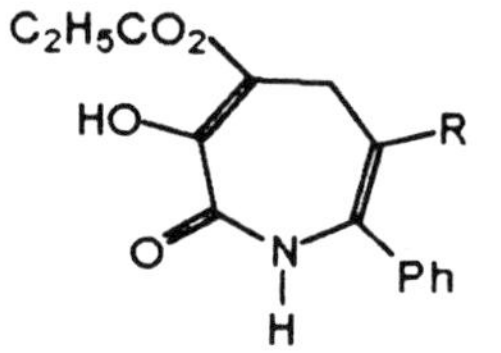

(36) 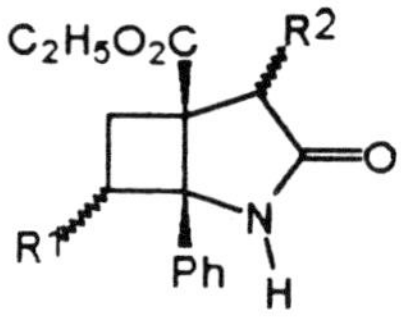 **(37)**

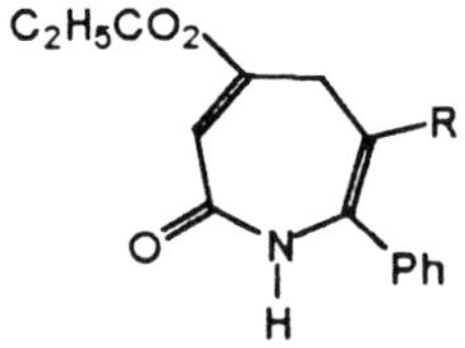

(38) 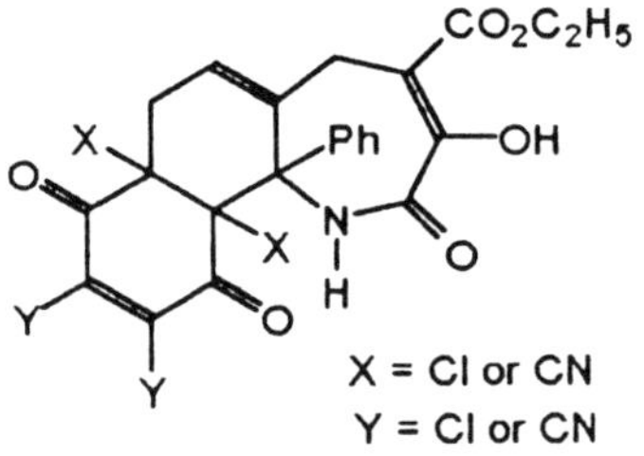 **(39)**

X = Cl or CN
Y = Cl or CN

The hydroxy group may be methylated with diazomethane.[42] In some cases[45] dehydrogenation (e.g. **36**, R = C_2H_5 or OAc) did not proceed cleanly since the products underwent ring contraction to 2-pyridones via a benzilic acid type of rearrangement. Further study of the chemistry of the 2-azabicyclo[3.2.0]heptane-3,4-diones revealed that thermal reactions could lead directly to pyridine derivatives[46,47] and that borohydride reduction gave 4-hydroxy compounds (**37**, R^2 = OH) whose acetates (**37**, R^2 = OAc) underwent eliminative ring expansion to 1,5-dihydro[2H]azepin-2-ones lacking the 3-hydroxy group (**38**, R = Ph, OEt, vinyl).[44] The vinyl compound **36** (R = vinyl) not surprisingly underwent Diels–Alder reaction with DDQ[45] to give **39**. An alternative method for obtaining analogues without 3-hydroxy substituents was reported as follows:[44] ^{n}Bu$_4$NBH$_4$ reduction of **36** (R = vinyl) gave **40** (R = H), whose monoacetate (**40**, R = Ac) reacted with DBU giving **38** (R = vinyl).

The main interest in 1,5-dihydro[2H]azepin-2-ones lies in their relationship with the so-called azatropolones and azatropones by dehydrogenation

Dihydroazepinones

(40), (41), (42)

referred to above.[45] Treatment with Meerwein's reagent converts them into [4H]azepines (e.g. 36 → 35).

All of the reported 1,5-dihydro[2H]azepin-2-ones are relatively stable solid materials, thoroughly analysed and authenticated spectroscopically (UV, IR, ^{1}H NMR) and, since a dehydrogenation product (41) has been confirmed by X-ray analysis,[48] this area can be said to have been fairly thoroughly explored. It is interesting to speculate whether the ring expansion of 2-azabicyclo[3.2.0]heptanes would be so facile in cases lacking a phenyl group at C-1 because an intermediate cation (42) seems likely especially in Lewis acid-promoted examples[41] and solvolyses (e.g. 37).

Table 2. 1,5-Dihydro[2H]azepin-2-ones

Structure	M.p. (°C)	UV: λ (nm) (ε (l mol^{-1} cm^{-1}))	IR (cm^{-1})	Ref.
(HO, C$_2$H$_5$CO$_2$, Ph, Ph)	226–236	226 (19 200), 269 (17 600)	3200, 1670, 1600	42, 45
(HO, C$_2$H$_5$CO$_2$, C$_2$H$_5$, Ph)	140–145	227 (13 000), 263 (13 500)	3200, 1665, 1600	42, 45
(HO, C$_2$H$_5$CO$_2$, OC$_2$H$_5$, Ph)	173–178	220 (13 600), 262 (17 400)	3200, 1620	42, 45

Table 2 (continued)

Structure	M.p. (°C)	UV: λ (nm) (ε (l mol^{-1} cm^{-1}))	IR (cm^{-1})	Ref.
C$_2$H$_5$CO$_2$, HO, OAc, O, N-H, Ph	192–194	220 (14 000), 258 (15 400)	3200, 1765, 1680, 1660	42, 45
C$_2$H$_5$CO$_2$, HO, SPh, O, N-H, Ph	178–180	220 (20 300), 263 (18 200)	3200, 3050, 1670, 1660	42, 45
C$_2$H$_5$CO$_2$, HO, H, O, N-H, Ph	194–197	220 (14 000), 257 (14 600)	3200, 1680, 1660, 1600	42, 45
C$_2$H$_5$CO$_2$, CH$_3$O, H, O, N-H, Ph	130–131	222 (14 000), 249 (12 800)	1690, 1660sh, 1640, 1620	45
C$_2$H$_5$CO$_2$, CH$_3$O, Ph, O, N-H, Ph	178–179	228 (18 800), 300sh (10 000)	1700, 1660, 1620	45
C$_2$H$_5$CO$_2$, CH$_3$O, C$_2$H$_5$, O, N-H, Ph	116–118	222 (13 300), 252 (12 200)	3160, 3050, 1700, 1660, 1620	45

Table 2 (continued)

Structure	M.p. (°C)	UV: λ (nm) (ε (l mol^{-1} cm^{-1}))	IR (cm^{-1})	Ref.
C$_2$H$_5$CO$_2$, CH$_3$O, OC$_2$H$_5$, Ph (dihydroazepinone)	97–99	222 (13 300), 252 (12 200)	1690, 1645, 1605	45
C$_2$H$_5$CO$_2$, CH$_3$O, OAc, Ph	136–138	220 (14 000), 253 (13 400)	1765, 1700, 1660, 1625	45
C$_2$H$_5$CO$_2$, CH$_3$O, SPh, Ph	136–141	220 (21 600), 260 (15 400), 315sh (7500)	1710, 1660, 1630, 1605	45
C$_2$H$_5$CO$_2$, HO, vinyl, Ph	162–167	233 (19 900), 263 (19 600)	3200, 1680, 1660, 1610	44, 45
C$_2$H$_5$CO$_2$, vinyl, Ph	154–155.5	238 (20 900), 279 (8300), 324 (9600)	1705, 1670	44
C$_2$H$_5$CO$_2$, Ph, Ph	163.5–165	238 (22 700), 330 (10 400)	3270, 1695, 1680	44

Table 2 (continued)

Structure	M.p. (°C)	UV: λ (nm) (ϵ (lmol^{-1}cm^{-1}))	IR (cm^{-1})	Ref.
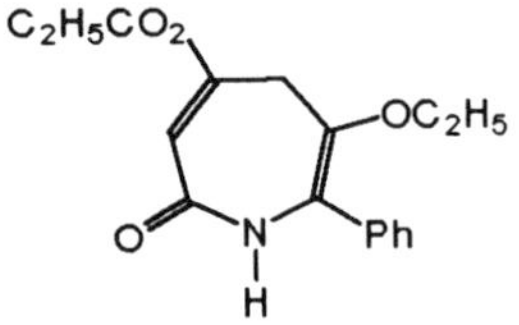 (structure: $C_2H_5CO_2$, OC_2H_5, Ph, N-H, C=O azepine)	100–102	340 (11 500)	3180, 1720, 1695, 1690	44

III. 1,7-DIHYDRO[2H]AZEPIN-2-ONES

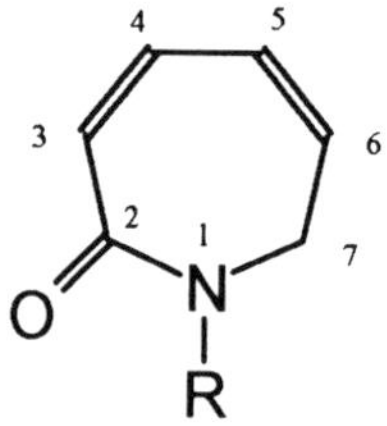

The only reference to this group involves ring expansion of enaminopyrrolidones with dimethyl acetylenedicarboxylate.[49] Thus the enamine **43** (R = H, PhCH$_2$) gave **44** (R = H, PhCH$_2$) in 65% and 58% yield, respectively. None of the other activated alkynes investigated gave any products. Dilute acid hydrolysis of **44** gave **45** (R = H, PhCH$_2$), which were comparatively stable

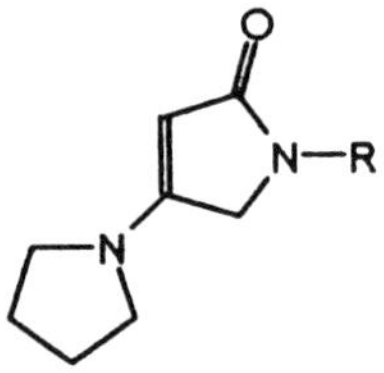

(43) 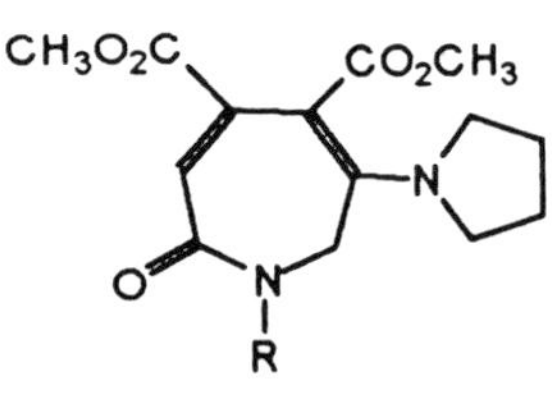   (44)

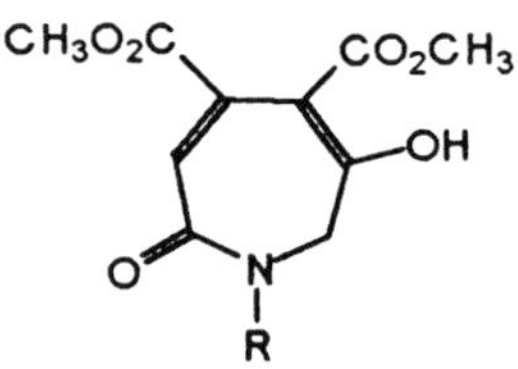

(45) 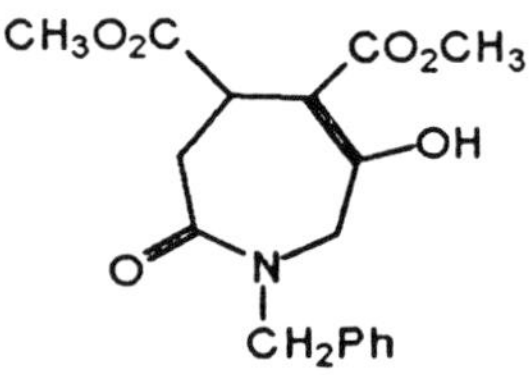   (46)

to mild hydrolysis. Compound **45** (R = PhCH$_2$) was *O*-methylated with diazomethane as one would expect; however, unexpectedly, one of the products obtained when **45** (R = PhCH$_2$) reacted with trimethylsilyl iodide proved to be the reduced compound **46**.

Reaction of the enols **45** (R = H, PhCH$_2$) with hydrazines gave pyrazolo[3,4-*c*] azepines (**47**, R^1 = H, PhCH$_2$; R^2 = H, Ph): these structures are tentative in so

(47)

far as one cannot be sure of the position of the phenyl group in the pyrazole ring. Hydroxylamine failed to react with **45** (R = H, PhCH$_2$).[49]

IV. 3,4-DIHYDRO[2*H*]AZEPIN-2-ONES

Uniquely,[37] compound **48** was obtained in 90% yield during catalytic hydrogenation of the azepin-2-one **49**. The latter was obtained by reaction of

(48)   (49)   (50)

triethyloxonium fluoroborate (Chapter 12) with the dione **50**, itself the product of the reaction of the appropriate benzoquinone with sodium azide in concentrated sulphuric acid (Chapter 10).

V. 3,6-DIHYDRO[2H]AZEPIN-2-ONES

Again, there is only one example. Cope rearrangement[50] of the isocyanate **51** gave **52** in 60% yield. It was hygroscopic and was characterised by spectroscopy

(51)   (52)   (53)

and chemical conversions to a variety of reduced structures (e.g. **53**, R = OH, OCH_3, SPh and H).

VII. 1,2-DIHYDRO[3H]AZEPIN-3-ONES

PREPARATION

The almost exclusive and very satisfactory method for these compounds depends upon the flash vacuum pyrolysis of certain dienamine derivatives of Meldrum's acid (**54a–d**)[51,52] at about 500 °C. In the case of **54** (R^1 = H, R^2 = ^{i}Pr), the products included both **55** (R^1 = H, R^2 = ^{i}Pr) and the fragmentation product **56**. All 1,2-dihydro[3H]azepin-3-ones (**55**, **56**) are yellow solids and X-ray data[52] indicated that although the seven-membered ring was

(54)

a; R1 = H, R2 = CH₃
b; R1 = H, R2 = Ph
c; R1 = CH₃, R2 = C₂H₅
d; R1 = R2 = Ph

distinctly non-planar, nevertheless the dienaminone portion of the molecule was nearly planar as is found in acyclic and other cyclic examples.

A plausible mechanism[51] for the production of these compounds involved hydrogen transfer from the methylene ketene **57** to an intermediate (**58**) capable

of undergoing either electrocyclisation to **55** or cycloaddition to **59**, which can fragment to cyclopentadienone and an imine. This is supported by the fact that the dimer of cyclopentadienone is generally isolated along with the imine; in the case of **54** ($R^1 = R^2 = Ph$) benzylidineaniline was identified. Recently, the synthesis was extended[53] to the 7-dimethylamino derivative **60**.

The Meldrum's acid derivatives (e.g. **54**) required for this protocol are in general fairly easily obtainable from Meldrum's acid (**61**) and the pyrolysis products are easily separated by chromatography on alumina.

There is only one other access to a 1,2-dihydro[3*H*]azepin-3-one, namely treatment of the pyridone **63** with LDA in THF, which gave the divinylogous amide **62**.[54]

PROPERTIES

The behaviour of 1,2-dihydro[3H]azepin-3-ones has been examined in admirable detail by McNab and co-workers. ^{1}H and ^{13}C NMR spectra of 1-substituted, 1,2-disubstituted and 1,2,2-trisubstituted examples were obtained[55] and analysed thoroughly. All parameters were assigned and discussed; this made it possible to state with some certainty the position of attack of reagents and structures of products obtained in later work.[56–59]

It was demonstrated[57] that O-protonation took place in trifluoroacetic acid since there were striking similarities between ^{1}H and ^{13}C NMR spectra of the protonated N-methyl compound (**55**; $R^1 = H$, $R^2 = CH_3$) and the undoubted O-ethyl compound (**64**) obtained by reaction with Meerwein's reagent. Deuterium exchange took place in [^{2}H]trifluoroacetic acid; the rates of exchange at indicated positions were 4 > 6 ≫ 2. Reaction of the N-methyl (**55**, $R^1 = H$, $R^2 = CH_3$) and N-phenyl (**55**, $R^1 = H$, $R^2 = Ph$) derivatives with one

equivalent of N-chlorosuccinimide in methanol at 0 °C gave the 4-chloro compounds (**65**; X = Cl, R = CH_3 and Ph) in tolerable yields. They proved to be stable in the solid state but not in solution, especially when heated. Further chlorination led to 4,6-dichloro compounds. N-Bromosuccinimide reacted with **55** ($R^1 = H$, $R^2 = CH_3$) giving the 4,6-dibromo compound **66** as the only clearly identified product irrespective of the proportions of reagent used; with one equivalent of NBS there appeared to be unreacted starting material, possibly contaminated with **65** (R = CH_3, X = Br). The structure of **65** (X = Cl, R = CH_3) was confirmed by X-ray analysis.[57]

Photolysis of **55** ($R^1 =$ H, $R^2 =$ Ph) gave the expected electrocyclisation product **67**; this reaction was thermally reversible.[58] Cycloadditions of 1,2-dihydro[3H]azepin-3-ones proceeded best with maleic anhydride to give **68** (R = CH_3 and Ph). Cycloaddition with DMAD and ethyl propiolate yielded benzenoid products, presumably by loss of fragments (CO + imine) as shown (**69**). Regiochemistry of addition was elucidated by reacting the dideuterio compound **70** with ethyl propiolate, in which case the product was the m-dideuterioethyl benzoate **71**.[58] Treatment of the compounds **55** ($R^1 = H$, $R^2 = CH_3$ or Ph) with sodium methoxide in methanol produced

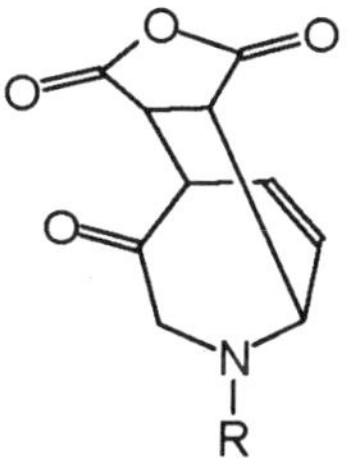

Dihydroazepinones

(68) 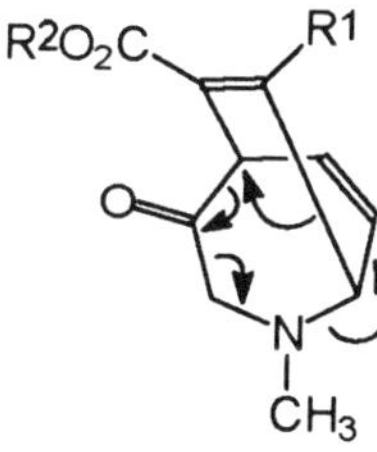 (69) 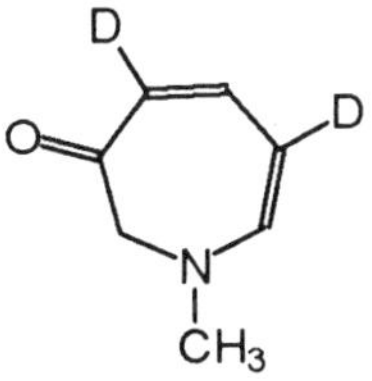 (70) 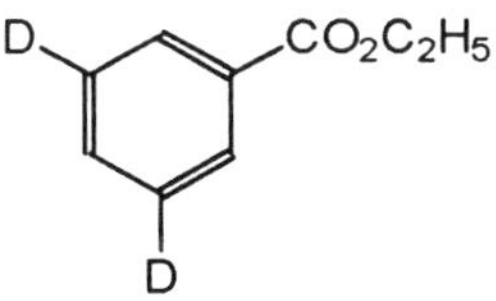 (71)

2-aminophenol derivatives; a plausible mechanism was proposed which involved an azanorcaradiene intermediate.[59]

A summary of the reactions undergone by 1,2-dihydro[3H]azepin-3-ones is shown in the scheme.[56]

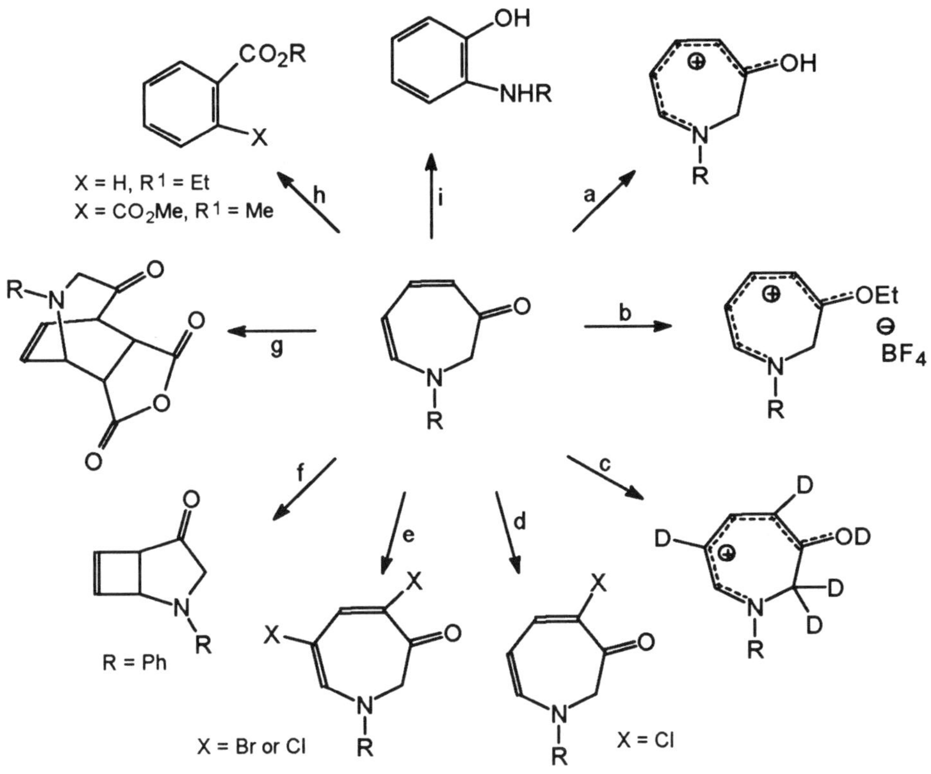

Reagents: a, CF_3CO_2H; b, Et_3O^+ BF_4^-; c, CF_3CO_2D; d, *N*-chlorosuccinimide; e, *N*-bromo- or *N*-chlorosuccinimide (2 equiv.); f, $h\nu$; g, maleic anhydride; h, $HC\equiv CCO_2Et$ or $MeO_2CC\equiv CCO_2Me$; i, NaOMe/MeOH, then H^+.

The 7-dimethylamino representative **60**, activated by the electron-donating substituent, behaved differently from those above.[53] Protonation of **60** gave a stable mixture of 22% *O*- and 78% 4*C*-protonated materials and cycloaddition with methyl propiolate yielded the 1:2 adduct (minus H_2O) (**72**).

(**72**)

XV. 2,3-DIHYDRO[4*H*]AZEPIN-4-ONES

There appears to be only one reference[60] in the literature to these derivatives. Thus, when 2-amino-1-azetines reacted in acetonitrile with 2,3-diphenylcyclopropenones, the products were azeto[1,2-*a*]pyrroles (e.g. **73**); the latter, on

(**73**)   (**74**)   (**75**)

hydrolysis with 2 M HCl in ethanol at room temperature, gave the enolic tautomers (**74**) of the tetrahydrodiones (**75**). In the case of **74** (R = Ph), hydrolysis yields of about 90% were seen, but with **74** (R = H), the yield was about 7%, along with **75** (R = H) in 30% yield.[60]

There are, therefore, representatives of only seven of the eighteen possible structures that can be drawn for dihydroazepinones. It seems surprising that 1,5-dihydro[4H]azepin-4-ones (**76**) and 1,7-dihydro[4H]azepin-4-ones (**77**) have not been reported since they contain some generally conjugative stabilising features.

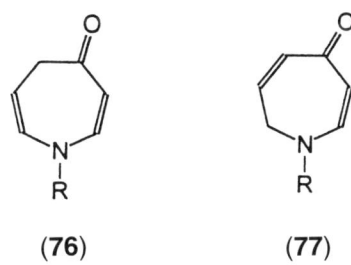

(**76**)   (**77**)

REFERENCES

1. L.A. Paquette, *J. Am. Chem. Soc.*, 1962, **84**, 4987.
2. L.A. Paquette, *J. Am. Chem. Soc.*, 1963, **85**, 3288.
3. W. Theilacker, K. Ebke, L. Seidl and S.Schwerin, *Angew. Chem.*, 1963, **75**, 208.
4. L.A. Paquette, *Org. Synth.*, 1964, **44**, 41.
5. L.A. Paquette, *US Pat.*, 3 239 505, *Chem. Abs.*, 1966, **64**, 19575f.
6. L.A. Paquette, *US Pat.*, 3 267 092, *Chem. Abs.*, 1966, **65**, 16952a.
7. L.A. Paquette, *US Pat.*, 3 275 653, *Chem. Abs.*, 1967, **66**, 2470q.
8. L.A. Paquette, *US Pat.*, 3 152 118, *Chem. Abs.*, 1964, **61**, 16054g.
9. L.A. Paquette, *US Pat.*, 3 158 600, *Chem. Abs.*, 1965, **62**, 5259a.
10. L.A. Paquette, *US Pat.*, 3 177 204, *Chem. Abs.*, 1965, **63**, 591c.
11. L.A. Paquette and W.C. Farley, *J. Am. Chem. Soc.*, 1967, **89**, 3595.
12. H.-D. Becker and K. Gustafsson, *Tetrahedron Lett.*, 1976, **20**, 1705.
13. H.-D. Becker and K. Gustafsson, *J. Org. Chem.*, 1977, **42**, 2966.
14. R.A. Abramovitch, S.R. Challand and E.F.V. Scriven, *J. Am. Chem. Soc.*, 1972, **94**, 1374.
15. H. Takeuchi and K. Koyama, *J. Chem. Soc., Chem. Commun.*, 1981, 202.
16. R. Purvis, R.K. Smalley, W.A. Strachan and H. Suschitzky, *J. Chem. Soc., Perkin Trans. 1*, 1978, 191.
17. R. Purvis, R.K. Smalley, H. Suschitzky and M.A. Aiknader, *J. Chem. Soc., Perkin Trans. 1*, 1984, 249.
18. (a) K. Lamara and R.K. Smalley, *Tetrahedron*, 1991, **47**, 2277; (b) K. Lamara, A.D. Redhouse, R.K. Smalley and J.R. Thompson, *Tetrahedron*, 1994, **50**, 5515.
19. W. Von E. Doering and R.A. Odum, *Tetrahedron*, 1966, **22**, 81.
20. F.R. Atherton and R.W. Lambert, *J. Chem. Soc., Perkin Trans. 1*, 1973, 1079.
21. M. Masaki, K. Fukui and J. Kita, *Bull. Chem. Soc. Jpn.*, 1977, **50**, 2013.
22. E. Vogel and R. Erb, *Angew. Chem.*, 1962, **72**, 76.
23. E. Vogel, R. Erb, G. Lenz and A.A. Bothner-by, *Liebigs Ann. Chem.*, 1965, **682**, 1.
24. L.A. Paquette, *Tetrahedron Lett.*, 1963, **29**, 2027.
25. L.A. Paquette, *J. Am. Chem. Soc.*, 1964, **86**, 4092.

26. L.A. Paquette, *J. Org. Chem.*, 1963, **28**, 3590.
27. L.A. Paquette, *J. Am. Chem. Soc.*, 1964, **86**, 4096.
28. L.A. Paquette, *J. Org. Chem.*, 1964, **29**, 3447.
29. L.A. Paquette and W.C. Farley, *J. Org. Chem.*, 1967, **32**, 2725.
30. L.A. Paquette, *J. Am. Chem. Soc.*, 1964, **86**, 500.
31. O.L. Chapman and E.D. Hoganson, *J. Am. Chem. Soc.*, 1964, **86**, 498.
32. J.W. Pavlik and C.A. Seymour, *Tetrahedron Lett.*, 1977, **30**, 2555.
33. H-D. Becker and A.B. Turner, *Tetrahedron Lett.*, 1979, **50**, 4871.
34. L.A. Paquette and J.K. Reed, *J. Med. Chem.*, 1963, **6**, 771.
35. T.B. Brown, P.R. Lowe, C.M. Schwalbe and M.F.G. Stevens, *J. Chem. Soc., Perkin Trans. 1*, 1983, 2485.
36. W. Heinzelmann and M. Marky, *Helv. Chim. Acta*, 1973, **56**, 1852.
37. E.J. Moriconi and I.A. Maniscalco, *J. Org. Chem.*, 1972, **37**, 208.
38. T. Sano and Y. Tsuda, *Heterocycles*, 1976, **4**, 1229.
39. T. Sano, Y. Tsuda, H. Ogura, K. Furumata and Y. Iitaka, *Heterocycles*, 1976, **4**, 1233.
40. T. Sano, Y. Horiguchi, T. Tsuda and Y. Itatani, *Heterocycles*, 1978, **9**, 161.
41. T. Sano, Y. Horiguchi, S. Kambe and Y. Tsuda, *Heterocycles*, 1981, **16**, 363.
42. T. Sano, Y. Horiguchi and Y. Tsuda, *Heterocycles*, 1979, **12**, 1427.
43. Y. Horiguchi, T. Sano and Y. Tsuda, *Heterocycles*, 1985, **23**, 1509.
44. T. Sano, Y. Horiguchi, K. Tanaka and Y. Tsuda, *Chem. Pharm. Bull.*, 1985, **33**, 5197.
45. T. Sano, Y. Horiguchi and Y. Tsuda, *Chem. Pharm. Bull.*, 1990, **38**, 3283.
46. T. Sano, Y. Horiguchi and Y. Tsuda, *Heterocycles*, 1981, **16**, 889.
47. T. Sano, Y. Horiguchi, K. Tanaka and Y. Tsuda, *Chem. Pharm. Bull.*, 1990, **38**, 36.
48. Y. Tsuda, M. Kaneda, T. Sano, Y. Horiguchi and Y. Iitaka, *Heterocycles*, 1979, **12**, 1423.
49. G.W. Heinicke, A.M. Morella, J. Orban, R.H. Prager and A.D. Ward, *Aust. J. Chem.*, 1985, **38**, 1847.
50. T. Sasaki, S. Eguchi and M. Ohno, *J. Org. Chem.*, 1972, **37**, 466.
51. H. McNab, L.C. Monahan and T. Gray, *J. Chem. Soc., Chem. Commun.*, 1987, 140.
52. A.J. Blake, H. McNab and L.C. Monahan, *J. Chem. Soc., Perkin Trans. 1*, 1989, 425.
53. E. Cartmell, J.E. Mayo, H. McNab and I.M. Sadler, *J. Chem. Soc., Chem. Commun.*, 1993, 1417.
54. A.R. Katritzky, J. Arrowsmith, Z. Bin Bahari, C. Jayaram, T. Siddiqui and S. Vassilatos, *J. Chem. Soc., Perkin Trans. 1*, 1980, 2851.
55. H. McNab and L.C. Monahan, *J. Chem. Soc., Perkin Trans. 1*, 1990, 3159.
56. H. McNab and L.C. Monahan, *J. Chem. Soc., Chem. Commun.*, 1987, 141.
57. H. McNab, L.C. Monahan and A.J. Blake, *J. Chem. Soc., Perkin Trans. 1*, 1990, 3163.
58. H. McNab and L.C. Monahan, *J. Chem. Soc., Perkin Trans. 1*, 1990, 3169.
59. H. McNab and L.C. Monahan, *J. Chem. Res. (S)*, 1990, 336.
60. F. Stierli, R. Prewo, J.H. Bieri and H. Heimgartner, *Helv. Chim. Acta*, 1983, **66**, 1366.

CHAPTER 10

Dihydroazepinediones

Of sixteen possible structures in this category, only three are known with certainty. It should be pointed out, however, that three of the apparently unknown dione (**1–3**) structures would be expected to exist in the corresponding enolic forms (**4–6**) which would be regarded as 'azatropolones' (Chapter 12).

In fact, structures related to **5** were claimed more than 30 years ago,[1] but later shown to be in error.[2] Furthermore, the tautomeric form (**6**) was indeed found to be the stable assembly in several examples (e.g. **7**; Chapter 12).[3]

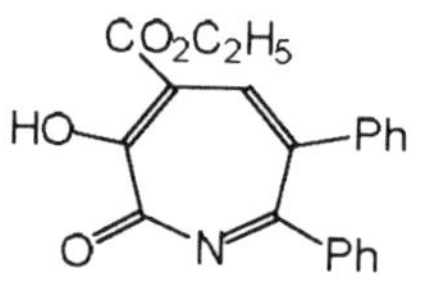

(7)

I. 2,3-DIHYDRO[1H]AZEPINE-2,3-DIONES

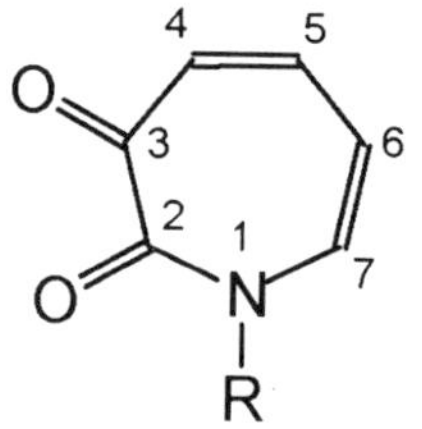

The thermal ring opening of **8** was shown to lead to the dione **9**,[6] which was obtained as yellow needles, m.p. 130–131 °C. The IR spectrum of the latter

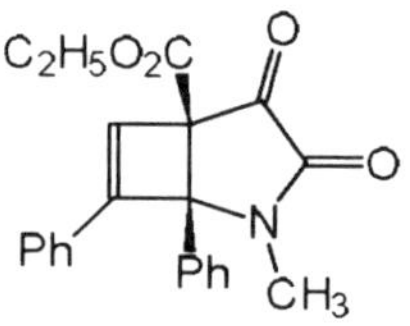

(8) 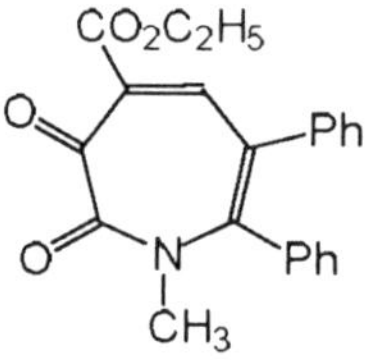   **(9)**

exhibited three distinct carbonyl stretching frequencies at 1730, 1685 and 1625 cm^{-1} as one might have expected. Both the ^{1}H and ^{13}C NMR spectra were consonant with structure **9**.

V. 2,5-DIHYDRO[1H]AZEPINE-2,5-DIONES

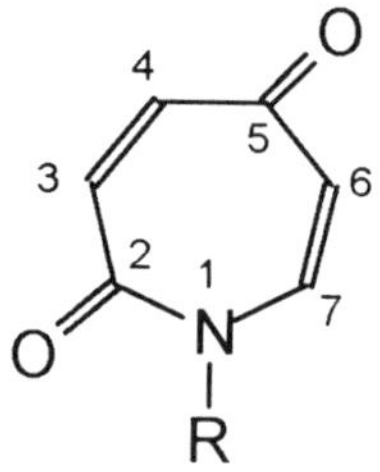

These have been obtained exclusively by application of the Schmidt reaction to 1,4-benzoquinones. The experimental conditions are crucial: the quinones were stirred at 0 °C in concentrated sulphuric acid while sodium azide was added slowly in portions.[7–10] Some structural errors[7,8] were later corrected,[9,10] partly by re-interpretation of ^{1}H NMR spectra and partly by hydrolysis of the diones and careful isolation and identification of steam-volatile ketones and accompanying acids. In particular, aldehydes were not formed after hydrolysis; this demanded that the carbon atom next to nitrogen in the heterocycle must carry an alkyl (methyl) substituent. Corrected structures are shown in Table 1.

Not all Schmidt reactions on 1,4-benzoquinones gave 2,5-dihydro[1H]-azepine-2,5-diones. Thus thymoquinone under slightly different conditions gave the amino lactone **10**[12] and hydroxynaphthoquinones also gave ring-contracted products[11] (e.g. **11**, R = H, X = NH, and R = CH$_3$, X = O). It has been shown that aqueous sodium hydroxide at reflux converted **12** into **13**;[13] seemingly the behaviour of the diones upon hydrolysis depends very much on the substituents present.

(10)   (11)   (12)   (13)

Table 1. 2,5-Dihydro[1H]azepine-2,5-diones

Structure	M.p. (°C)	Spectroscopic evidence	Ref.
(CH$_3$, CH$_3$ azepinedione)	216–217	^{1}H NMR, UV, IR	8, 9
(iC$_3$H$_7$, CH$_3$ azepinedione)	169–170	^{1}H NMR, UV, IR	8–10
(H$_3$C, CH$_3$, CH$_3$ azepinedione)	199–200	^{1}H NMR, UV, IR	8, 9

Table 1 (continued)

Structure	M.p. (°C)	Spectroscopic evidence	Ref.
	201–202 214–215	^{1}H NMR, UV, IR	9 8
	253–255	IR, MS, ^{1}H NMR	13
	285–287	IR, MS, ^{1}H NMR	13
	226–226.5	IR, MS, ^{1}H NMR	13
	250–251	IR, MS, ^{1}H NMR	13
	111–112	IR, MS, ^{1}H NMR	13
	197.5–199	IR, MS, ^{1}H NMR	13

IX. 2,7-DIHYDRO[1H]AZEPINE-2,7-DIONES

Two examples of the 2,7-dihydro[1H]azepine-2,7-dione system have been reported in the literature. The 1-methyl derivative (**14**) was synthesised[14] as

(**14**) CH$_3$NHCO(CH$_2$)$_4$CO$_2$H (**15**) (**16**) (**17**)

follows: thionyl chloride treatment of N-methyladipamic acid (**15**) gave the imide **16**, which could be brominated (CuBr$_2$) and dehydrobrominated (Et$_3$N) to **14**. The mass spectrum of the latter was discussed in some detail; it is interesting that all the major peaks are identical with those of 1-methyl-2-pyridone.

On bromination (ice-cold hexane) **14** yielded trans-3,4-dihydro-3,4-dibromo-1-methylazepine-2,7-dione (**17**) and the cycloaddition reaction with cyclopentadiene gave the exo and endo products (**18** and **19**) wherein the heterocycle had reacted as a dienophile. It did not react with maleic anhydride or with TCE, that

(**18**)   (**19**)   (**20**)

is, it lacks diene properties in the Diels–Alder sense. Ethanolysis of **14** gave ethyl N-methyl-cis,cis-muconamate (**20**).[14]

The parent compound (**22**) has recently been prepared in poor yield by the UV irradiation of the benzofuroxan **21**; a mechanism for the reaction was postulated involving ring opening to the unstable di(nitrile oxide) 23 followed by rearrangement to the azepinedione 22.[15]

(21)　　　　　(22)　　　　　(23)

The physical data reported for the two 2,7-dihydro[1H]azepine-2,7-diones 14 and 22 are given in Table 2.

Table 2. 2,7-Dihydro[1H]azepine-2,7-diones

Structure	M.p. (°C)	Spectroscopic evidence	Ref.
	76–77.8	UV, IR, MS, ^{1}H NMR	15
	167–168	UV, IR, MS, ^{1}H, ^{13}C NMR	14

REFERENCES

1. W. Triebs and A. Lange, *J. Prakt. Chem.*, 1961, **14**, 208.
2. A. H. Rees, *J. Chem. Soc.*, 1965, 1749.
3. T. Sano, Y. Horiguchi and Y. Tsuda, *Heterocycles*, 1978, **9**, 731.
4. T. Sano, Y. Horiguchi and Y. Tsuda, *Heterocycles*, 1979, **12**, 1427.
5. T. Sano, Y. Horiguchi and Y. Tsuda, *Heterocycles*, 1985, **23**, 1509.
6. T. Sano, Y. Horiguchi and Y. Tsuda, *Chem. Pharm. Bull.*, 1990, **38**, 3283.
7. D. Misiti, H. W. Moore and K. Folkers, *Tetrahedron Lett.*, 1965, 1071.
8. D. Misiti, H. W. Moore and K. Folkers, *Tetrahedron*, 1966, 1201.
9. R. W. Rickards and R. M. Smith, *Tetrahedron Lett.*, 1966, 2361.
10. G. R. Bedford, G. Jones, and B. R. Webster, *Tetrahedron Lett.*, 1966, 2367.
11. H. W. Moore and H. R. Sheldon, *J. Org. Chem.*, 1967, **32**, 3603.
12. A. H. Rees, *J. Chem. Soc.*, 1962, 3097.
13. A. H. Rees. *Can. J. Chem.*, 1974, **52**, 3327.
14. R. Shapiro and S. Nesnow, *J. Org. Chem.*, 1969, **34**, 1695.
15. M. Hasegawa and T. Takabatake, *J. Heterocycl. Chem.*, 1991, **28**, 1079.

CHAPTER 11

Azepines

The fully unsaturated azepine system can exist in four isomeric forms, [1H]-, [2H]-, [3H]- and [4H]azepines (**1, 2, 3** and **4**, respectively), and stable derivatives of each isomer have been reported in the literature.

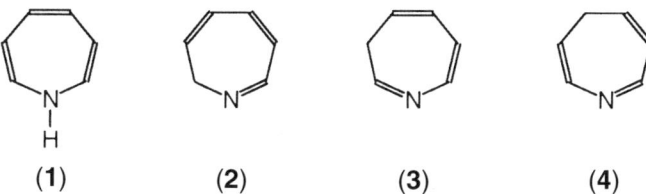

In early reports there was apparently some confusion in identifying the correct structures and many of the erroneous assignments were only corrected with the introduction of ^{1}H NMR spectroscopy as a routine analytical technique. For example, dibenzamil, isolated by Wolff in 1912,[1] was not recognised as an azepine until 1958 when Huisgen and co-workers[2,3] identified the seven-membered ring system, but preferred the [2H] or [4H] structures **5** or **6** on the basis of the UV and IR spectra. The correct identification was finally achieved by Doering and Odum,[4] who proposed the [3H] structure **7** on the basis of the ^{1}H NMR spectrum. A later review,[5] in which the alicyclic cycloheptatriene **8** was compared with the heterocyclic analogues confirmed the assignment and clarified the situation.

I. [1H]AZEPINES

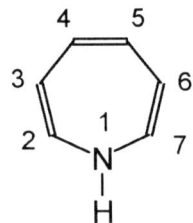

PREPARATION

The parent [1H]azepine **1** was first generated by Hafner in 1963[6] by hydrolysis of ethyl [1H]azepine-N-carboxylate (**9**).[6–13] Owing to its high reactivity, however (tautomerism to the [3H]azepine **3** and polymerisation), it was not characterised until 1980 when Vogel and his co-workers[8] obtained the ^{1}H and ^{13}C NMR spectra of the pure compound in solution before it decomposed. At $-78\,°$C the compound is stable for only a few hours in solution and it has not been isolated.

[1H]Azepines substituted on the N atom, such as the 1-carbethoxy derivative **9**, have been prepared by the method of Hafner and König[7] by insertion of a nitrene into a six-membered ring. The nitrene was generated by UV irradiation of the corresponding azide (**10**). The presumed bicyclic [4.1.0]azaheptene **11** has not been isolated, but was proposed to explain the mechanism of the insertion[7,8,10] (Scheme 1) (see, however, section III for an alternative explanation for the mechanism of the reaction).

Scheme 1

Azepines 489

This method is of general application and has been used with various azides, e.g. the azidonitrile **12** (R = CN) and substituted benzenes **13** (X = H, alkyl, halogen, etc.) to give a range of 1-cyano[1H]azepines (**14**, X = H, alkyl, halogen, etc.)[14,15] (see Table 1 for details).

N_3—R

(**12**)   (**13**)   (**14**)   (**15**)

A range of substituted benzenesulphonyl analogues (**15**, X = H, Cl, Br, NO_2, CH_3, OCH_3) has been made from the corresponding benzenesulphonyl azides (**12**, R = $SO_2C_6H_4X$, X = H, Cl, Br, NO_2, CH_3, OCH_3) with benzene or substituted benzenes, thus illustrating further the generality of the method.[16–18] The use of a phase-transfer catalyst has resulted in improved yields and shorter reaction times,[19] and increasing the reaction pressure has been shown to have a similar effect.[20]

The reaction of a disubstituted benzene such as **16** (R^1 = CH_3, R^2 = Cl, Br) with the azido ester **12** (R = $CO_2C_2H_5$)[21] or treatment of **16** (R^1 = R^2 = CO_2CH_3) with p-toluenesulphonyl azide (**12**, R = $SO_2C_6H_4$-p-CH_3)[22] resulted in the formation of 2,5-disubstituted azepines (**17**, R^1 = CH_3, R^2 = Cl, Br and **18**, R^1 = R^2 = CO_2CH_3, respectively).

(**16**)   (**17**)   (**18**)

(**19**)   (**20**)

In an exceptional case, di-*tert*-butylbenzene (**16**, R^1 = R^2 = ^{t}Bu) when reacted with a range of azido esters (**19**, R = C_2H_5, $CH_2^cC_3H_5$, ^{t}Bu,

C_6H_4-p-OCH_3) gave the 3,6-disubstituted [1H]azepines **20** (R = C_2H_5, $CH_2{}^cC_3H_5$, ^tBu, C_6H_4-p-OCH_3) as the main products.[23]

A systematic study of the reactions of o-, m- and p-xylene (**16**, $R^1 = R^2 = CH_3$) with the azido ester **19** (R = C_2H_5) gave mixtures of isomers of dimethyl[1H]azepines. Of the four possible products from o-xylene, only the 4,5- and the 3,4-dimethyl[1H]azepines (**21** and **22**) were formed. Similarly, m-xylene gave the 3,5- and 2,4-dimethyl[1H]azepines (**23** and **24**) and p-xylene (**16**, $R^1 = R^2 = CH_3$) gave both the 2,5- and the 3,6-dimethyl[1H]azepines (**17**, $R^1 = R^2 = CH_3$, and **25**), although only the former compound could be isolated from the 66:34 mixture.[24]

A variety of other substituted [1H]azepines have been prepared in an analogous manner from appropriately substituted benzenes and azides.[22,25,26a,27–33,71] In one case the nitrene from the azide **26** has been induced to insert intramolecularly to give the intermediate **27**, which yielded the [1H]]azepine **28** on treatment with ethylamine. By contrast, acid hydrolysis of the intermediate **27** gave the [3H]azepine **29**.[34]

Although the bicyclic structure **11** does not appear to have been isolated,[7,8,10] Paquette and his co-workers have isolated a series of analogues (**31**, R = H, or R = CH$_3$, n = 1, 2) which they prepared from the corresponding 1,4-cyclohexadienes (**30**, R = H or CH$_3$, n = 1, 2) using iodine isocyanate in acetic acid or methanol, and reacting the intermediate iodoamide (**32**, R = H or CH$_3$, n = 1, 2) with NaOCH$_3$ in THF. Bromination of **31** to the dibromo compound **33** and dehydrobromination/ring opening with NaOCH$_3$ in THF gave the corresponding [1H]azepines **34** (R = H or CH$_3$, n = 1, 2) as shown in Scheme 2.[35,36]

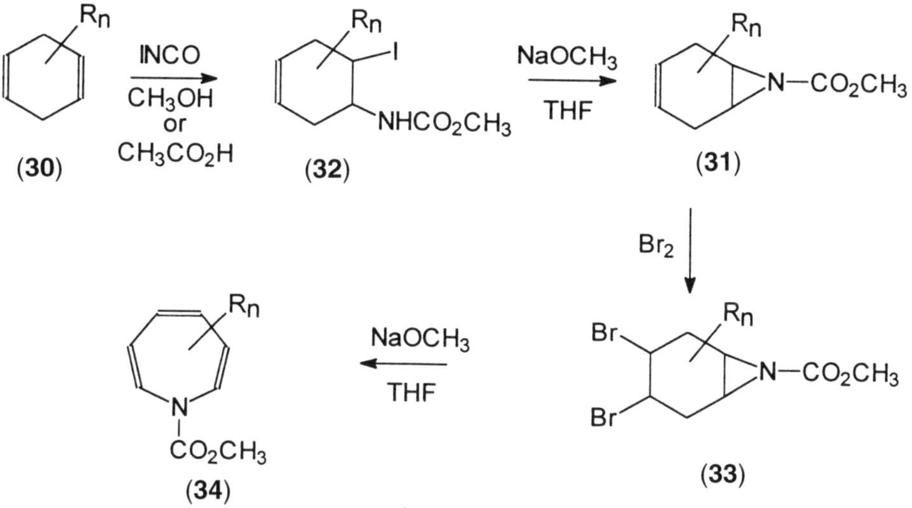

Scheme 2

N-Substituted pyrroles, on treatment with acetylenes under Diels–Alder conditions, form [2.2.1]bicyclic systems which rearrange on UV irradiation to azaquadricyclanes. Thermal rearrangement of these azaquadricyclanes generates N-substituted [1H]azepines. This sequence of events has been used by Bansal et al.[37] and Prinzbach et al.[38] to produce [1H]azepines (**39**, R = CO$_2$CH$_3$ and R = COCH$_3$, SO$_2$C$_6$H$_4$-p-CH$_3$) from the pyrroles **35** (R = CO$_2$CH$_3$ and R = COCH$_3$, SO$_2$C$_6$H$_4$-p-CH$_3$, respectively) on treatment with dimethyl acetylenedicarboxylate (**36**) according to Scheme 3.[37,38]

This reaction sequence has been extensively studied and exploited by Prinzbach and co-workers in the period 1968–86,[38–43] during which time they produced a wide range of azepines (**40**, R^1 = COCH$_3$, CO$_2$CH$_3$, CONH$_2$, SO$_2$CH$_3$, SO$_2$C$_6$H$_4$-p-CH$_3$, C$_6$H$_4$-p-NO$_2$, R^2 = H, CH=CH$_2$, R^3 = H, Cl, Ph, R^4 = H, CF$_3$, CO$_2$CH$_3$) from the corresponding pyrroles (**41**, R^1 = COCH$_3$, CO$_2$CH$_3$, CONH$_2$, SO$_2$CH$_3$, SO$_2$C$_6$H$_4$-p-CH$_3$, C$_6$H$_4$-p-NO$_2$, R^2 = H, CH=CH$_2$, R^3 = H, Cl, Ph) and appropriately substituted acetylenes (**42**, R^4 = H, CF$_3$, CO$_2$CH$_3$) via the corresponding azaquadricyclane intermediates

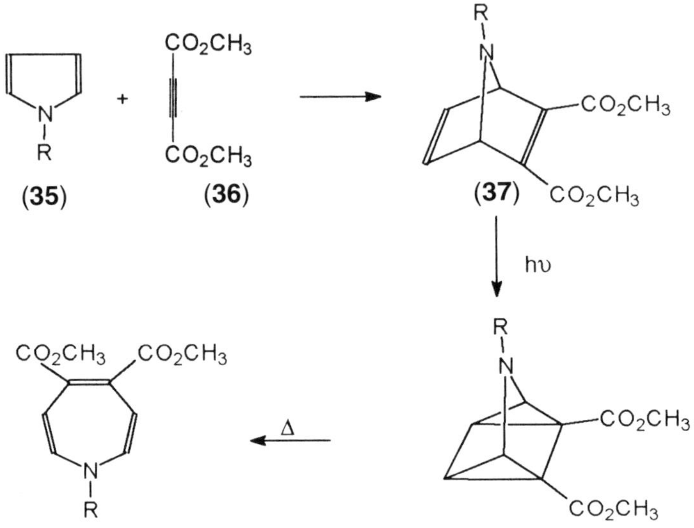

(39) (38)

Scheme 3

(**43**, R^1 = $COCH_3$, CO_2CH_3, $CONH_2$, SO_2CH_3, $SO_2C_6H_4$-*p*-CH_3, C_6H_4-*p*-NO_2, R^2 = H, CH=CH_2, R^3 = H, Cl, Ph, R^4 = H, CF_3, CO_2CH_3)[38–43] (see Table 1 for details).

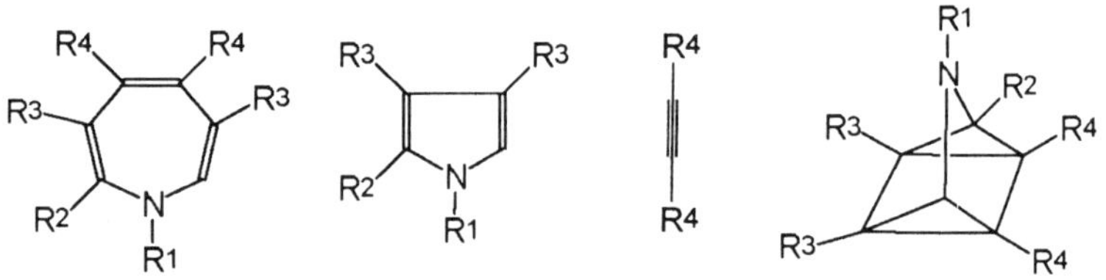

(**40**) (**41**) (**42**) (**43**)

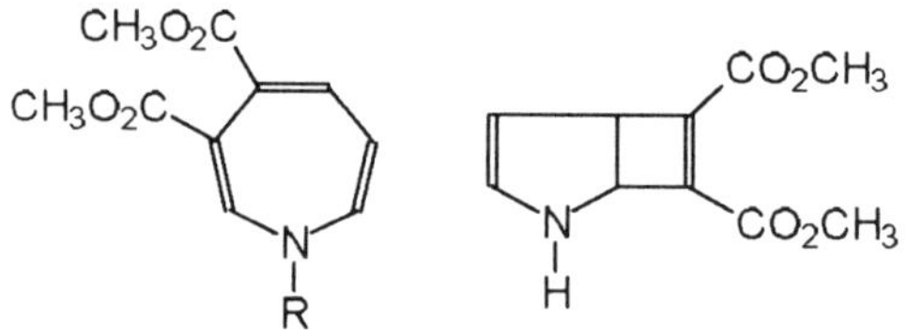

(**44**) (**45**)

Azepines

In contrast to these results,[37-40] it has also been reported that UV irradiation of the unsubstituted pyrrole **35** (R = H) led to the 3,4-dicarbomethoxy-[1H]azepine **44** rather than the 4,5-analogue **39** (R = H) as might be expected. In this case, the alternative product **39** (R = H) was excluded on the basis of the proton NMR spectrum, and the intermediate proposed to explain the results was the bicyclo[3.2.1]azaheptadiene **45** rather than the azaquadricyclane **38** (R = H).[44]

Other ring expansion reactions have also been used to prepare [1H]azepines. For example, the N-phenylaziridine **47**, prepared from the triazoline **46**, on heating to 150 °C, gave 1-phenyl[1H]azepine (**48**), but unfortunately with substituents on the N atom other than phenyl, thermolysis does not result in the corresponding azepines.[45]

(46) (47) (48)

A similar sequence of reactions from hexafluorobenzene (**49**) and either phenyl azide (**12**, R = Ph) or ethyl azidoformate (**10**) gave the corresponding triazolines **50** (R = Ph, $CO_2C_2H_5$), which in turn were pyrolysed to the hexafluoro[1H]azepines **51** (R = Ph, $CO_2C_2H_5$).[46]

(49) (50) (51)

The thermally induced ring expansion of the *trans*-aziridine **52** has been reported to proceed at 100 °C to give the 1-*tert*-butylazepine **53**, albeit in low yield,[47] and the 4-chloromethyldihydropyridines **54** ($R^1 = CH_3$, $R^2 = H$, $R^3 = CO_2CH_3$, and $R^1 = CH_3$, Ph, p-$CH_3C_4H_6$, $R^2 = CH_3$, $R^3 = CN$) have been converted into the corresponding azepines **55** ($R^1 = CH_3$, $R^2 = H$, $R^3 = CO_2CH_3$, and $R^1 = CH_3$, Ph, p-$CH_3C_4H_6$, $R^2 = CH_3$, $R^3 = CN$) by treatment with KO^{t}Bu or KOH, respectively, at room temperature.[48,49]

In a novel sequence of reactions (see Scheme 4), the 2,4,6-triphenylanilides **56** (R = CH_3, Ph, p-$CH_3C_6H_4$, p-$CH_3OC_6H_4$, p-BrC_6H_4) were oxidised with lead tetraacetate in methanol to give the corresponding cyclohexa-2,5-diene-1-ylidenecarboxamides **57** (R = CH_3, Ph, p-$CH_3C_6H_4$, p-$CH_3OC_6H_4$, p-BrC_6H_4), which in turn were alkylated using methyllithium to produce the cyclohexa-2,5-dienamides **58** (R = CH_3, Ph, p-$CH_3C_6H_4$, p-$CH_3OC_6H_4$, p-BrC_6H_4). The latter compounds on subsequent treatment with acid gave the bicyclic oxazoline derivatives **59** (R = CH_3, Ph, p-$CH_3C_6H_4$, p-$CH_3OC_6H_4$, p-BrC_6H_4), which on heating to 180 °C were converted smoothly into the corresponding N-acyl[1H]azepines **60** (R = CH_3, Ph, p-$CH_3C_6H_4$, p-$CH_3OC_6H_4$, p-BrC_6H_4).[50]

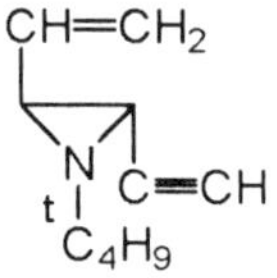

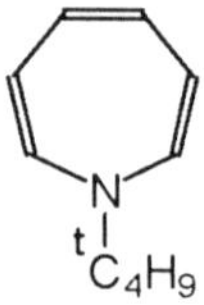

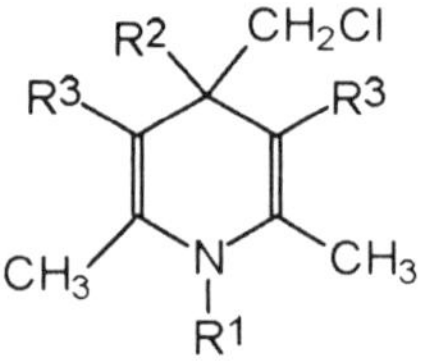

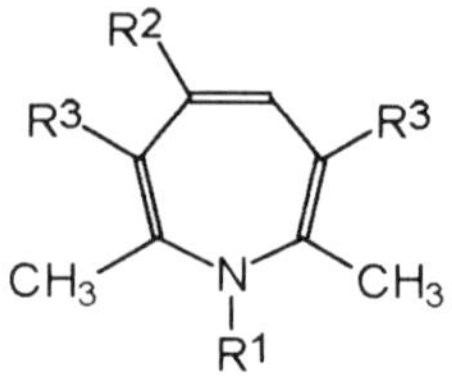

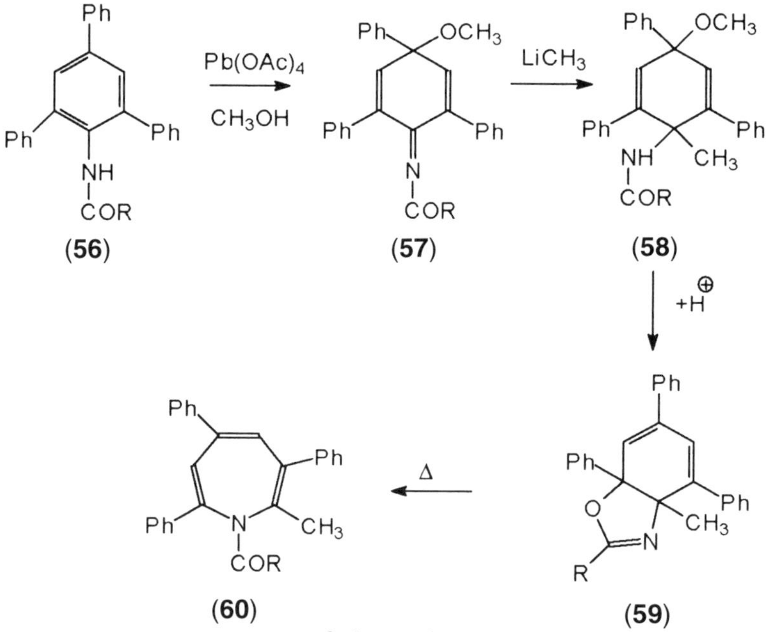

Scheme 4

The N-tosyl analogue **61** was prepared in a similar manner from the corresponding p-toluenesulphonamide **62**, but in this case the reaction proceeded without isolation of an analogous oxathiazoline intermediate.[50]

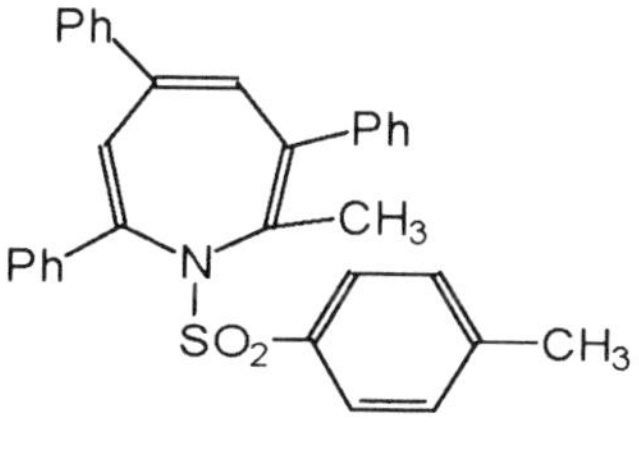

Azepines

(61)

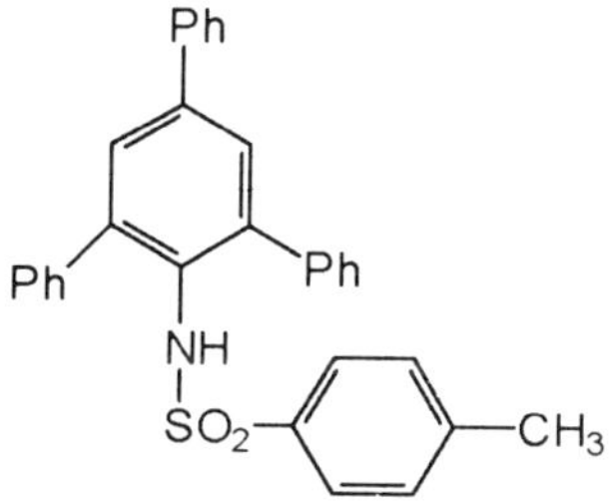

(62)

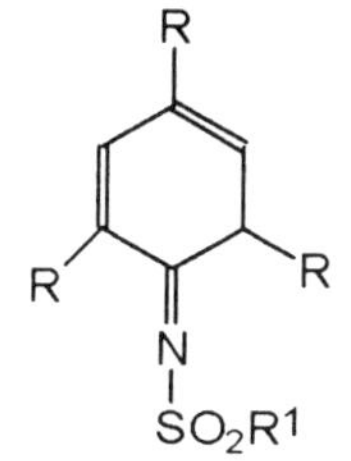

(63)

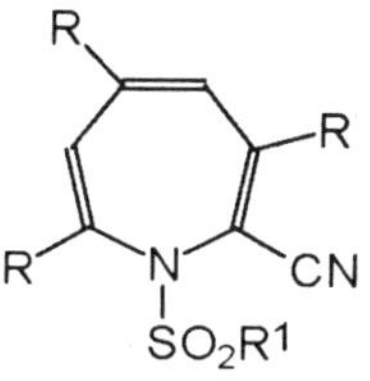

(64)

A similar ring-expansion reaction was subsequently reported, resulting from the action of potassium cyanide on the cyclohexa-2,4-diene-1-ylidene-sulphonamides **63** (R = CH_3, R^1 = CH_3, Ph, p-$CH_3C_6H_4$ and R = Ph, R^1 = p-$CH_3C_6H_4$) to yield a series of 1-sulphonyl-2-cyano[1H]azepines (**64**, R = CH_3, R^1 = CH_3, Ph, p-$CH_3C_6H_4$ and R = Ph, R^1 = p-$CH_3C_6H_4$).[51]

Other heterocyclic and benzenoid rings have been successfully expanded to give the [1H]azepine ring system.[52-55] These include the benzofuran **65**[52] and the azirine aldehyde **66**,[53] which gave the azepines **67** and **68**, respectively on UV irradiation, the cyclopropylbenzene **69**, which was elaborated to the

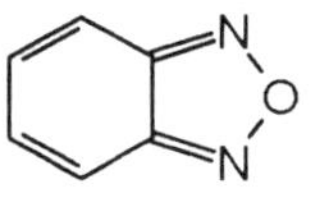

(65)

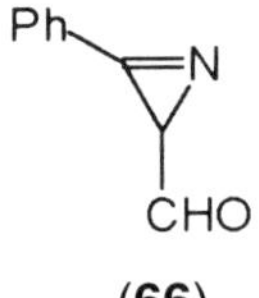

(66)

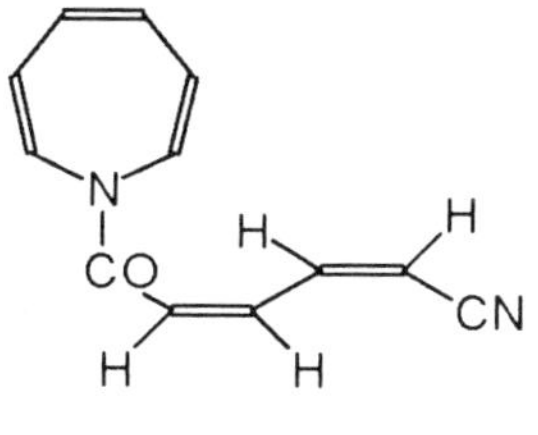

(67)

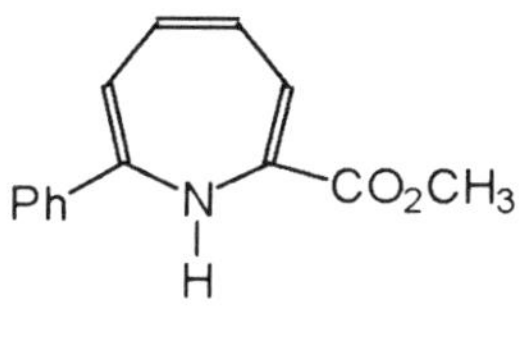

(68)

[1H]azepine **70** in the presence of diethylamine,[54] and the bicyclic compound **71** (R = ^{s}Bu), which on treatment with dimethyl acetylenedicarboxylate at 100 °C was converted into the tetramethyl[1H]azepine-3,4,5,6-tetracarboxylate **72**.[55]

(69) (70)

(71) (72)

A full list of [1H]azepines and their physical constants is given in Table 1.

PROPERTIES

Most of the substituted [1H]azepines which have been studied are solids with a clearly defined melting point, although a few have only been isolated as oils. The [1H]azepine ring system is stable to a variety of chemical reagents, for example, withstanding acid conditions which are strong enough to hydrolyse a 1-nitrile to a 1-carboxamide,[56] or treatment with Raney nickel to remove a 1-tosylate group.[57] Even at elevated temperatures many of the compounds are stable, as witnessed by their formation under pyrolytic conditions.[37,38,45–47,50] In some cases, however, pyrolytic rearrangement or dimerisation can occur depending on the pattern of substitution on the [1H]azepine ring. For example, the 2-methyl derivative **73** undergoes rearrangement at 200 °C to give a bicyclic ring system (**74**),[58] whereas the unsubstituted esters [**75**, R = CH$_3$, C$_2$H$_5$, C(CH$_3$)$_3$] dimerise under the same conditions to give the bridged system **76** [R = CH$_3$, C$_2$H$_5$, C(CH$_3$)$_3$)].[59] No dimerisation takes place in the presence of a 2-, 4- or 7-methyl group.[58]

By contrast, the highly substituted [1H]azepine **77** rearranges in boiling toluene to give the fulvene **78**.[60]

Photochemically induced rearrangement using UV light of a series of 2-, 3- and 4-methyl[1H]azepines resulted in the formation of [3.2.0] bicyclic ring

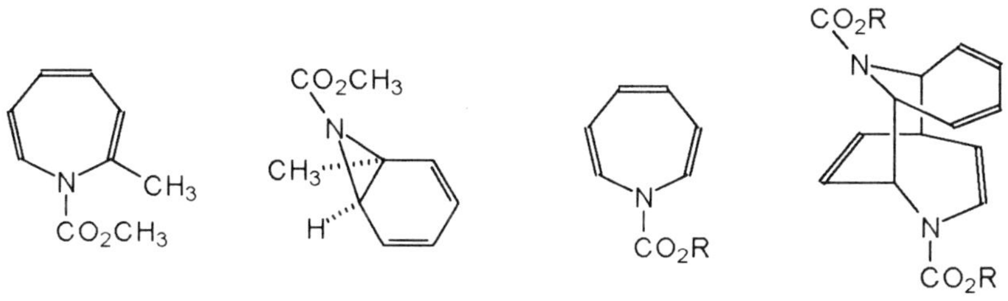

Azepines

(73)   (74)   (75)   (76)

systems. For example, the 2-methyl derivative **73** gave a 14:1 isomeric mixture of **79** and **80**.[61] The products arise from the formation of a new bond from C-2

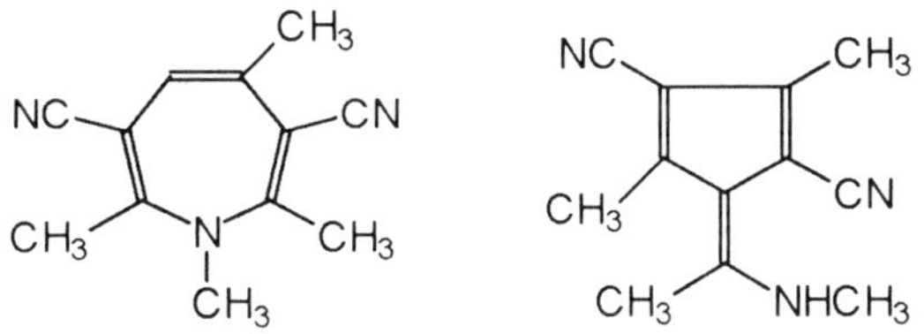

(77)   (78)

to C-5 or from C-4 to C-7 of the original [1H]azepine. The 3- and 4-methyl[1H]azepines **81** and **84** behave analogously but the products, **82, 83** and **85, 86**, respectively are formed in approximately 1:1 ratios.[61]

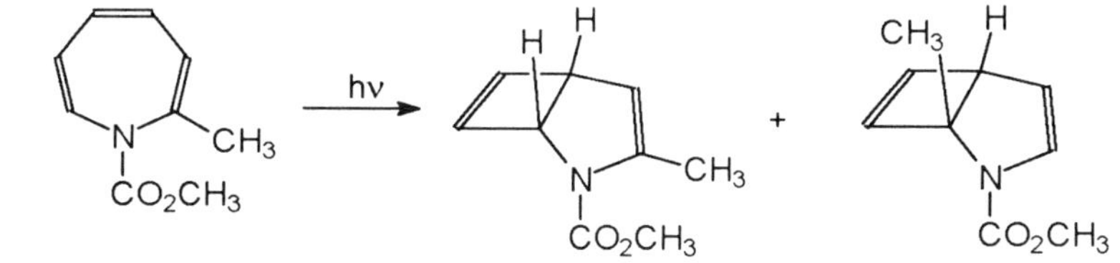

(73)   (79)   (80)

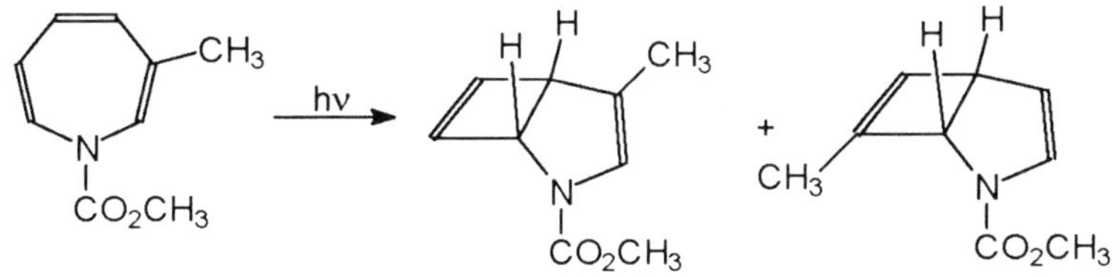

(81)   (82)   (83)

(84) —hv→ (85) + (86)

1,2-Addition to both the 2,3- and the 4,5-double bonds of the [1H]azepine system has been reported. For example, at room temperature in the presence of hexamethylphosphoric acid triamide (HMPA) and magnesium, 1-carbethoxy[1H]azepine (9) reacted with the chlorosilane derivatives (87, R = CH_3, Ph, and 88) to give the corresponding 4,5-*trans*-silyl derivatives 89 [R = $Si(CH_3)_2OH$, $SiPh_2OH$, $Si(CH_3)_3$, respectively].[62,63] On heating to 120 °C, the hydroxysilyl derivatives 89 [R = $Si(CH_3)_2OH$, $SiPh_2OH$] afforded

(9) Cl_2SiR_2 (87) $Cl_3Si(CH_3)$ (88)

(89)   (90)   (91)

the bicyclic compounds 90 (R = CH_3, Ph) in almost quantitative yield.[63] This addition reaction only occurred on the 4,5-double bond of the [1H]azepine ring system. By contrast, addition of acetoxy groups to the 2,3-double bond of 1-carbethoxy[1H]azepine (9) was reported to take place using $Pd(OAc)_2$ to give the 2,3-diacetoxy derivative (91). The yield, however, was only 19%.[64]

Cycloaddition reactions of *N*-substituted azepines to various dienophiles and dienes have been studied.[65–70] For example, tetracyanoethylene (92), when reacted with a range of 4-substituted [1H]azepines (84) and 93 (R = H, Br, Cl, F, OCH_3), gave addition across the 2,5-double bond and formed the corresponding [3.2.2] bicyclic products 94 (R = H, CH_3, Br, Cl, F, OCH_3).[65]

Azepines

The structure of the bromo compound **94** (R = Br) has been confirmed by X-ray crystallography.[66] The reaction has also been extended to the 2-methyl- (**73**)[65,67] and 3-methyl derivatives (**81**).[67] With the 2-methyl derivative (**73**), addition occurred over the 4,7-position to give **95**, whereas with the 3-methyl derivative (**81**) a mixture of isomers (**96** and **97**) resulted from 2,5- and 4,7-addition.[67]

N-Nitrosobenzene (**98**) has been reported to form cyclo-adducts with 1-carbethoxy[1H]azepine (**9**), to give a single adduct (**99**) in one case[68] and two isomeric adducts (**99** and **100**) in another.[69] By contrast, N,α-diphenyl nitrone (**101**) added across the 4,5-double bond of 1-carbomethoxy[1H]azepine (**93**, R = H) to give the bicyclic oxazolidine **102**[69] and, in a similar manner, tetrachlorocyclopentadiene dimethyl acetal (**103**) added to the carbethoxy analogue **9** to give the bridged ring system **104**.[70]

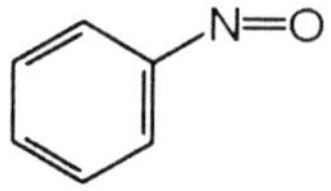

(98) 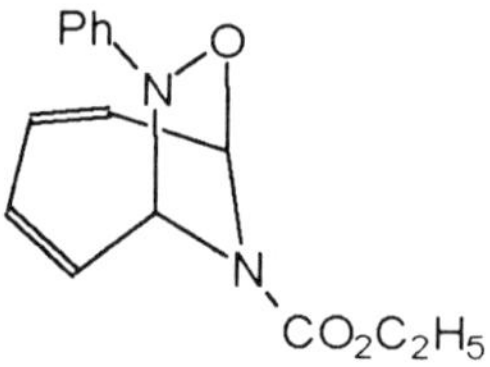 (99) 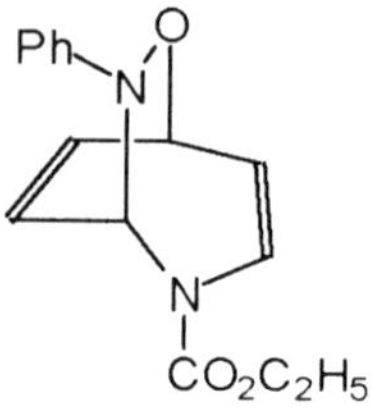 (100)

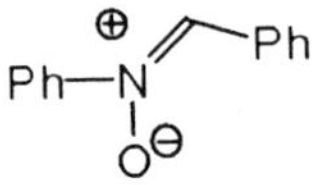

(101) 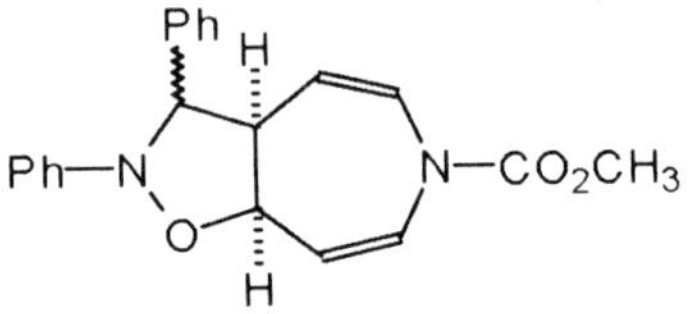 (102) 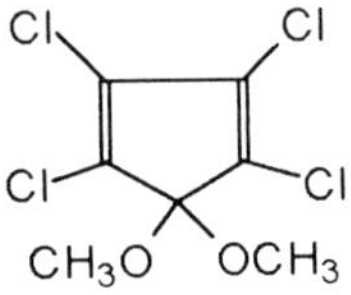 (103)

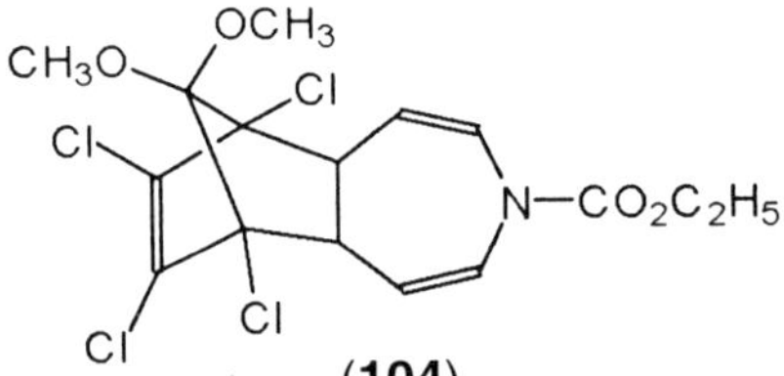

(104)

Table 1 gives details of the [1H]azepines described.

Table 1. [1H]Azepines

Structure	M.p. (°C)/b.p. (°C/mmHg)	Spectroscopic evidence/data	Ref.
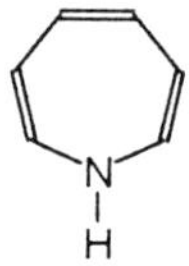 azepine N-H		^{1}H, ^{13}C NMR	6 8
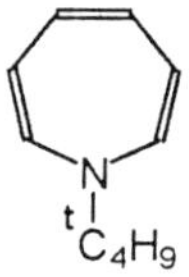 azepine N-^{t}C$_4$H$_9$		^{1}H NMR ^{1}H, ^{13}C NMR	47b 47a
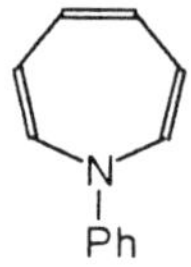 azepine N-Ph	49–49.5	UV, ^{1}H NMR	28 45

Table 1 (continued)

Structure	M.p. (°C)/b.p. (°C/mmHg)	Spectroscopic evidence/data	Ref.
(azepine-N-triazine with two OCH₃)	105–106	UV, MS, ^{1}H NMR	30
(azepine-N-CO₂H)		^{1}H, ^{13}C NMR	8
(azepine-N-CO₂CH₃)	59–61/0.2 62–63/0.05	n_D^{25} 1.5379, IR, ^{1}H NMR, UV, MS ^{13}C NMR	36 11
(azepine-N-CO₂C₂H₅)	130/20	IR, ^{1}H NMR UV, ^{1}H NMR UV, IR ^{13}C NMR MS, ^{1}H NMR claimed	33 7 10 11 13 9, 12, 26b, 26c, 32
(azepine-N-CO₂Si(CH₃)₃)	63/0.02		8
(azepine-N-C(OC₂H₅)=NCN)		MS, ^{1}H NMR claimed	13
(azepine-N-CO-CH=CH-CH=CH-CN)	121–122	UV, IR, MS, ^{1}H NMR	52

Table 1 (continued)

Structure	M.p. (°C)/b.p. (°C/mmHg)	Spectroscopic evidence/data	Ref.
1-(CON₃)-azepine			12
1-(CONH₂)-azepine			12
1-(CONHCH₂Ph)-azepine		^{1}H NMR	12
1-(CN)-azepine	49–50/2	IR, UV, MS, ^{1}H NMR $n_D^{25.2}$ 1.5520, IR, ^{1}H NMR	14 15
1-(SO₂CH₃)-azepine	91.5–92.5	IR, ^{1}H NMR, UV, MS ^{1}H, ^{13}C NMR	36 8
1-(SO₂Ph)-azepine	132–133 132–133 133	 IR, ^{1}H NMR, UV, MS IR, ^{1}H NMR, MS	18 36 19
1-(SO₂CH₂CH₂Ph)-azepine	35–36	IR, ^{1}H NMR, MS	27

Table 1 (continued)

Structure	M.p. (°C)/b.p. (°C/mmHg)	Spectroscopic evidence/data	Ref.
N-SO₂-C₆H₄-CH₃ (p-tolyl)	169 (d) 167 (d) 167 169 167	¹H NMR IR, ¹H NMR, UV, MS IR, ¹H NMR, MS IR, UV, ¹H NMR	43 20a 19 39 18, 20b
N-SO₂-C₆H₄-OCH₃	111–112 110–111	IR, ¹H NMR, MS IR, UV, ¹H NMR, MS	19 18
N-SO₂-C₆H₄-F	Semi-solid	IR, ¹H NMR, MS	19
N-SO₂-C₆H₄-Cl	122–123 121–122	IR, ¹H NMR, MS IR, UV, ¹H NMR, MS	19 18
N-SO₂-C₆H₄-Br	132.5–134 132 132	IR, ¹H NMR, UV IR, ¹H NMR, MS	36 19 18
N-SO₂-C₆H₄-NO₂	175* 145* (*Different m.p. quoted by same authors)	IR, ¹H NMR, MS IR, UV, ¹H NMR, MS	19 18

Table 1 (continued)

Structure	M.p. (°C)/b.p. (°C/mmHg)	Spectroscopic evidence/data	Ref.
1-(4-aminophenylsulfonyl)azepine	170–171	IR, UV, ^{1}H NMR, MS	18
1-trimethylsilylazepine		^{1}H, ^{13}C NMR	8
1-(diphenylphosphinoyl)azepine	151.5–153	IR, ^{1}H NMR, UV, MS	36
2-acetoxy-1H-azepine	144	UV, IR, ^{1}H NMR	57
2-methyl-1-(methoxycarbonyl)azepine	62–64/0.1	UV, ^{1}H NMR n_D^{25} 1.5264, IR, ^{1}H NMR, UV, MS	35 36
2-methyl-1-(ethoxycarbonyl)azepine		^{1}H NMR, C,H,N analysis IR, ^{1}H NMR	21 33 32
2-ethyl-1-(ethoxycarbonyl)azepine		^{1}H NMR, C,H,N analysis	21

Table 1 (continued)

Structure	M.p. (°C)/b.p. (°C/mmHg)	Spectroscopic evidence/data	Ref.
n-C_3H_7 azepine, N-$CO_2C_2H_5$		^{1}H NMR, C,H,N analysis	21
i-C_3H_7 azepine, N-$CO_2C_2H_5$		^{1}H NMR, C,H,N analysis	21
n-C_4H_9 azepine, N-$CO_2C_2H_5$		^{1}H NMR, C,H,N analysis	21
2-(2-hydroxyphenyl)azepine, N-$CONHC_2H_5$	121–122	IR, ^{1}H NMR	34
2-CH_3 azepine, N-SO_2-C$_6$H$_4$-Br	94.5–95.5	IR, ^{1}H NMR, UV	36
2-CH_3O azepine, N-SO_2-C$_6$H$_4$-CH_3	125	IR, ^{1}H NMR, MS, UV	17

Table 1 (continued)

Structure	M.p. (°C)/b.p. (°C/mmHg)	Spectroscopic evidence/data	Ref.
(1-tosyl-2-methoxycarbonyl azepine)	92	IR, UV, MS, ^{1}H NMR	16
(1-tosyl-2-nitro azepine)		IR, ^{1}H NMR, MS, UV claimed	17
(3-methyl-1-methoxycarbonyl azepine)	62–65/0.15–0.25	UV, ^{1}H NMR IR, ^{1}H NMR, UV, MS	35 36
(3-methyl-1-ethoxycarbonyl azepine)		IR, ^{1}H NMR	33 32
(4-methyl-1-methoxycarbonyl azepine)	62–65/0.2	UV, ^{1}H NMR n_D^{25} 1.5338, IR, ^{1}H NMR UV, MS	35 36
(4-methyl-1-ethoxycarbonyl azepine)		IR, ^{1}H NMR	33 32

Azepines 507

Table 1 (continued)

Structure	M.p. (°C)/b.p. (°C/mmHg)	Spectroscopic evidence/data	Ref.
4-methyl-1-(2-allyl-4-methylbenzoyloxy)azepine		^{1}H, ^{13}C NMR, IR, MS	29
4-methyl-1-(2-allyl-6-methylbenzoyloxy)azepine		^{1}H, ^{13}C NMR, IR, MS	29
Ethyl-ethoxycarbonyl-azepine (Isomeric mixture)		GC/MS	26a
Fluoro-cyano-azepine (Isomeric mixture)	25/0.03–0.07	IR, ^{1}H, ^{19}F NMR	15 14
Methyl-cyano-azepine (Isomeric mixture)		^{1}H NMR	15 14

Table 1 (continued)

Structure	M.p. (°C)/b.p. (°C/mmHg)	Spectroscopic evidence/data	Ref.
CF$_3$-substituted azepine with N-CN; Isomeric mixture	30–40/0.02	IR, MS	15, 14
CO$_2$CH$_3$-substituted azepine with N-CN; Isomeric mixture	55/0.1	IR, ^{1}H NMR	15, 14
CH$_3$-substituted azepine with N-(dimethoxytriazinyl); Isomeric mixture	101–102	UV, MS, ^{1}H NMR	30
C$_2$H$_5$-substituted azepine with N-(dimethoxytriazinyl); Isomeric mixture	87–88	UV, MS, ^{1}H NMR	30
CH$_2$CN-substituted azepine with N-(dimethoxytriazinyl); Isomeric mixture	107–108	UV, MS, ^{1}H NMR	30

Table 1 (continued)

Structure	M.p. (°C)/b.p. (°C/mmHg)	Spectroscopic evidence/data	Ref.
CF$_3$-substituted azepine with 4,6-dimethoxy-1,3,5-triazinyl group (Isomeric mixture)	109–110	UV, MS, ^{1}H NMR	30
F-substituted azepine with 4,6-dimethoxy-1,3,5-triazinyl group (Isomeric mixture)	117–118	UV, MS, ^{1}H NMR	30
Cl-substituted azepine with 4,6-dimethoxy-1,3,5-triazinyl group (Isomeric mixture)	159–160	UV, MS, ^{1}H NMR	30
Cyclopropyl-substituted azepine with N(C$_2$H$_5$)$_2$ group		n_D^{25} 1.5295, IR, ^{1}H NMR	54
Azepine with CO$_2$CH$_3$ and CH$_3$O$_2$C substituents	112	IR, UV, ^{1}H NMR	57

Table 1 (continued)

Structure	M.p. (°C)/b.p. (°C/mmHg)	Spectroscopic evidence/data	Ref.
CH₃O₂C–[azepine]–Ph	156–157	IR, UV, MS, ¹H NMR	53
CH₃O₂C–[azepine(CO₂CH₃)]–NH	104	IR, UV, MS, ¹H NMR	44
[4-CH₃, 2-CH₃ azepine N-CO₂C₂H₅]		¹H NMR	24
[4-iC₃H₇, 2-iC₃H₇ azepine N-CO₂CH₃]	Pale yellow oil	IR, UV, ¹H, ¹³C NMR, MS	71c
[4-tC₄H₉, 2-tC₄H₉ azepine N-CO₂CH₃]	56–57	IR, UV, ¹H, ¹³C NMR, MS	71b
[4-CH₃, 2-CH₃ azepine N-CO₂C₂H₅]		¹H NMR	24
[4-Cl, 2-CH₃ azepine N-CO₂C₂H₅]		¹H NMR, C,H,N,Cl analysis	21

Table 1 (continued)

Structure	M.p. (°C)/b.p. (°C/mmHg)	Spectroscopic evidence/data	Ref.
4-Br, 7-CH$_3$, N-CO$_2$C$_2$H$_5$ azepine		^{1}H NMR, C,H,N,Br analysis	21
4-CO$_2$CH$_3$, 7-CO$_2$CH$_3$, N-CO$_2$C$_2$H$_5$ azepine	94–95 93	IR, MS, ^{1}H NMR IR, UV, MS, ^{1}H NMR	22 25
4-CO$_2$H, 7-CO$_2$H, N-SO$_2$-C$_6$H$_4$-CH$_3$ azepine	253	MS, IR	16
4-CO$_2$CH$_3$, 7-CO$_2$CH$_3$, N-SO$_2$-C$_6$H$_4$-CH$_3$ azepine	127 126	IR, ^{1}H NMR IR, UV, MS, ^{1}H NMR	22 16
4-CO$_2$C$_2$H$_5$, 7-CO$_2$C$_2$H$_5$, N-SO$_2$-C$_6$H$_4$-CH$_3$ azepine	69	IR, UV, MS, ^{1}H NMR	33
4-CO$_2$CH$_3$, 7-CO$_2$CH$_3$, N-SO$_2$-C$_6$H$_4$-N=N-C$_6$H$_3$(OH)(CH$_3$) azepine	191 (d)	IR, ^{1}H NMR, UV, MS	31

Table 1 (continued)

Structure	M.p. (°C)/b.p. (°C/mmHg)	Spectroscopic evidence/data	Ref.
2,7-dimethyl-1-methyl-azepine	Decomp.	IR, ^{1}H NMR, UV	36
2,7-dimethyl-1-(CO$_2$CH$_3$)-azepine	53–54	IR, ^{1}H NMR, UV, MS	36
3,4-dimethyl-1-(CO$_2$C$_2$H$_5$)-azepine		^{1}H NMR	24
3,5-dimethyl-1-(CO$_2$C$_2$H$_5$)-azepine		^{1}H NMR	24
3,6-dimethyl-1-(CO$_2$CH$_3$)-azepine	29–31	IR, ^{1}H NMR, UV, MS ^{13}C NMR	36 11
3,6-di-iC$_3$H$_7$-1-(CO$_2$CH$_3$)-azepine	Pale yellow oil	^{13}C NMR IR, ^{1}H NMR, UV, MS	11 71c
3,6-di-tC$_4$H$_9$-1-(CO$_2$CH$_3$)-azepine	Yellow oil	^{13}C NMR IR, ^{1}H NMR, UV, MS	11 71b

Table 1 (continued)

Structure	M.p. (°C)/b.p. (°C/mmHg)	Spectroscopic evidence/data	Ref.
3,6-dimethyl-1-(ethoxycarbonyl)azepine			24
3,6-di-t-butyl-1-(ethoxycarbonyl)azepine	67–69	UV, ^{1}H NMR	23
3,6-di-t-butyl-1-(cyclopropylmethoxycarbonyl)azepine	41–43	UV, ^{1}H NMR	23
3,6-di-t-butyl-1-(t-butoxycarbonyl)azepine	100–102	UV, ^{1}H NMR	23
3,6-di-t-butyl-1-(4-methoxyphenoxycarbonyl)azepine	142–144	UV, ^{1}H NMR	23
3,6-bis(trifluoromethyl)-1-(methoxycarbonyl)azepine	Yellow oil	IR, UV, ^{1}H NMR	40
3,6-bis(trifluoromethyl)-1-(4-methylphenylsulfonyl)azepine	137–138	IR, UV, ^{1}H NMR	40, 39

Table 1 (continued)

Structure	M.p. (°C)/b.p. (°C/mmHg)	Spectroscopic evidence/data	Ref.
3,6-dichloro-1-(p-tolylsulfonyl)azepine	136	^{1}H, ^{13}C NMR ^{1}H, ^{13}C NMR, IR, UV	42 39
4,5-dimethyl-1-(methoxycarbonyl)azepine	60–61	IR, ^{1}H NMR, UV, MS	36
4,5-bis(methoxycarbonyl)-1-acetylazepine	Yellow oil	UV, ^{1}H NMR UV, IR, MS, ^{1}H NMR	38 40 39, 41
4,5-bis(methoxycarbonyl)-1-(methoxycarbonyl)azepine		UV, ^{1}H NMR n_D^{25} 1.5112, MS, UV, ^{1}H NMR	40 37
4,5-dimethyl-1-(ethoxycarbonyl)azepine		^{1}H NMR	24
4,5-bis(methoxycarbonyl)-1-(allyloxycarbonyl)azepine	Oil	IR, ^{1}H NMR	39

Table 1 (continued)

Structure	M.p. (°C)/b.p. (°C/mmHg)	Spectroscopic evidence/data	Ref.
4,5-bis(CO₂CH₃), N-CONH₂ azepine	Yellow oil	^{1}H NMR	40
4,5-bis(CO₂CH₃), N-(4-nitrophenyl) azepine	174–175	UV, IR, ^{1}H NMR	40
4,5-bis(CO₂CH₃), N-SO₂-CH₃ azepine	90–91	UV, IR, ^{1}H NMR	40
4,5-bis(CO₂CH₃), N-SO₂-(4-tolyl) azepine	111 111	UV, MS, ^{1}H NMR UV, IR, ^{1}H NMR UV, ^{1}H NMR	38 40 16 39
4,5-bis(CF₃), N-CO₂CH₃ azepine	Yellow oil	UV, IR, ^{1}H NMR	40

Table 1 (continued)

Structure	M.p. (°C)/b.p. (°C/mmHg)	Spectroscopic evidence/data	Ref.
CF$_3$, CF$_3$ azepine N-SO$_2$-C$_6$H$_4$-CH$_3$	111.5–112.5	UV, IR, ^{1}H NMR	40
CH$_3$, CH$_3$ azepine N-CN (Isomeric mixture)		^{1}H NMR	15, 14
CF$_3$, CF$_3$ azepine N-CN (Isomeric mixture)	35–40/0.02	$n_D^{25.2}$ 1.4307–1.4320, IR, ^{1}H, ^{19}F NMR	15, 14
Ph, Ph, Ph, CH$_3$ azepine N-COCH$_3$	134–135	IR, ^{1}H, ^{13}C NMR	50
Ph, Ph, Ph, CH$_3$ azepine N-COPh	149–150	IR, ^{1}H, ^{13}C NMR	50

Table 1 (continued)

Structure	M.p. (°C)/b.p. (°C/mmHg)	Spectroscopic evidence/data	Ref.
2,3,5-triphenyl-azepine with N-CO-C₆H₄-CH₃ (p-tolyl), 7-Ph, 2-CH₃	145–146	IR, ^{1}H, ^{13}C NMR	50
2,3,5-triphenyl-azepine with N-CO-C₆H₄-OCH₃, 7-Ph, 2-CH₃	154–155	IR, ^{1}H, ^{13}C NMR	50
2,3,5-triphenyl-azepine with N-CO-C₆H₄-Br, 7-Ph, 2-CH₃	138–139	IR, ^{1}H, ^{13}C NMR, X-ray	50
2,3,5-triphenyl-azepine with N-SO₂-C₆H₄-CH₃, 7-Ph, 2-CH₃	182–184	IR, ^{1}H, ^{13}C NMR	50
3,4-dimethyl-5-methyl-7-methyl-2-cyano azepine with N-SO₂-CH₃	200	IR, MS, ^{1}H, ^{13}C NMR	51

Table 1 (continued)

Structure	M.p. (°C)/b.p. (°C/mmHg)	Spectroscopic evidence/data	Ref.
(3,5-dimethyl-7-methyl-2-cyano-1-phenylsulfonyl azepine)	161	IR, MS, ^{1}H, ^{13}C NMR	51
(3,5-dimethyl-7-methyl-2-cyano-1-(p-tolylsulfonyl) azepine)	141	IR, MS, ^{1}H, ^{13}C NMR	51
(3,5,7-triphenyl-2-cyano-1-(p-tolylsulfonyl) azepine)	134	IR, MS, ^{1}H, ^{13}C NMR	51
(4,5-bis(methoxycarbonyl)-2-(propargyloxymethyl)-1-(p-tolylsulfonyl) azepine)		^{1}H NMR	39
(3,5-bis(methoxycarbonyl)-2,6-dimethyl-1-methyl azepine)	105–107	IR, UV, ^{1}H NMR	48b, 48a

Table 1 (continued)

Structure	M.p. (°C)/b.p. (°C/mmHg)	Spectroscopic evidence/data	Ref.
	156–157	IR, UV, ^{1}H NMR	40
	142–143	MS, UV, ^{1}H NMR	40
		^{1}H, ^{13}C NMR ^{1}H, ^{13}C NMR, UV, IR, X-ray	42 39
	135–137	IR, MS, ^{1}H NMR	55
	97–99.5	IR, UV, ^{1}H NMR	49
	110–111	IR, UV, ^{1}H NMR	49

Table 1 (continued)

Structure	M.p. (°C)/b.p. (°C/mmHg)	Spectroscopic evidence/data	Ref.
[1-(4-methoxyphenyl)-2,5-dimethyl-3,6-dicyano-4-methyl azepine]		^{1}H NMR	49
[1-methyl-2,5-dimethyl-3,6-dicyano-4-(chloromethyl) azepine]		^{1}H NMR	49
[perfluoro-1-phenyl azepine]	64–65	UV, ^{19}F NMR	46
[perfluoro-1-cyano azepine]	51–52 51–52	IR, UV, MS, ^{19}F NMR	15 14
[perfluoro-1-(ethoxycarbonyl) azepine]	58/0.05	IR, UV, ^{1}H, ^{19}F NMR	46

Table 1 (continued)

Structure	M.p. (°C)/b.p. (°C/mmHg)	Spectroscopic evidence/data	Ref.
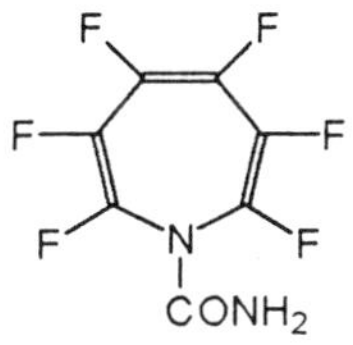 (perfluoro-azepine with CONH$_2$ on N)	102–103	IR, UV, MS, ^{19}F NMR	56

II. [2H]AZEPINES

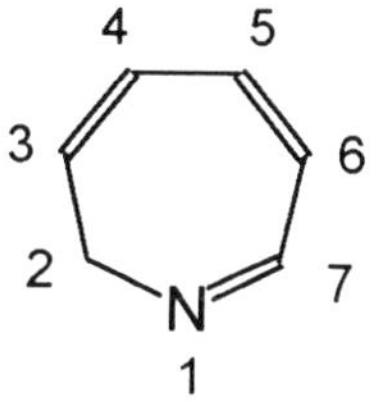

The parent compound in this series has not been prepared. 3,5-Di-*tert*-butyl[2H]azepine (106) has been formed, along with the [3H] and [4H] isomers, by heating the corresponding [1H]azepine (105) with DBU in xylene. The isomers were separated and characterised by NMR.[71a,c]

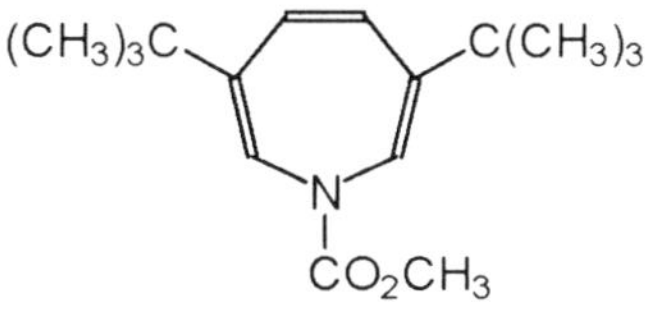

(105) 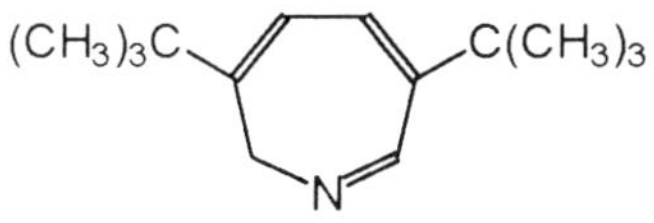   (106)

Two pungent compounds isolated from the mushroom *Chalciporus piperatus* have been identified as the substituted [2H]azepines chalciporone (107) and

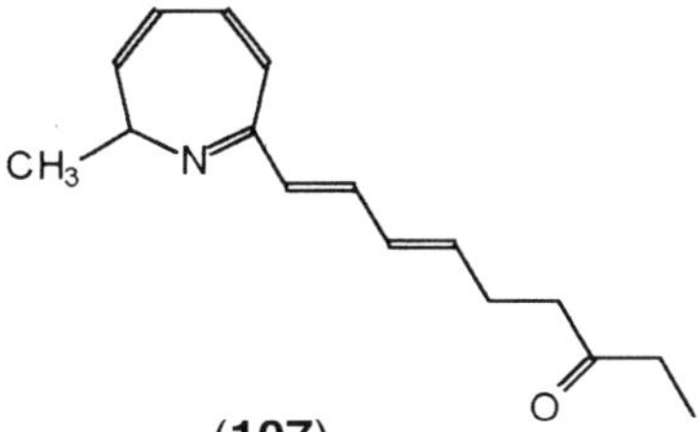

(107) 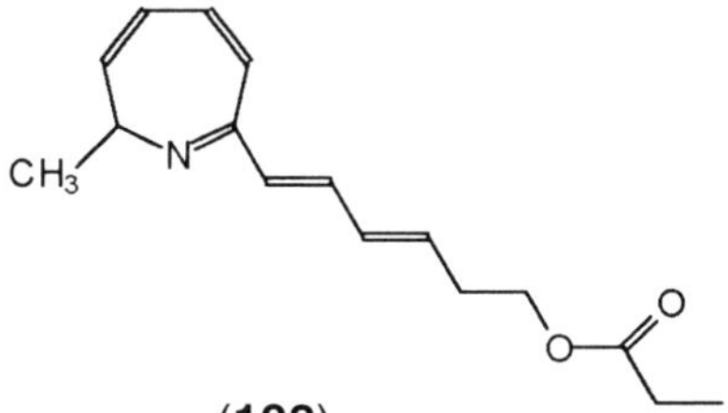   (108)

norchalciporyl propionate (**108**).⁷² Physical constants of these and the reduced derivative **109** are given in Table 2. Other than the 2-keto derivatives discussed in Chapter 12, these are the only [2H]azepines described in the literature.

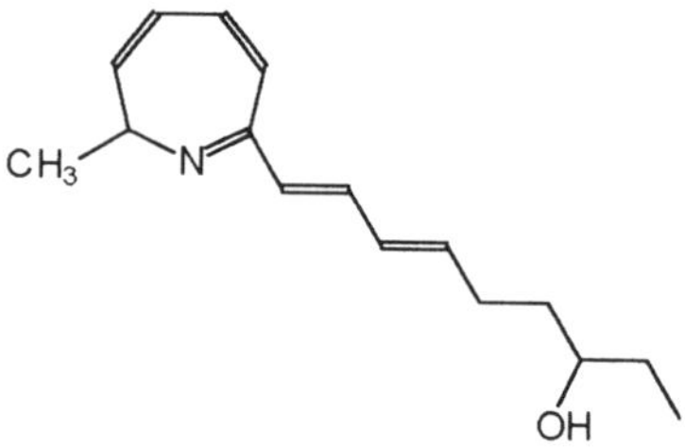

(**109**)

Table 2. [2H]Azepines

Structure	M.p. (°C)	Spectroscopic evidence/data	Ref.
(CH₃)₃C / C(CH₃)₃ azepine	68.5–69	¹H, ¹³C NMR IR, UV, ¹H,¹³C NMR	71a 71c
CH₃-azepine-dienone		¹H, ¹³C NMR, IR, UV, MS, $[\alpha]_D^{22}$ −452°	72
CH₃-azepine-dienyl ester		¹H NMR, IR, UV, MS, $[\alpha]_D^{22}$ −290°	72
CH₃-azepine-dienyl-OH		¹H NMR, IR, UV, MS, $[\alpha]_D^{22}$ −351°	72

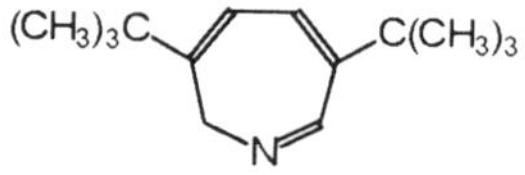

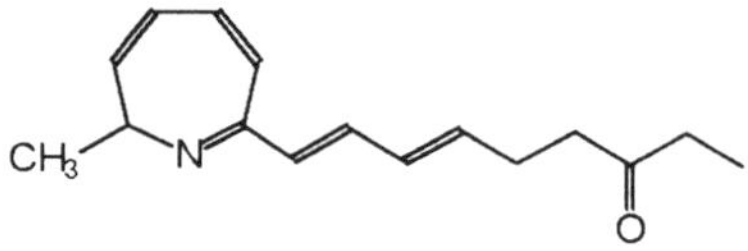

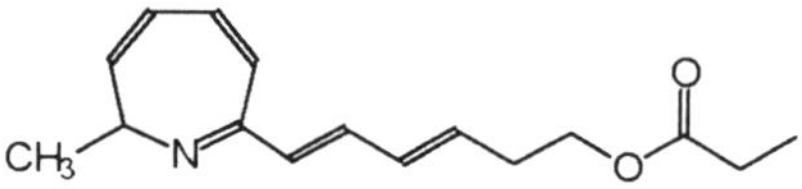

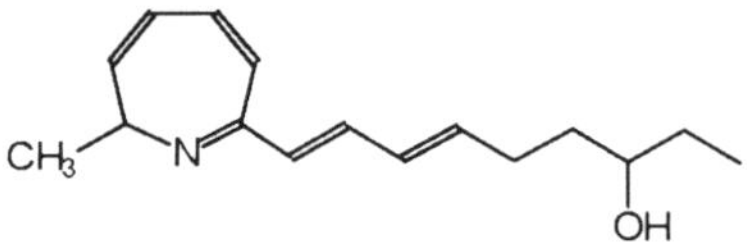

III. [3H]AZEPINES

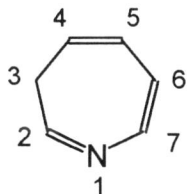

PREPARATION

Evidence for the transient existence of the parent [3H]azepine (**3**) has been presented by Schaden,[73-75] who demonstrated its formation from the alcohol (**110**) by pyrolysis at temperatures ranging from 150 to 900 °C. Using Curie point pyrolysis at the latter temperature, he showed the presence of [3H]azepine (**3**) by mass spectrometry which gave a band corresponding to $M = 93$.

The most frequently used technique for the preparation of the [3H]azepine system involves the ring expansion of a nitrene (**111**) either thermally or utilising UV irradiation and trapping the intermediate with a suitable nucleophilic reagent. Various methods have been used to generate the nitrene **111**, the most popular being the decomposition of the corresponding azide **112**. The earliest reported use of this procedure by Wolff in 1912 was not recognised as producing an azepine,[1] but when it was applied by Huisgen and co-workers in 1958, the thermal decomposition of the azides **113** (R = H, CH$_3$) in the presence of aniline gave the 2-phenylamino[3H] azepines **114** (R = H, CH$_3$).[2] Unfortunately the authors were unable to distinguish between the different possible isomeric [2H]-, [3H]- or [4H]azepines and the products were only confirmed as the [3H] isomers using ^{1}H NMR by Doering and Odum some 8 years later.[4]

(**111**) (**112**) (**113**) (**114**)

Further work demonstrated that treatment of phenyl azide (**113**, R = H) with 3-substituted anilines and aliphatic amines gave analogous products (**115**, R^1 = 3-$CH_3C_6H_4$, ^{i}Pr, $R^2 = R^3$ = H)[76] and it was later shown that irradiation at UV wavelengths rather than thermolysis could be used as an alternative means of generating the nitrene from the azide **113** (R = H) to give the corresponding [3H]azepine **114** (R = H).[77]

(**115**)　　　　　(**116**)

Subsequent investigators confirmed the general nature of the reaction, using a variety of phenylnitrenes (**111**, R = H, alkyl, aryl, acyl, halogen, etc.) generated from phenyl azides (**112**) or from substituted bicyclic isoxazoles (**117**), and a selection of amines such as ammonia,[4] diethylamine,[4,78–93,155] dimethylamine,[94] n-butylamine,[78,95] di-n-butylamine,[82–84,87] piperidine[96–98] and morpholine.[98] In general, it was found that if the phenyl nitrene **111** was substituted in the 4-position, then the corresponding 5-substituted [3H]azepine was formed,[88,90,92,94,98] the 2-substituted phenylnitrenes (**111**) gave 3-substituted [3H]azepines[79,84,155] and 3-substituted phenylnitrenes (**111**) gave mixtures of 4- and 6-substituted [3H]azepines.[89,98] With multiple substitution the pattern was less clearly defined.[91,93] Full details of the structures obtained (**115**) are given in Table 3.

(**117**)

Treatment of phenylnitrenes **111** (R = H, alkyl, aryl, acyl, halogen, etc.), formed from either the corresponding azides (**112**) or from the bicyclic isoxazoles (**117**), with an alcohol in place of an amine to terminate the reaction sequence led to the corresponding 2-alkoxy- or 2-aryloxy[3H]azepines. For example, 2-alkoxy[3H]azepines (**116**, R^1 = alkyl) were obtained using methanol,[78,86,96,98–108] ethanol,[86,108] n-propanol[86] and cyclohexanol,[86] and 2-aryloxy[3H]azepines (**116**, R^1 = Ph, β-naphthyl) have been prepared using phenol[86,109] or 2-methoxynaphthalene.[154] *Ortho-*, *meta-* and *para*-substituted

Azepines

phenyl azides (112) gave substituted 2-alkoxy[3H]azepines (116) in which the substitution patterns resembled that of the 2-amino[3H]azepines discussed above.[86,98,100–108] Full structural details and physical constants of these products are also listed in Table 3.

There has been considerable discussion in the literature[87] regarding the nature of the intermediate formed by thermolysis or irradiation of a nitrene (111). Some authors claim evidence in support of a bicyclic intermediate (118)[2,4,81,82,85,99,101] whereas others prefer the dehydroazepine structure (119).[79,88,93] Neither intermediate, however, has been isolated and arguments in support of the involvement of both structures have been made.[87]

(118)   (119)   (120)   (121)

Nitrosobenzene (120) has been deoxygenated with triphenyl- or tri-n-butylphosphine as an alternative procedure for generating phenylnitrene (111) and in the presence of an appropriate amine has yielded the 2-substituted amino[3H]azepines 115 ($R^1 = CH_3$, $R^2 = H$, CH_3, $R^3 = H$; $R^1 = {}^nBu$, $R^2 = R^3 = H$; $R^1 = R^2 = C_2H_5$, $R^3 = H$).[110] Similar deoxygenation of nitrobenzene (121, R^1 to $R^5 = H$) with tri-n-butylphosphine in an alcohol has yielded a range of 2-alkoxy[3H]azepines (116, $R^1 = C_1–C_7$ alkyl, $R^2 = H$).[111] Nitrobenzenes (121, R^1 to $R^5 = H$;[112–114] $R^1 = CH_3$ or $R^2 = CH_3$ or $R^1 = R^3 = R^5 = CH_3$, other values of $R = H$;[112] $R^3 = CH_3$ or $R^2 = R^3 = CH_3$, other values of $R = H$;[114] and $R^1 = CH_3$ or $R^3 = CH_3$, Br, $CO_2C_2H_5$, other values of $R = H$[115,116]) on deoxygenation in a similar manner in the presence of diethylamine using $P(OC_2H_5)_3$, $CH_3P(OC_2H_5)_2$ or $P(NEt_2)_3$ yield the corresponding 2-ethylamino[3H]azepines 115 ($R^1 = R^2 = C_2H_5$). In the absence of an amine, deoxygenation of substituted nitrobenzenes with trimethyl or triethyl orthophosphate gave poor yields of the corresponding 7-phosphate esters (122, $R^1 = H$, CH_3, C_2H_5, $R^2 = CH_3$, C_2H_5).[116–118]

(122)   (123)

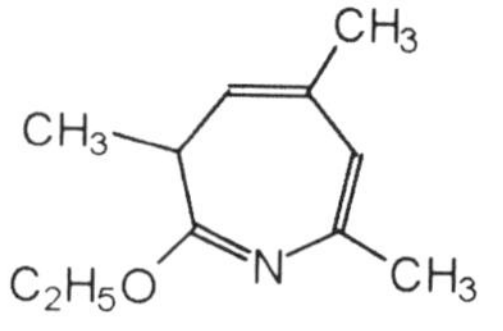

(124) 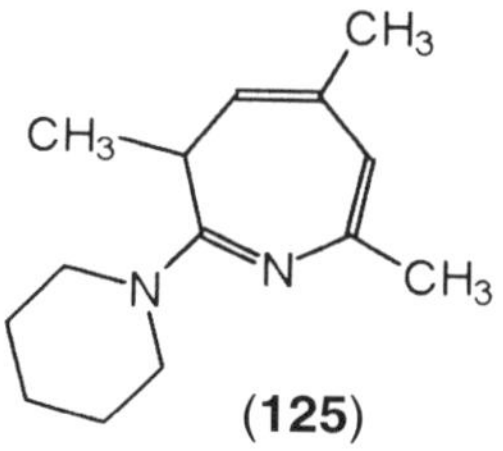 (125)

Since 2-alkoxy- and 2-amino[3H]azepines (**116** and **117**) are enol ethers and enamines, respectively, of 1,3-dihydro[2H]azepin-2-ones (**123**), this offers an alternative means of access to 2-substituted [3H]azepines. The unsubstituted lactam **123** (R = H) has been treated with $(CH_3)_3O^+ BF_4^-$ to give 2-methoxy-[3H]azepine (**116**, $R^1 = CH_3$, $R^2 = H$)[119] and 3,5,7-trimethyl-1,3-dihydro[2H]-azepin-2-one (**123**, R = CH_3) has been treated in a similar manner with $(C_2H_5)_3O^+ BF_4^-$ to give the corresponding 2-ethoxy-3,5,7-trimethyl[3H]-azepine (**124**).[120,121] Treatment of the 2-methoxy[3H]azepine **116** ($R^1 = CH_3$, $R^2 = H$) with aniline and the 2-n-butoxy[3H]azepine **116** ($R^1 = {}^nBu$, $R^2 = H$) with pyrrolidine, piperidine and morpholine led to the corresponding amines **115** ($R^1 = Ph$, $R^2 = R^3 = H$)[119] and R^1 and R^2 together = $-C_4H_8-$, $-C_5H_{10}-$ and $-C_2H_4OC_2H_4-$, $R^3 = H$,[122] respectively). Similar treatment of the 3,5,7-trimethyl analogue **124** with piperidine gave the 2-piperidyl derivative **125**.[120,121]

Other ring-expansion procedures have been used to produce [3H]azepines. For example, Nair in 1972[123] and Hassner and Anderson from 1971 to 1974[124–127] reacted a series of 3,4-diphenylcyclopentadienones (**126**, $R^1 = CH_3$, C_2H_5, Ph) with a range of substituted azirines (**127**, $R^2 = Ph$, CH_2Ph, C_2H_5; $R^3 = H$, CH_3, C_2H_5, Ph)[123,124,126,127] or with the azides **128** ($R^3 = Ph$, ${}^tBu$)[125] to give the corresponding substituted [3H]azepines **129** ($R^1 = CH_3$, C_2H_5, Ph; $R^2 = Ph$, CH_2Ph, C_2H_5; $R^3 = H$, CH_3, C_2H_5, Ph, ${}^tBu$) as shown in Scheme 5.

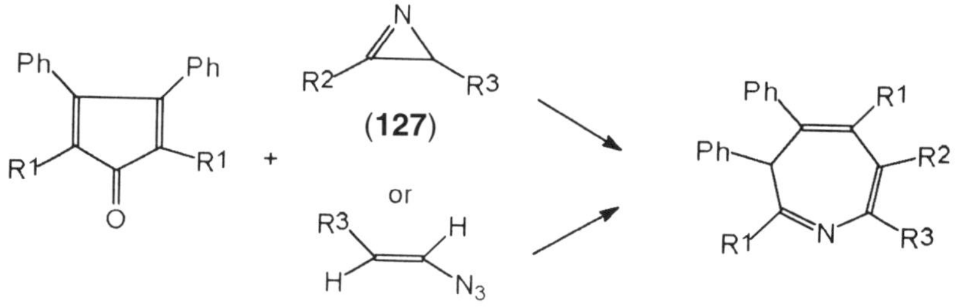

(126)   (128)   (129)

Scheme 5

Ring expansion of the 1,4-dihydropyridines **130** (R = CH_3, C_2H_5) using sodium ethoxide in ethanol gave initially the dihydro[1*H*]azepines **131** (R = CH_3, C_2H_5) and the [4*H*]azepines **132** (R = CH_3, C_2H_5), but at reflux temperature the products rearranged to give the corresponding [3*H*]azepines **133** (R = CH_3, C_2H_5).[128] In a similar manner, the 1,2-dihydropyridine **134** gave the [3*H*]azepine **135** on treatment with pyridine at 100 °C.[129] In this case the formation of a [4*H*]azepine did not appear to be involved.

(130)   (131)   (132)

(133)   (134)   (135)

A range of 1,2,4-triazines (**138**, R^1 = CH_3, Ph, R^2 = H, CH_3, Ph), prepared by condensation of the corresponding diketones (**137**, R^1 = CH_3, Ph, R^2 = H, CH_3, Ph) with the amidrazone **136**, was used by Sauer and his co-workers[130] in a series of [4 + 2] cycloaddition reactions with the cyclopropenes **139** (R = H, Ph). The reaction sequence is shown in Scheme 6. The initial products obtained were the [4*H*]azepines **140** (R = H, Ph, R^1 = CH_3, Ph, R^2 = H, CH_3, Ph), which were converted on treatment with base into the corresponding substituted [3*H*]azepines **141** (R = H, Ph, R^1 = CH_3, Ph, R^2 = H, CH_3, Ph). The reaction was found to be of general application and was used to synthesise 22 substituted [4*H*]azepines, details of which are given in Section IV.

Other successful rearrangements which have been reported include the treatment of 2,4,6-triphenyl-*N*-benzylpyridinium tetrafluoroborate (**142**) with NaH in refluxing toluene to give 2,4,6,7-tetraphenyl[3*H*]azepine (**143**)[131] and the rearrangement of the di-*tert*-butyl[1*H*]azepines **144** [R^1 = H, R^2 = $C(CH_3)_3$ and R^1 = $C(CH_3)_3$, R^2 = H] to give the corresponding [3*H*]azepines **145** [R^1 = H, R^2 = $C(CH_3)_3$ and R^1 = $C(CH_3)_3$, R^2 = H], using UV irradiation[132] or DBU in refluxing xylene.[71] In the case of the UV-induced rearrangement,[132] the reaction proceeds via the bicyclic intermediates **146** [R^1 = H, R^2 = $C(CH_3)_3$ and R^1 = $C(CH_3)_3$, R^2 = H], which are hydrolysed and decarboxylated to **147** [R^1 = H, R^2 = $C(CH_3)_3$ and R^1 = $C(CH_3)_3$, R^2 = H]. The final rearrangement to the [3*H*]azepines **145** is achieved by heating **147** in

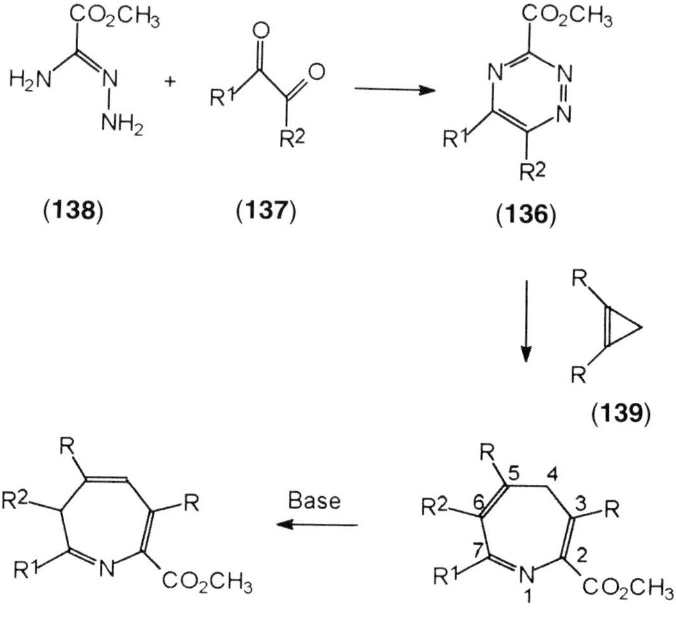

(138) (137) (136) (139) (141) (140)

Scheme 6

benzene. The related thermal rearrangement[71] is less specific, leading to mixtures of di-*tert*-butyl-[2*H*]-, -[3*H*]- and -[4*H*]azepines. In a similar manner, the

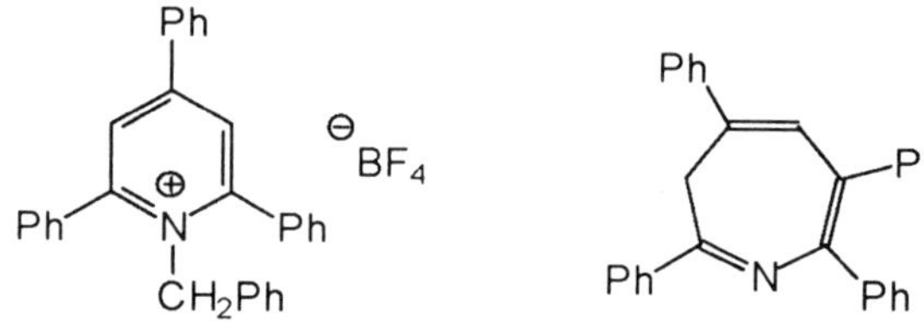

(142) (143)

2,5- and 3,6-dimethyl[3*H*]azepines **145** (R^1 = H, R^2 = CH_3 and R^1 = CH_3, R^2 = H) and the 2,5- and 3,6-diisopropyl[3*H*]azepines **145** [R^1 = H, R^2 = $CH(CH_3)_2$ and R^1 = $CH(CH_3)_2$, R^2 = H] were formed from the corresponding dialkyl[1*H*]azepine analogues **144**;[71c] in these latter examples, however, no [2*H*]- or [4*H*]azepines were obtained.

There is only one reference in the literature to the formation of [3*H*]azepines by a ring-closure procedure. The dinitriles **148** (R^1 = H, R^2 = CH_3, Ph, C_6H_4Ph; R^1 = R^2 = Ph) have been cyclised using HBr to yield the

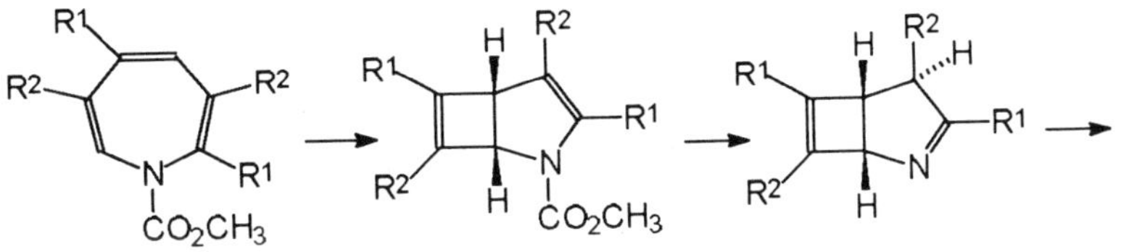

Azepines

(144) (146) (147)

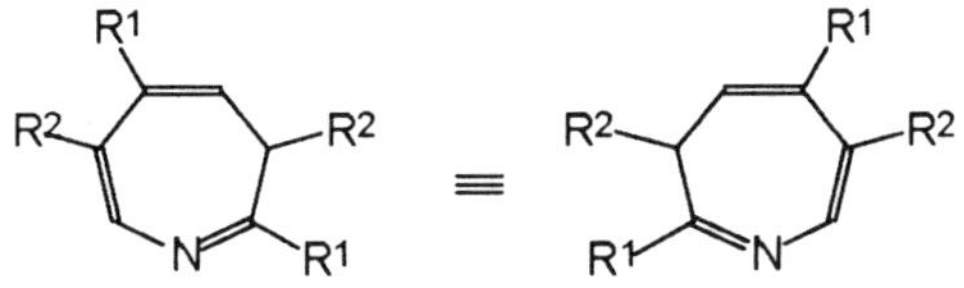

(145)

2-amino[3H]azepines **149** ($R^1 = H$, $R^2 = CH_3$, Ph, C_6H_4Ph; $R^1 = R^2 = Ph$).[133]

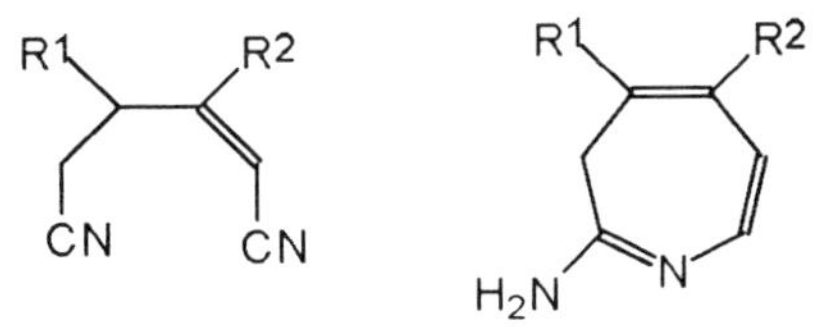

(148) (149)

2-Diethylamino-5-phenyl[3H]azepine (**150**, R = H) has been alkylated successfully at the 3-position, using base and an appropriate halide, to yield a range of 3-substituted products (**150**, R = alkyl, aryl, carbethoxy, etc.). Using this technique, a series of 13 3-substituted derivatives have been prepared,[134,135] Treatment of the halides **151** (R = Cl, F) with ^{t}BuLi gives a mixture of the 4- and 5-alkylated products [**152** and **151**, R = $C(CH_3)_3$].[136] Details of these compounds and other related [3H]azepines are given in Table 3.[137–142]

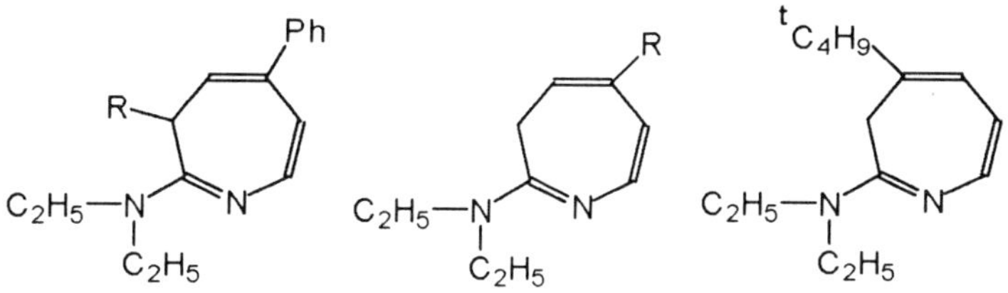

(150) (151) (152)

PROPERTIES

The [3H]azepines are stable oils or crystalline solids. With the obvious exception of the 2-substituted-amino and 2-alkoxy derivatives such as 115 and 116, they are able to withstand strong acid and alkaline conditions and will undergo alkylation with base and alkyl halides as discussed above.[134,135]

(115)   (116)

Rearrangement reactions occur occasionally either thermally or utilising UV irradiation. For example, the amines 115 ($R^1 = R^2 = H$, CH_3, $R^3 = H$) and the ether 116 ($R^1 = CH_3$, $R^2 = H$) rearranged under UV irradiation and led to the 2-azabicyclo[3.2.0]hepta-2,6-dienes 153 [$R = N(CH_3)_2$, NH_2, OC_2H_5],[143] and ring contraction of the ester 154 occurred on methylation in the presence of lithium 2,2,6,6-tetramethylpiperidide, resulting in the formation of the bicyclic compound 155.[144]

(153)   (154)   (155)

Treatment of the diester 133 ($R = CH_3$) with ethanolic HCl gave a mixture of three products, the main one being the pyridine ester 156 and two by-products, one of which was identified as the chloromethyldihydropyridine 130 ($R = CH_3$), indicating that the base-catalysed ring expansion discussed earlier[128] is reversed in the presence of acid. The second by-product of the

(130)   (133)

Azepines 531

acid treatment in this reaction was formed by the 1,6-addition of the elements of water to the [3H]azepine ester **133** (R = CH$_3$), giving the 3,4-dihydro-[2H]azepine **157**, the structure of which was confirmed by ^{1}H NMR spectroscopy.[145]

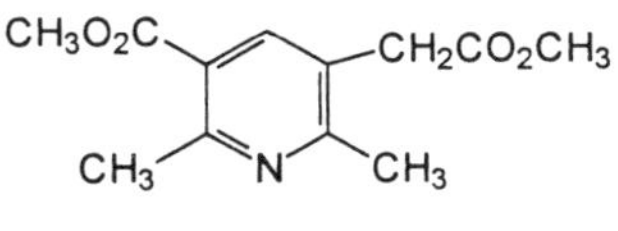

(**156**) 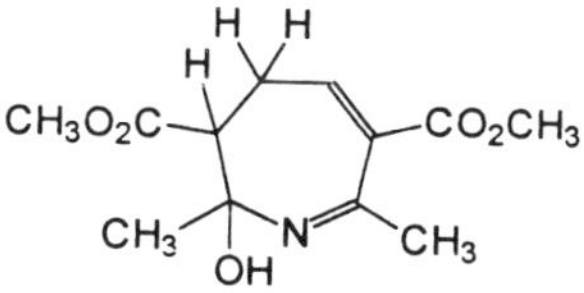   (**157**)

Table 3. [3H]Azepines

Structure	M.p. (°C)/b.p. (°C/mmHg)	Derivatives and m.p. (°C)	Spectroscopic evidence/data	Ref.
	25/0.1		^{1}H, ^{13}C NMR MS	8 74, 75
			MS	73
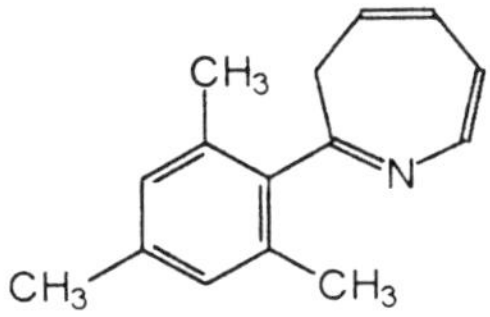	Oil		MS, ^{1}H NMR	138
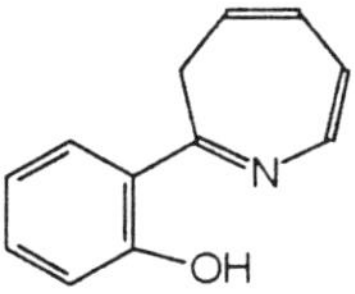	Yellow liquid		IR, ^{1}H NMR	34
	52/0.11 74–76/32		n_D^{20} 1.5203, IR, UV, ^{1}H NMR IR, MS, ^{1}H NMR	119 111
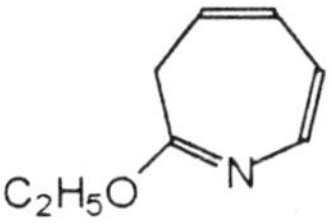	105/78		IR, MS, ^{1}H NMR	111

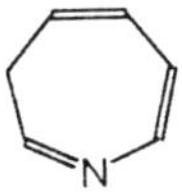

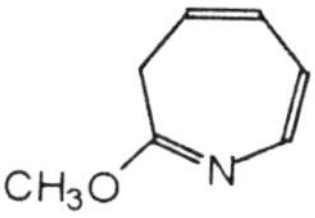

Table 3 (continued)

Structure	M.p. (°C)/b.p. (°C/mmHg)	Derivatives and m.p. (°C)	Spectroscopic evidence/data	Ref.
^{i}C$_3$H$_7$O-azepine	89/31		IR, MS, ^{1}H NMR	111
^{n}C$_4$H$_9$O-azepine	126/49		IR, MS, ^{1}H NMR	111
^{n}C$_6$H$_{13}$O-azepine	98/3.5		IR, MS, ^{1}H NMR	111
^{c}C$_6$H$_{11}$O-azepine	104–106/5.5		IR, MS, ^{1}H NMR	111
^{n}C$_7$H$_{15}$O-azepine	86/1		IR, ^{1}H NMR	111
PhCH$_2$O-azepine	173/27		IR, MS, ^{1}H NMR	111
CF$_3$CH$_2$O-azepine	69/27		IR, MS, ^{1}H NMR	111
PhO-azepine			IR, MS, ^{1}H NMR	109

Table 3 (continued)

Structure	M.p. (°C)/b.p. (°C/mmHg)	Derivatives and m.p. (°C)	Spectroscopic evidence/data	Ref.
naphthyloxy-azepine				154
2-amino-azepine (NH₂)	90.0–91.0	Picrate 200–201 (d)	¹H NMR	4
2-(methylamino)-azepine (H–N–CH₃)			IR, ¹H NMR	110
2-(isopropylamino)-azepine (H–N–iC₃H₇)		Oxalate 96–97 (d)		114
2-(n-butylamino)-azepine (H–N–nC₄H₉)			IR, ¹H NMR	110
2-(benzylamino)-azepine (CH₂NH–)	52.5–53.5		UV	2

Table 3 (continued)

Structure	M.p. (°C)/b.p. (°C/mmHg)	Derivatives and m.p. (°C)	Spectroscopic evidence/data	Ref.
[structure: 2-(phenylamino)azepine]	151 151–152 150–151 151–152.5 94–95/0.12	HI salt 160–161	IR, UV ^{1}H NMR n_D^{20} 1.402, IR, UV, ^{1}H NMR ^{1}H NMR	3 1 2 77 4 119 141
[structure: 2-(cyclohexylamino)azepine]	123–124 118–119		^{1}H NMR	2 83
[structure: 2-(o-tolylamino)azepine]	98–99		UV	76
[structure: 2-(p-tolylamino)azepine]	116 115–115.5			1 2

Table 3 (continued)

Structure	M.p. (°C)/b.p. (°C/mmHg)	Derivatives and m.p. (°C)	Spectroscopic evidence/data	Ref.
(2-(4-ethoxyphenylamino)-3H-azepine)	90–91		UV	2
(2-dimethylamino-3H-azepine)			IR, ^{1}H NMR	110
(2-[N-methyl-N-(2-dimethylaminoethyl)amino]-3H-azepine)			IR, MS, ^{1}H NMR	85
(2-diethylamino-3H-azepine)	85/1		IR, ^{1}H NMR	86a
	59–61/0.1		IR, MS, ^{1}H NMR	85
	65–80/2	Picrate 99–100	IR, ^{1}H NMR	110
			$n_D^{25.5}$ 1.5513, d^{31} 0.9632, IR, UV, ^{1}H NMR	4
			^{1}H NMR	80, 116
	50–60/0.05		^{1}H NMR	83
			n_D^{25} 1.5509, ^{1}H NMR	113
			UV, ^{1}H NMR	112
			UV	82
			MS, ^{1}H NMR	90
				79, 87
(2-diisopropylamino-3H-azepine)	80–90/0.001	Picrate 140–142	UV	76

Table 3 (continued)

Structure	M.p. (°C)/b.p. (°C/mmHg)	Derivatives and m.p. (°C)	Spectroscopic evidence/data	Ref.
(n-C₄H₉)-N-azepine with n-C₄H₉			UV ¹H NMR IR, MS, ¹H NMR	82 83 87
CH₃-N(Ph)-azepine	95–97/0.001	Picrate 158–159	IR	2
pyrrolidinyl-azepine	90–95/0.1		IR, UV, ¹H NMR	122
piperidinyl-azepine	114–116/0.7 139–144/12	Oxalate 155	IR, UV, ¹H NMR	114 122 97
morpholinyl-azepine	127–130/0.1		IR, UV, ¹H NMR	122
PhSO₂-N(Ph)-azepine	158–159			2

Table 3 (continued)

Structure	M.p. (°C)/b.p. (°C/mmHg)	Derivatives and m.p. (°C)	Spectroscopic evidence/data	Ref.
[pyrazole-azepine with CH₃, Ph, OC₂H₅]	73.5–75	HBF₄ salt 158–159, ¹H NMR	¹H NMR	122
[2-(phenylthio)azepine]	117–118		UV	76
[cycloheptatrienyl-azepine]	Yellow oil		¹H, ¹³C NMR	140
[cycloheptatrienyl-azepine isomer]	Pale yellow oil		¹H, ¹³C NMR	140
[azepine-P(O)(OC₂H₅)₂]			¹H NMR	116
[CF₃, CH₃-substituted mesityl azepine]	Oil		¹H NMR	138

Table 3 (continued)

Structure	M.p. (°C)/b.p. (°C/mmHg)	Derivatives and m.p. (°C)	Spectroscopic evidence/data	Ref.
[structure: 2-(2,4,6-trimethylphenyl)-azepine with CO₂CH₃]			MS, ¹H NMR	138
[structure: 3-CHO, 2-OCH₃ azepine]	100/2		IR, UV, ¹H NMR	78
[structure: 3-OCOCH₃, 2-OCH₃ azepine]	115–116/13 117–118/11		IR, UV, ¹H NMR IR, UV, MS, ¹H NMR IR	78 99 96
[structure: 3-OCOPh, 2-OCH₃ azepine]	65–66		IR, UV, ¹H NMR	78
[structure: 3-(2-furoyloxy), 2-OCH₃ azepine]	73		IR, UV, ¹H NMR	78
[structure: 3-(2-thienoyloxy), 2-OCH₃ azepine]	160/0.3		IR, UV, ¹H NMR	78

Azepines 539

Table 3 (continued)

Structure	M.p. (°C)/b.p. (°C/mmHg)	Derivatives and m.p. (°C)	Spectroscopic evidence/data	Ref.
CH₃O₂C-, CH₃O- azepine	75/0.5		MS, UV ¹H NMR	86a 86b
PhO₂C-, CH₃O- azepine	115/0.3 115/0.3		MS ¹H NMR	86a 86b
HO-CH₂CH₂-O₂C-, CH₃O- azepine	120/0.2		¹H, ¹³C NMR, MS	107
H-(O-)₂ ester, CH₃O- azepine	163/0.2		¹H, ¹³C NMR, MS	107
H-(O-)₃ ester, CH₃O- azepine	179/0.2		¹H, ¹³C NMR, MS	107
NC-, CH₃O- azepine			IR, ¹H, ¹³C NMR	106
H₂NCO-O-, CH₃O- azepine	162–163		IR, UV, MS, ¹H NMR MS, ¹H NMR	101 100, 104

Table 3 (continued)

Structure	M.p. (°C)/b.p. (°C/mmHg)	Derivatives and m.p. (°C)	Spectroscopic evidence/data	Ref.
CH₃NHCO–, CH₃O– azepine	128–139		IR, UV, MS, ^{1}H NMR	101
			MS, ^{1}H, NMR	100
C₂H₅NHCO–, CH₃O– azepine	111–112		UV, MS, ^{1}H NMR	101
			MS, ^{1}H, NMR	100
PhC₂H₄NHCO–, CH₃O– azepine	100–101		IR, UV, MS ^{1}H NMR	101, 100
PhCH₂CH(CO₂C₂H₅)NHCO–, CH₃O– azepine	114–116		IR, UV ^{1}H NMR	101, 100
PhHNCO–, CH₃O– azepine	159		MS, UV, ^{1}H NMR	86a
Cl-C₆H₄-CH=NNHCO–, CH₃O– azepine	226			106
CF₃–, CH₃O– azepine	80/0.6		^{1}H, ^{13}C, ^{19}F NMR	106

Table 3 (continued)

Structure	M.p. (°C)/b.p. (°C/mmHg)	Derivatives and m.p. (°C)	Spectroscopic evidence/data	Ref.
H$_2$NCO–, C$_2$H$_5$O– azepine	153–155 153		MS, ^{1}H NMR MS, UV, ^{1}H NMR	104 86a
PhHNCO–, C$_2$H$_5$O– azepine	111		MS, ^{1}H NMR	86a
(o-CH$_3$-C$_6$H$_4$)HNCO–, C$_2$H$_5$O– azepine	150		MS, ^{1}H NMR	86a
CH$_3$O$_2$C–, C$_2$H$_5$O– azepine	120/4		MS ^{1}H NMR	86a 86b
CH$_3$O$_2$C–, n-C$_3$H$_7$O– azepine	130/1		^{1}H NMR	86b
CH$_3$O$_2$C–, i-C$_3$H$_7$O– azepine	100/3		^{1}H NMR	86b
H$_2$NCO–, n-C$_3$H$_7$O– azepine	124–125 122		MS MS, ^{1}H NMR	104 86a

Table 3 (continued)

Structure	M.p. (°C)/b.p. (°C/mmHg)	Derivatives and m.p. (°C)	Spectroscopic evidence/data	Ref.
CH₃O₂C–, ⁿC₄H₉O– azepine	110/0.1		¹H NMR	86b
H₂NCO–, ⁿC₄H₉O– azepine	109–110		MS	104
CH₃O₂C–, cyclohexyl-O– azepine	127/0.1		UV, MS, ¹H NMR	86a
CH₃O₂C–, C₂H₅OCH₂CH₂O– azepine	130/0.4		¹H, ¹³C NMR, MS	107
CH₃O₂C–, CH₃O(CH₂)₂O(CH₂)₂O– azepine	150/0.4		¹H, ¹³C NMR, MS	107
C₂H₅OCH₂CH₂O-C(O)–, C₂H₅OCH₂CH₂O– azepine	140/0.4		¹H, NMR, MS	107
CH₃(OCH₂CH₂)₂O-C(O)–, CH₃(OCH₂CH₂)₂O– azepine	170/0.4		¹H, ¹³C NMR, MS	107

Table 3 (continued)

Structure	M.p. (°C)/b.p. (°C/mmHg)	Derivatives and m.p. (°C)	Spectroscopic evidence/data	Ref.
3-CH₃, 2-N(C₂H₅)₂ azepine	70–75/0.01		^{1}H NMR UV, ^{1}H NMR MS, ^{1}H NMR	80, 116 112 115
3-C₂H₅, 2-N(C₂H₅)₂ azepine			^{1}H NMR	80
3-^{i}C₃H₇, 2-N(C₂H₅)₂ azepine			^{1}H NMR	80
3-Ph(CH₂)$_n$, 2-N(C₂H₅)₂ azepine; n = 1, 2, 3 and 4	Oils		^{1}H NMR	155
3-Ph, 2-N(C₂H₅)₂ azepine	110–112/0.1		^{1}H NMR n_D^{25} 1.5904, IR, UV ^{1}H NMR, MS, UV	116 113 81 79
3-(4-CH₃-C₆H₄), 2-N(C₂H₅)₂ azepine			IR, ^{1}H NMR	81

Table 3 (continued)

Structure	M.p. (°C)/b.p. (°C/mmHg)	Derivatives and m.p. (°C)	Spectroscopic evidence/data	Ref.
3-(4-CF$_3$-C$_6$H$_4$)-2-N(C$_2$H$_5$)$_2$ azepine	Yellow oil		IR, ^{1}H NMR	81
3-(4-CH$_3$O$_2$C-C$_6$H$_4$)-2-N(C$_2$H$_5$)$_2$ azepine	171–172		IR, ^{1}H NMR	81
3-(2-pyridyl)-2-N(C$_2$H$_5$)$_2$ azepine	135/0.3		UV, IR, MS, ^{1}H NMR	84
3-(2-pyridyl)-2-N(n-C$_4$H$_9$)$_2$ azepine	155/0.3		UV, IR, MS, ^{1}H NMR	84
3-CH$_3$CO-2-piperidino azepine	67–68		UV, IR, MS, ^{1}H NMR	96

Table 3 (continued)

Structure	M.p. (°C)/b.p. (°C/mmHg)	Derivatives and m.p. (°C)	Spectroscopic evidence/data	Ref.
4-CH₃CO, 2-CH₃O azepine	Pale yellow oil		UV, IR, ¹H NMR	98, 105
4-PhCO, 2-CH₃O azepine	Yellow oil		UV, IR, ¹H NMR	98
4-CH₃O₂C, 2-CH₃O azepine	Yellow oil		UV, IR, ¹H NMR	98
4-O₂N, 2-CH₃O azepine			UV, IR, ¹H NMR	98
4-Cl, 2-NHPh azepine	165–166			77
4-CH₃, 2-N(C₂H₅)₂ azepine		Oxalate 139		114

Table 3 (continued)

Structure	M.p. (°C)/b.p. (°C/mmHg)	Derivatives and m.p. (°C)	Spectroscopic evidence/data	Ref.
4-^{t}Bu, 2-NEt$_2$ azepine		Picrate 121–122	MS, ^{1}H NMR	136
4-CH$_3$O, 2-NEt$_2$ azepine		Oxalate 125–127 CH$_3$I 146–148	UV, IR, ^{1}H NMR	114
4-CH$_3$CO, 2-NEt$_2$ azepine		Oxalate 115–117 CH$_3$I 154–156	UV, IR, ^{1}H NMR	114
4-Cl, 2-NEt$_2$ azepine		Oxalate 108–110	UV, IR, ^{1}H NMR	114
4-Ph, 2-NEt$_2$ azepine		Oxalate 171–172	UV, IR, ^{1}H NMR	114

Table 3 (continued)

Structure	M.p. (°C)/b.p. (°C/mmHg)	Derivatives and m.p. (°C)	Spectroscopic evidence/data	Ref.
	76–78	Oxalate 165–172	UV, IR, ^{1}H NMR	114
			IR, MS, ^{1}H NMR	89
	Pale yellow oil		IR, UV, ^{1}H, ^{13}C NMR	71c
	Pale yellow oil		IR, UV, ^{1}H, ^{13}C NMR	71c
	Yellow oil 20.5–21 20.5–21		^{1}H, ^{13}C NMR ^{1}H, ^{13}C NMR IR, UV, ^{1}H, ^{13}C NMR	71a 132 71b 71c
	Yellow oil		MS, ^{1}H NMR UV, IR, ^{1}H NMR	90 98

Table 3 (continued)

Structure	M.p. (°C)/b.p. (°C/mmHg)	Derivatives and m.p. (°C)	Spectroscopic evidence/data	Ref.
4-COPh, 2-OCH₃ azepine	Yellow oil		UV, IR, ¹H NMR	98
4-CO₂CH₃, 2-OCH₃ azepine	77–87/0.5		IR, UV, MS, ¹H, ¹³C NMR	106
	Yellow oil		UV, IR, ¹H NMR	98
4-CONH₂, 2-OCH₃ azepine	147–148		MS, ¹H NMR	104
4-CON(CH₃)₂, 2-OCH₃ azepine			MS, ¹H NMR	108
4-SO₂NH₂, 2-OCH₃ azepine	130–131		MS, ¹H NMR	104, 102
4-CON(CH₃)₂, 2-OC₂H₅ azepine			MS, ¹H NMR	108

Table 3 (continued)

Structure	M.p. (°C)/b.p. (°C/mmHg)	Derivatives and m.p. (°C)	Spectroscopic evidence/data	Ref.
2-methoxy-5-cyano azepine (CH₃O, CN substituents)	56		IR, UV, MS, ^{1}H, ^{13}C NMR	106
n-C$_4$H$_9$NH, COCH$_3$ azepine			^{1}H NMR claimed	95
n-C$_4$H$_9$NH, CN azepine			^{1}H NMR claimed	95
n-C$_4$H$_9$NH, OCOCH$_3$ azepine			^{1}H NMR claimed	95
PhNH, CH$_3$ azepine	157–158			2
(CH$_3$)$_2$N, CH$_3$ azepine	68/0.9	Oxalate 158–160	n_D^{20} 1.5573, IR, ^{1}H NMR	114

Table 3 (continued)

Structure	M.p. (°C)/b.p. (°C/mmHg)	Derivatives and m.p. (°C)	Spectroscopic evidence/data	Ref.
[4-CN, 2-NMe₂ azepine]			IR, ¹H NMR	94
[4-CH₃, 2-NEt₂ azepine]	68–70/0.2 120/0.1		¹H NMR ¹H NMR MS, ¹H NMR	80 112 115
[4-tBu, 2-NEt₂ azepine]		Picrate 112–113	MS, ¹H NMR	136
[4-Ph, 2-NEt₂ azepine]	78–79 77–78		UV, IR, ¹H NMR ¹H, ¹³C NMR MS, ¹H NMR	114 135 90
[4-(4-piperidyl), 2-NEt₂ azepine]	62–63		UV, IR, ¹H NMR	114

Table 3 (continued)

Structure	M.p. (°C)/b.p. (°C/mmHg)	Derivatives and m.p. (°C)	Spectroscopic evidence/data	Ref.
4-COCH₃, 2-N(C₂H₅)₂ azepine			MS, ¹H NMR	90
4-CO₂H, 2-N(C₂H₅)₂ azepine			MS, ¹H NMR	90
4-CO₂C₂H₅, 2-N(C₂H₅)₂ azepine			MS, ¹H NMR	115
4-CON(CH₃)₂, 2-N(C₂H₅)₂ azepine			MS, ¹H NMR	88
4-CONHCH(CO₂H)CH₂(3-I-4-HO-C₆H₃), 2-N(C₂H₅)₂ azepine			MS, ¹H NMR	92

Table 3 (continued)

Structure	M.p. (°C)/b.p. (°C/mmHg)	Derivatives and m.p. (°C)	Spectroscopic evidence/data	Ref.
[HO, I-substituted phenyl-CH₂-CONHCHCO₂H azepine with N(C₂H₅)₂]			MS, ^{1}H NMR	92
[CN-substituted azepine with N(C₂H₅)₂]			MS, ^{1}H NMR	90
[F-substituted azepine with N(C₂H₅)₂]	78–79/0.3		n_D^{20} 1.5294, MS, ^{1}H NMR	136
[Cl-substituted azepine with N(C₂H₅)₂]	126–127.5/0.7 96.5–97.5/0.2 39–40	Oxalate 107.0–109.0	MS, ^{1}H NMR n_D^{20} 1.5718, ^{1}H NMR	90 136
[Br-substituted azepine with N(C₂H₅)₂]	80/0.15		MS, ^{1}H NMR	90, 115

Table 3 (continued)

Structure	M.p. (°C)/b.p. (°C/mmHg)	Derivatives and m.p. (°C)	Spectroscopic evidence/data	Ref.
(4-iodo-2-(N,N-diethylamino)-3H-azepine)			MS, ^{1}H NMR	90
(4-methylthio-2-(N,N-diethylamino)-3H-azepine)			MS, ^{1}H NMR	90
(4-carboxy-2-(N-methyl-N-butanoyl amino)-3H-azepine)			No data	137
(4-acetyl-2-piperidino-3H-azepine)		Brown oil	UV, IR, ^{1}H NMR	98
(4-carboxamido-2-piperidino-3H-azepine)		156–157	MS	104

Table 3 (continued)

Structure	M.p. (°C)/b.p. (°C/mmHg)	Derivatives and m.p. (°C)	Spectroscopic evidence/data	Ref.
(morpholino-substituted azepine with COCH₃)	Yellow oil		UV, IR, MS, ^{1}H NMR	98
(CH₃O-substituted azepine with COCH₃)	Yellow oil Pale yellow oil		UV, IR, ^{1}H NMR UV, IR, ^{1}H NMR	98 105
(CH₃O-substituted azepine with COPh)	Yellow oil		UV, IR, ^{1}H NMR	98
(CH₃O-substituted azepine with CO₂CH₃)	Yellow oil		IR, UV, MS, ^{1}H, ^{13}C NMR UV, IR, ^{1}H NMR	106 98
(Et₂N-substituted azepine with CH₃)		Oxalate 120–125	UV, IR, ^{1}H NMR	114
(Et₂N-substituted azepine with COCH₃)		CH₃I 134–137	UV, IR, ^{1}H NMR	114
(Et₂N-substituted azepine with Ph)	75–76	Oxalate 137–138	UV, IR, ^{1}H NMR	114

Table 3 (continued)

Structure	M.p. (°C)/b.p. (°C/mmHg)	Derivatives and m.p. (°C)	Spectroscopic evidence/data	Ref.
	97–99		UV, IR, ^{1}H NMR	114
		Oxalate 119–120	UV, IR, ^{1}H NMR	114
	Yellow oil		IR, MS, ^{1}H NMR UV, IR, ^{1}H NMR	89 98
	235–236		IR, UV, ^{1}H NMR	122
	224–225		^{1}H NMR	122
	Brown oil		UV, IR, MS, ^{1}H NMR	98

Table 3 (continued)

Structure	M.p. (°C)/b.p. (°C/mmHg)	Derivatives and m.p. (°C)	Spectroscopic evidence/data	Ref.
(piperidine-azepine-COPh)	Yellow oil		UV, IR, ^{1}H NMR	98
(morpholine-azepine-CPh$_3$)	227–229		^{1}H NMR	122
(morpholine-azepine-COCH$_3$)	Yellow oil		UV, IR, ^{1}H NMR	98
(morpholine-azepine-COPh)	Brown oil		UV, IR, MS, ^{1}H NMR	98
(pyrrolidine-azepine-SCH$_3$)	Pale yellow oil		IR, ^{1}H NMR	122
(piperidine-azepine-SCH$_3$)	Pale yellow oil		^{1}H NMR	122

Table 3 (continued)

Structure	M.p. (°C)/b.p. (°C/mmHg)	Derivatives and m.p. (°C)	Spectroscopic evidence/data	Ref.
morpholine-SCH₃ azepine	Pale yellow oil		¹H NMR	122
CH₃S, SCH₃ azepine	Pale yellow oil, 150/0.01		IR, ¹H NMR	122
CH₃, P(O)(OCH₃)₂ azepine	105/0.04		¹H NMR	118, 116
CH₃, P(O)(OC₂H₅)₂ azepine	115–120/0.02		n_D^{20} 1.5040 ¹H NMR	118, 116
C₂H₅, P(O)(OC₂H₅)₂ azepine	103/0.02		n_D^{20} 1.5066 ¹H NMR	118, 116
Ph, CO₂CH₃ azepine			IR, UV, ¹H, ¹³C NMR	130
ketone-alkenyl-CH₃ azepine			IR, UV, MS, ¹H, ¹³C NMR	72

Table 3 (continued)

Structure	M.p. (°C)/b.p. (°C/mmHg)	Derivatives and m.p. (°C)	Spectroscopic evidence/data	Ref.
			IR, UV, MS, ^{1}H, ^{13}C NMR	72
			^{1}H NMR	105
			IR, ^{1}H NMR	96
			^{1}H, ^{13}C NMR	106
	100/0.8		IR, ^{1}H, ^{13}C NMR	106
			^{1}H NMR	139
			^{1}H NMR	116

Table 3 (continued)

Structure	M.p. (°C)/b.p. (°C/mmHg)	Derivatives and m.p. (°C)	Spectroscopic evidence/data	Ref.
[structure: C₂H₅–N(C₂H₅)–azepine–S–C₆H₄–tC₄H₉]			¹H NMR	116
[structure: piperidino-azepine-COCH₃]			IR, UV, MS, ¹H NMR	96
[structure: 3,6-dimethyl-azepine]	Pale yellow oil		IR, UV, ¹H, ¹³C NMR	71c
[structure: 3,6-diisopropyl-azepine]	Pale yellow oil		IR, UV, ¹H, ¹³C NMR	71c
[structure: 3,6-di-tert-butyl-azepine]	26–27 57.5–58.5 57.5–58.5		¹H, ¹³C NMR ¹H, ¹³C NMR IR, ¹H, ¹³C NMR, MS	71a 132 71b 71c
[structure: Ph₃C-azepine-pyrrolidinyl]	153–154		IR, UV, ¹H NMR	122
[structure: Ph₃C-azepine-piperidinyl]	151–152		¹H NMR	122

Table 3 (continued)

Structure	M.p. (°C)/b.p. (°C/mmHg)	Derivatives and m.p. (°C)	Spectroscopic evidence/data	Ref.
Ph₃C-azepine-morpholine	153–154		^{1}H NMR	122
^{i}C₃H₇-azepine-^{i}C₃H₇	Pale yellow oil		IR, UV, ^{1}H, ^{13}C NMR	71c
^{t}C₄H₉-azepine-^{t}C₄H₉	Pale yellow oil		^{1}H, ^{13}C NMR IR, UV, ^{1}H, ^{13}C NMR	71a 71c
CH₃-azepine-P(O)(OCH₃)₂	110/0.05		^{1}H NMR	118 116
CH₃-azepine-P(O)(OC₂H₅)₂			No data	117
OCH₃-azepine-P(O)(OCH₃)₂	125–130/0.05		^{1}H NMR	118 116

Table 3 (continued)

Structure	M.p. (°C)/b.p. (°C/mmHg)	Derivatives and m.p. (°C)	Spectroscopic evidence/data	Ref.
OCH₃ / NH₂CO / CH₃O azepine	196		UV, MS, ¹H NMR	86a
CH₃, CH₃ / C₂H₅—N(C₂H₅) azepine			¹H NMR	80
Cl / PhCO / CH₃O azepine	83.5–84.5		IR, UV, MS, ¹H NMR	78, 99
OCH₃ / CH₃O₂C / CH₃O azepine	77		UV, MS, ¹H NMR	86a
Cl / CH₃O₂C / CH₃O azepine	44 / 90/0.01		UV, MS, ¹H NMR	86a
Br / CH₃O₂C / CH₃O azepine	42 / 90/0.02		UV, MS, ¹H NMR	86a

Table 3 (continued)

Structure	M.p. (°C)/b.p. (°C/mmHg)	Derivatives and m.p. (°C)	Spectroscopic evidence/data	Ref.
PhCO, Cl, n-C₄H₉NH azepine	105–107		IR, UV, ¹H NMR	78
CH₃, CH₃, C₂H₅-N(C₂H₅) azepine		Oxalate 85–86	UV, IR, ¹H NMR ¹H NMR	114 80
CH₃, Ph, C₂H₅-N(C₂H₅) azepine	88.5–89.5 88.5–89.5		¹H NMR MS, ¹H, ¹³C NMR	134 135
C₂H₅, Ph, C₂H₅-N(C₂H₅) azepine	71.5–72.5		MS, ¹H, ¹³C NMR	135
iC₃H₇, Ph, C₂H₅-N(C₂H₅) azepine	80.5–81.5		MS, X-ray, ¹H, ¹³C NMR ¹H NMR	135 142
Ph, Ph, C₂H₅-N(C₂H₅) azepine	82.5–93.5		MS, ¹H, ¹³C NMR	135

Table 3 (continued)

Structure	M.p. (°C)/b.p. (°C/mmHg)	Derivatives and m.p. (°C)	Spectroscopic evidence/data	Ref.
PhCH₂-, Ph, C₂H₅-N(C₂H₅) azepine	124–125 125–126		¹H NMR MS, ¹H, ¹³C NMR	134 135
OH, PhCH-, Ph, C₂H₅-N(C₂H₅) azepine	172–173.5		IR, MS, ¹H NMR	135
CH₃S-, Ph, C₂H₅-N(C₂H₅) azepine	109.5–110.5 109.5–110.5		¹H NMR MS, ¹H, ¹³C NMR	134 135
PhS-, Ph, C₂H₅-N(C₂H₅) azepine	91.5–92.5		MS, ¹H, ¹³C NMR	135
ClCH=C(Cl)-, Ph, C₂H₅-N(C₂H₅) azepine	64–65		MS, ¹H NMR	135

Table 3 (continued)

Structure	M.p. (°C)/b.p. (°C/mmHg)	Derivatives and m.p. (°C)	Spectroscopic evidence/data	Ref.
(3-CO₂C₂H₅, 5-Ph, 2-N(C₂H₅)₂ azepine)	72.5–73.5		IR, MS, ^{1}H, ^{13}C NMR	135, 144
(3-OCONHPh, 5-Ph, 2-N(C₂H₅)₂ azepine)	152.5–153.5		IR, MS, ^{1}H, ^{13}C NMR	135
(3-CON(C₂H₅)₂, 5-Ph, 2-N(C₂H₅)₂ azepine)	127.5–128		IR, MS, ^{1}H, ^{13}C NMR	135
(3-C(=NC₆H₁₁)NHC₆H₁₁, 5-Ph, 2-N(C₂H₅)₂ azepine)	101.5–102.5		IR, MS	135
(3-OCOPh, 5-Cl, 2-N(C₂H₅)₂ azepine)	131–132		IR, UV	78

Table 3 (continued)

Structure	M.p. (°C)/b.p. (°C/mmHg)	Derivatives and m.p. (°C)	Spectroscopic evidence/data	Ref.
PhCO, Cl, PhNH (azepine)	129–131		IR, UV, ^{1}H NMR	78
OHC, Cl, CH$_3$O (azepine)	70–80/6 78/6		IR, UV, ^{1}H NMR IR, UV, MS, ^{1}H NMR	78 99
CH$_3$O$_2$C, OCH$_3$, CH$_3$O (azepine)	62		UV, MS, ^{1}H NMR	86a
CH$_3$O$_2$C, CO$_2$CH$_3$, CH$_3$O (azepine)	120/0.1		UV, MS, ^{1}H NMR	86a
CH$_3$O$_2$C, Cl, CH$_3$O (azepine)	40 80/0.02		UV, MS, ^{1}H NMR	86a
PhCO, Cl, CH$_3$O (azepine)	93.5–95.5		IR, UV, ^{1}H NMR	78
CH$_3$, CH$_3$, C$_2$H$_5$–N(C$_2$H$_5$) (azepine)			^{1}H NMR	80

Table 3 (continued)

Structure	M.p. (°C)/b.p. (°C/mmHg)	Derivatives and m.p. (°C)	Spectroscopic evidence/data	Ref.
3-Ph, 2-Ph, 7-CO₂CH₃ azepine			IR, UV, ^{1}H, ^{13}C NMR	130
3-CH₃, 2-N(C₂H₅)₂, 7-CH₃ azepine			^{1}H NMR	80
3-CH₃, 2-N(C₂H₅)₂, 7-^{i}C₃H₇ azepine			^{1}H NMR	80
3-^{i}C₃H₇, 2-N(C₂H₅)₂, 7-CH₃ azepine			^{1}H NMR	80
4-CH₃O, 5-CO₂CH₃, 2-CH₃O azepine			IR, UV, MS, ^{1}H NMR	103
4-CH₃, 5-CH₃, 2-N(C₂H₅)₂ azepine		Oxalate 119–120	IR, ^{1}H NMR	114

Azepines

Table 3 (continued)

Structure	M.p. (°C)/b.p. (°C/mmHg)	Derivatives and m.p. (°C)	Spectroscopic evidence/data	Ref.
[structure: 4-CH₃CONH, 5-NHCOCF₃, 2-N(C₂H₅)₂ azepine]			MS, ^{1}H, ^{13}C NMR	93
[structure: 4-CH₃O₂C, 5-F, 2-N(C₂H₅)₂ azepine]			MS, ^{1}H, ^{19}F NMR	91a
[structure: 4-F, 5-CO₂CH₃, 2-N(C₂H₅)₂ azepine]			MS, ^{1}H, ^{19}F NMR	91a
[structure: 4,6-dimethyl-2-N(C₂H₅)₂ azepine]	96–99/1.3		UV, IR, ^{1}H NMR	114
[structure: 4-CH₃O, 2-CH₃O, 7-phthalimido azepine]			^{1}H NMR	139

Table 3 (continued)

Structure	M.p. (°C)/b.p. (°C/mmHg)	Derivatives and m.p. (°C)	Spectroscopic evidence/data	Ref.
(3H-azepine with 5-CH₃, 7-mesityl, 2-N(C₂H₅)₂)			UV, ¹H NMR	80
(3H-azepine with 5-NHCOCF₃, 6-NHCOCH₃, 2-N(C₂H₅)₂)			MS, ¹H, ¹³C NMR	93
(3H-azepine with 5-CO₂CH₃, 6-F, 2-N(C₂H₅)₂)			MS, ¹H, ¹⁹F NMR	91a
(3H-azepine with 5-F, 6-CO₂CH₃, 2-N(C₂H₅)₂)			MS, ¹H, ¹⁹F NMR	91a
(3H-azepine with 5-CH₃, 7-Br, 2-NH₂)	160–165		IR, ¹H NMR	133
(3H-azepine with 5-Ph, 7-Br, 2-NH₂)	173–175 (d)	HBr, 210–230 IR, ¹H NMR	IR	133

Table 3 (continued)

Structure	M.p. (°C)/b.p. (°C/mmHg)	Derivatives and m.p. (°C)	Spectroscopic evidence/data	Ref.
(2-amino-7-bromo-4-(4-phenylphenyl)-azepine)	195–200 (d)	HBr, 245–250 IR	IR	133
(N,N-diethyl-5-methyl-7-methyl-azepin-2-amine)		Oxalate 140–141	UV, IR, ^{1}H NMR	114
(methyl ester, difluoro, N,N-diethylamino azepine)			MS, ^{1}H, ^{19}F NMR	91a
(methyl ester, fluoro, bis(N,N-diethylamino) azepine)			MS, ^{1}H, ^{19}F NMR	91a
(methyl 5-methyl-6,7-diphenyl-azepine-2-carboxylate)			IR, UV, ^{1}H, ^{13}C NMR	130

Table 3 (continued)

Structure	M.p. (°C)/b.p. (°C/mmHg)	Derivatives and m.p. (°C)	Spectroscopic evidence/data	Ref.
(3-CH$_3$, 4-CH$_3$, 7-CH$_3$, 2-OC$_2$H$_5$ azepine)	57/1.1		IR, UV, ^{1}H NMR n_D^{26} 1.4946, IR, UV, ^{1}H NMR	120 121
(3-CH$_3$, 4-CH$_3$, 7-CH$_3$, 2-SC$_2$H$_5$ azepine)	76/0.25 76/0.25		^{1}H NMR n_D^{26} 1.4498, UV, ^{1}H NMR	120 121
(3-CH$_3$, 4-CH$_3$, 7-CH$_3$, 2-N(C$_2$H$_5$)$_2$ azepine)	Yellow liquid		^{1}H NMR ^{1}H NMR	80 112
(3-CH$_3$, 4-CH$_3$, 7-CH$_3$, 2-piperidino azepine)	105–112/1.3	HClO$_4$ salt 156–157 HClO$_4$ salt 156–157		120 121
(3-CO$_2$CH$_3$, 6-CO$_2$CH$_3$, 2-CH$_3$, 7-CH$_3$ azepine)	130–132/1.5		IR, ^{1}H NMR	128b 128a
(3-CO$_2$C$_2$H$_5$, 6-CO$_2$C$_2$H$_5$, 2-CH$_3$, 7-CH$_3$ azepine)	122–123/0.2		IR, UV, ^{1}H NMR	128b 128a

Table 3 (continued)

Structure	M.p. (°C)/b.p. (°C/mmHg)	Derivatives and m.p. (°C)	Spectroscopic evidence/data	Ref.
(Ph, Ph, H₂N, Br azepine)	212–215 (d)	HBr 275 (d), IR	IR	133
(CH₃O₂C, F, (C₂H₅)₂N, F azepine)			MS, ^{1}H, ^{19}F NMR	91a
(CH₃, CH₃, CH₃, P(O)(OC₂H₅)₂ azepine)			IR, UV, ^{1}H NMR	112
(CH₃O₂C, CO₂CH₃, CH₃, CH₃ azepine)	Oil		IR, UV, ^{1}H NMR	129
(CH₃O₂C, CO₂CH₃, PhCH=CH, CH₃ azepine)	122–123.5		IR, UV, ^{1}H NMR	129
(Ph, Ph, Ph, CO₂CH₃ azepine)			IR, UV, ^{1}H, ^{13}C NMR	130

Table 3 (continued)

Structure	M.p. (°C)/b.p. (°C/mmHg)	Derivatives and m.p. (°C)	Spectroscopic evidence/data	Ref.
(Ph, Ph, Ph, Ph azepine)	214–215		IR, UV, MS, ^{1}H, ^{13}C NMR X-ray	131
(CH₃O, CH₃, CH₃O, phthalimido azepine)			^{1}H NMR	139
(CH₃O, CH₂CN, CH₃O, phthalimido azepine)			^{1}H NMR	139
(CH₃O, OCH₃, CH₃O, phthalimido azepine)			^{1}H NMR	139
(F, CO₂CH₃, C₂H₅–N(C₂H₅), F azepine)			MS, ^{1}H, ^{19}F NMR	91a
(Ph, CH₃, Ph, Ph, CH₃ azepine)	123 133–134		^{1}H NMR ^{1}H NMR UV, ^{1}H NMR	126 124 123

Table 3 (continued)

Structure	M.p. (°C)/b.p. (°C/mmHg)	Derivatives and m.p. (°C)	Spectroscopic evidence/data	Ref.
Ph, CH₃, Ph, CH₂Ph, CH₃, N (azepine)	90		¹H NMR	126
Ph, Ph, Ph, Ph, CH₃, N	174		¹H NMR	126
Ph, Ph, Ph, C₆H₄OCH₃, CH₃, N	197		¹H NMR	126
Ph, C₂H₅, Ph, Ph, C₂H₅, N	Oil Oil	Picrate 169, Picrate 169, ¹H NMR	¹H NMR	127 126
Ph, CH₃, Ph, Ph, Ph, N	170		¹H NMR	126
Ph, CH₃, Ph, C₆H₄OCH₃, Ph, N	192		¹H NMR	126
Ph, Ph, Ph, Ph, Ph, N	212 217–218		¹H NMR ¹H NMR UV, ¹H NMR	126 124 123

Table 3 (continued)

Structure	M.p. (°C)/b.p. (°C/mmHg)	Derivatives and m.p. (°C)	Spectroscopic evidence/data	Ref.
(Ph, Ph, Ph, Ph, CH₂Ph azepine)	161		^{1}H NMR ^{1}H NMR	126 124
(Ph, CH₃, Ph, CH₃, ^{t}C₄H₉ azepine)	97		MS, ^{1}H NMR	125
(Ph, CH₃, Ph, CH₃, Ph azepine)	108		MS, ^{1}H NMR	125
(Ph, Ph, CH₃, CH₃, CO₂CH₃ azepine)			^{1}H, ^{13}C NMR	130
(Ph, Ph, CH₃, Ph, CO₂CH₃ azepine)			IR, UV, ^{1}H, ^{13}C NMR	130
(Ph, Ph, Ph, Ph, CO₂CH₃ azepine)			IR, UV, ^{1}H, ^{13}C NMR	130

Table 3 (continued)

Structure	M.p. (°C)/b.p. (°C/mmHg)	Derivatives and m.p. (°C)	Spectroscopic evidence/data	Ref.
[3H-azepine with 4-F-C6H4, Ph, Ph, 4-F-C6H4, CO2CH3, N]			IR, UV, ^{1}H, ^{13}C NMR	130
[3H-azepine with Ph, CH3, Ph, Ph, CH3, CH3, N]	182–183 182		UV, ^{1}H NMR ^{1}H NMR	123 126
[3H-azepine with Ph, CH3, Ph, CH3, CH3, Ph, N]	181		^{1}H NMR	126
[3H-azepine with Ph, CH3, Ph, Ph, CH3, Ph, N]	189 186–188		^{1}H NMR ^{1}H NMR UV, ^{1}H NMR	126 124 123
[3H-azepine with Ph, Ph, Ph, Ph, CH3, CH3, N]	164		^{1}H NMR ^{1}H NMR	126 124
[3H-azepine with Ph, Ph, Ph, Ph, CH3, Ph, N]	206		^{1}H NMR	126

Table 3 (continued)

Structure	M.p. (°C)/b.p. (°C/mmHg)	Derivatives and m.p. (°C)	Spectroscopic evidence/data	Ref.
(Ph, Ph, Ph, Ph, C₂H₅, C₂H₅ azepine)	127 128		¹H NMR ¹H NMR	126 124
(Ph, Ph, Ph, C₂H₅, C₂H₅, CH₃ azepine)	113		¹H NMR	126
(Ph, Ph, Ph, Ph, CH₃, CH₃ azepine)	165		¹H NMR	126
(Ph, Ph, Ph, Ph, Ph, CH₃ azepine)	187		¹H NMR	126
(Ph, Ph, Ph, Ph, Ph, CH₃ azepine)	177		¹H NMR	126
(Ph, Ph, Ph, Ph, C₂H₅, C₂H₅ azepine)	151 151		¹H NMR ¹H NMR	126 124
(Ph, Ph, Ph, Ph, Ph, CH₃ azepine)	212 208		¹H NMR UV, ¹H NMR	124, 126 123

Table 3 (continued)

Structure	M.p. (°C)/b.p. (°C/mmHg)	Derivatives and m.p. (°C)	Spectroscopic evidence/data	Ref.
Ph, Ph, Ph, Ph, Ph, N, CH₂OH (azepine)	213		¹H NMR	126
Ph, Ph, Ph, Ph, Ph, N, Ph (azepine)	227 227		UV, ¹H NMR ¹H NMR ¹H NMR	123 126 124
Ph, CH₃, Ph, Ph, PhCH=CH, N, CH₃ (azepine)	161–163		UV, ¹H NMR	123

IV. [4H]AZEPINES

PREPARATION

Evidence for the existence of the parent [4H]azepine (**4**) has not been presented. Substituted derivatives are, however, fairly widely represented in the literature.

As discussed earlier (Section III), ring-expansion procedures have been used to prepare [3H]azepines, and some of these methods have also been usefully applied with suitable modifications to give the [4H] isomers. Thus Gregory and Johnson and their co-workers treated the dihydropyridines **158** (R = CH₃, C₂H₅) with potassium cyanide[128,146–148] and with sodium ethoxide at room temperature[128] to achieve the 4,5-dihydro[1H]azepines **159** (R = CH$_3$, C$_2$H$_5$, R^1 = CN, OC$_2$H$_5$) (see also Chapter 8, Section IV). At elevated temperatures

(4)   (158)   (159)

(160)   (161)

these latter compounds were converted into the corresponding [4H]azepines **160** (R = CH$_3$, C$_2$H$_5$) with loss of HCN or C$_2$H$_5$OH, respectively.[128,148] Elimination of ethanol occurred readily at temperatures up to the boiling point of ethanol, whereas eliminating HCN required fairly forcing conditions (265 °C). Even higher temperatures (320 °C) resulted in further rearrangement to the [3H]azepines **161** (R = CH$_3$, C$_2$H$_5$).[128] The [4H]azepine **163** was obtained directly from the dihydropyridine **162** using either potassium carbonate in dimethyl sulphoxide at room temperature or sodium ethoxide in ethanol.[148]

(162)   (163)

Photolysis of aromatic azides, which proved to be a successful approach to [3H]azepines[79,80] (Section III), has also been shown to give [4H]azepines although usually in poor yields as the minor products of the reaction. For example, the [4H]azepines **165** (R = H, CH$_3$) were separated as the minor components from the photolysis of the biphenyl azides **164** (R = H, CH$_3$) in diethylamine, the corresponding [3H]azepines being the other components.[79,80]

(164)   (165)

[4*H*]Azepines were also formed, in this case as the main products of the reaction, by the [4 + 2] cycloaddition of 1,2,4-triazines (**136**, R^1 = H, CH_3, Ph, *p*-anisyl, *p*-FC_6H_4, R^2 = H, CH_3, Ph, *p*-FC_6H_4 or R^1 = R^2 = H, Ph, CO_2CH_3)[130,149] and cyclopropenes (**166**, R^3 to R^6 = H, CH_3, Ph). Using this procedure, Sauer and his co-workers[130,149] produced 28 examples of substituted [4*H*]azepines (see Scheme 6, Section III); details of the products (**167**, R^1, R^2 = H, CH_3, Ph, *p*-anisyl, *p*-FC_6H_4, CO_2CH_3; R^3 to R^6 = H, CH_3, Ph) are given in Table 4.

(136)   (166)   (167)

Sano and his co-workers have investigated the rearrangement of bicycloheptenes[150] and -heptanes[151,152] as a method for the preparation of a series of 'azatropolones'. In the course of their investigations they found that treatment of the azabicyclo[3.2.0]heptane-3,4-diones **168** (R = Ph, OC_2H_5, $OCOCH_3$, SPh, C_2H_5) with DBU led to the corresponding 1,5-dihydro-[2*H*]azepine-2-ones **169** (R = Ph, OC_2H_5, $OCOCH_3$), which on treatment with $(C_2H_5)_3O^+$ BF_4^- gave the 7-ethoxy[4*H*]azepines **170** (R = Ph, OC_2H_5, $OCOCH_3$).[151,152] An alternative procedure, treating the biyclic enol ethers **171** (R = Ph, OC_2H_5, $OCOCH_3$) with $SnCl_4$, led directly to the same products. Treatment of **168** (R = SPh) with Raney Ni gave the dihydro[2*H*]azepin-2-one **169** (R = H), which was similarly converted into the 7-ethoxy[4*H*]-azepine **170** (R = H) with $(C_2H_5)_3O^+$ BF_4^-.[151,152]

An unusual ring-expansion reaction was shown to occur on treatment of 3-methyl-3-vinyl-2-dimethylaminoazirine (**172**) with dimethyl acetylenedicarboxylate (**173**), which led to the formation of the [4*H*]azepine **174** by incorporation of both the acetylene and the vinyl substituents into the ring system. The structure of **174** was confirmed by X-ray analysis.[153]

[4H]Azepines (**176**, R = CH₃, CH₂Ph) were formed by alkylation of 5-phenyl[3H]azepine (**175**, X = SCH₃), using methyl iodide and benzyl chloride, respectively, in the presence of KNH₂/liquid NH₃. Similarly, alkylation of the [3H]azepine **175** (X = H) with 2-bromopropane in the presence of lithium diisopropylamide gave the 4-phenyl-4-isopropyl[4H]azepine **177**.[142]

Physical data for the [4H]azepines are detailed in Table 4.

Azepines 581

Table 4. [4H]Azepines

Structure	M.p. (°C)/b.p. (°C/mmHg)	Derivatives and m.p. (°C)	Spectroscopic evidence	Ref.
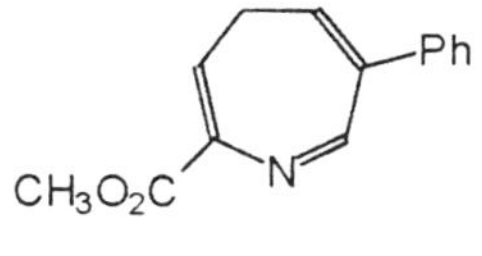 (CH₃O₂C, Ph substituted [4H]azepine)			IR, UV, ^{1}H, ^{13}C NMR	130
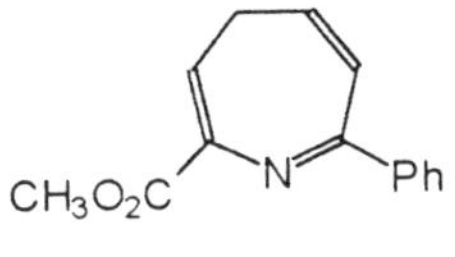 (CH₃O₂C, Ph substituted [4H]azepine)			IR, UV, ^{1}H, ^{13}C NMR	130
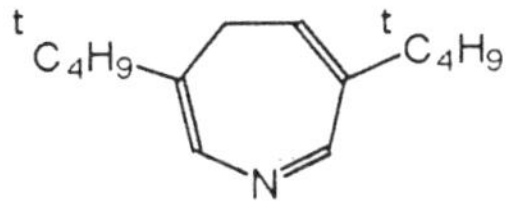 (di-^{t}C₄H₉ substituted [4H]azepine)			^{1}H, ^{13}C NMR	71
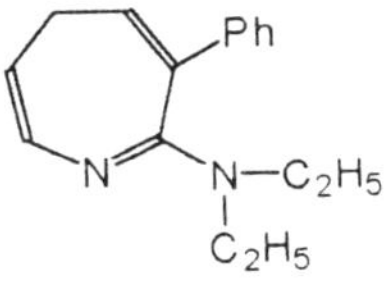 (Ph, N(C₂H₅)₂ substituted [4H]azepine)			UV, ^{1}H NMR	79
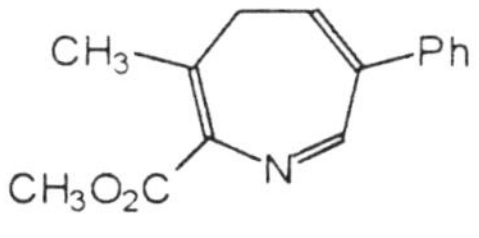 (CH₃, Ph, CH₃O₂C substituted [4H]azepine)			^{1}H, ^{13}C NMR	130
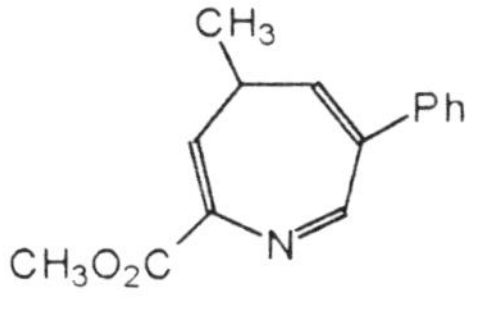 (CH₃, Ph, CH₃O₂C substituted [4H]azepine)			^{1}H, ^{13}C NMR	130
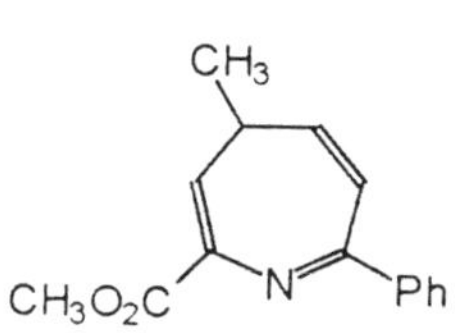 (CH₃, Ph, CH₃O₂C substituted [4H]azepine)			IR, UV, ^{1}H, ^{13}C NMR	130

Table 4 (continued)

Structure	M.p. (°C)/b.p. (°C/mmHg)	Derivatives and m.p. (°C)	Spectroscopic evidence	Ref.
(4-CH₃, 3-Ph, 7-CH₃O₂C azepine)			^{1}H, ^{13}C NMR	130
(3-CH₃, 2-CH₃, 7-CH₃O₂C azepine)			^{1}H, ^{13}C NMR	130
(4-CH₃, 3-Ph... 7-CH₃O₂C azepine)			IR, UV, ^{1}H, ^{13}C NMR	130
(3-Ph, 2-Ph, 7-CH₃O₂C azepine)			IR, UV, ^{1}H, ^{13}C NMR	130
(3-CO₂CH₃, 2-CO₂CH₃, 7-CH₃O₂C azepine)	95–96		^{1}H NMR	149
(mesityl-substituted N,N-diethylamino azepine)		Picrate 179–179.5	^{1}H NMR	80
(Ph, iC₃H₇, N(C₂H₅)₂ azepine)	48–49		MS, ^{1}H, ^{13}C NMR	142

Azepines 583

Table 4 (continued)

Structure	M.p. (°C)/b.p. (°C/mmHg)	Derivatives and m.p. (°C)	Spectroscopic evidence	Ref.
3,5-Ph, 4-Ph, 2-CO₂CH₃ azepine			IR, UV, ^{1}H, ^{13}C NMR	130
3-Ph, 4-Ph, 7-Ph, 2-CO₂CH₃ azepine			IR, UV, ^{1}H, ^{13}C NMR	130
3-Ph, 4-Ph, 7-(4-OCH₃-C₆H₄), 2-CO₂CH₃ azepine			IR, UV, ^{1}H, ^{13}C NMR	130
3,6-(CO₂CH₃)₂, 2,7-(CH₃)₂ azepine	76.5–77		IR, UV, ^{1}H NMR	128b, 128a
3,6-(CO₂C₂H₅)₂, 2,7-(CH₃)₂ azepine	142–145/0.1		IR, UV, ^{1}H NMR	128b, 128a
7-CH₃, 3-CO₂CH₃, 2,2-(CO₂CH₃)₂ azepine	82–83		^{1}H NMR	149
7-CH₃, 3-Ph, 2-Ph, 2-CO₂CH₃ azepine	113–114		^{1}H NMR; IR, UV, ^{1}H, ^{13}C NMR	149, 130

Table 4 (continued)

Structure	M.p. (°C)/b.p. (°C/mmHg)	Derivatives and m.p. (°C)	Spectroscopic evidence	Ref.
			X-ray	153
			IR, UV, ^{1}H, ^{13}C NMR	130
			IR, UV, ^{1}H, ^{13}C NMR	130
			IR, UV, ^{1}H, ^{13}C NMR	130
	139–141		^{1}H NMR	149
	144–145		^{1}H NMR	149

Table 4 (continued)

Structure	M.p. (°C)/b.p. (°C/mmHg)	Derivatives and m.p. (°C)	Spectroscopic evidence	Ref.
(Ph, CO₂C₂H₅, OH, OC₂H₅ azepine)	Gum		IR, UV, MS, ^{1}H NMR	151, 152
(Ph, CH₃, SCH₃, N(C₂H₅)₂ azepine)	36–37		MS, X-ray, ^{1}H, ^{13}C NMR	142
(Ph, CH₂Ph, SCH₃, N(C₂H₅)₂ azepine)	Oil		MS, ^{1}H, ^{13}C NMR	142
(Ph, Ph, Ph, Ph, CH₃O₂C azepine)			IR, UV, ^{1}H, ^{13}C NMR	130
(CH₃O₂C, CH₃, CO₂CH₃, CH₃, CH₃ azepine)	110–115/0.03		IR, UV, MS, ^{1}H NMR	148
(Ph, Ph, CH₃, CH₃, CH₃O₂C azepine)			IR, UV, ^{1}H, ^{13}C NMR	130

Table 4 (continued)

Structure	M.p. (°C)/b.p. (°C/mmHg)	Derivatives and m.p. (°C)	Spectroscopic evidence	Ref.
(Ph, Ph, CH₃, Ph, CH₃O₂C azepine)			^{1}H, ^{13}C NMR	130
(Ph, Ph, Ph, CH₃, CH₃O₂C azepine)			IR, UV, ^{1}H, ^{13}C NMR	130
(Ph, Ph, Ph, Ph, CH₃O₂C azepine)			IR, UV, ^{1}H, ^{13}C NMR	130
(Ph, 4-F-C₆H₄, Ph, 4-F-C₆H₄, CH₃O₂C azepine)			IR, UV, ^{1}H, ^{13}C NMR	130
(Ph, CO₂C₂H₅, OH, OC₂H₅, Ph azepine)		114–115	IR, UV, MS, ^{1}H NMR	151, 152
(C₂H₅O, CO₂C₂H₅, OH, OC₂H₅, Ph azepine)		Gum	IR, UV, MS, ^{1}H NMR	151, 152

Table 4 (continued)

Structure	M.p. (°C)/b.p. (°C/mmHg)	Derivatives and m.p. (°C)	Spectroscopic evidence	Ref.
[structure: azepine with CH$_3$CO$_2$, CO$_2$C$_2$H$_5$, OH, Ph, N, OC$_2$H$_5$ substituents]	73–75		IR, UV, MS, ^{1}H NMR	151, 152

REFERENCES

1. L. Wolff, *Liebigs Ann. Chem.*, 1912, **394**, 59.
2. R. Huisgen, D. Vossius and M. Appl, *Chem. Ber.*, 1958, **91**, 1.
3. R. Huisgen and M. Appl, *Chem. Ber.*, 1958, **91**, 12.
4. W. von E. Doering and R.A. Odum, *Tetrahedron*, 1966, **22**, 81.
5. G. Maier, *Angew. Chem., Int. Ed. Engl.*, 1976, **6**, 402.
6. K. Hafner, *Angew. Chem.*, 1963, **75**, 1041.
7. K. Hafner and C. König, *Angew. Chem.*, 1963, **75**, 89.
8. E. Vogel, H.J. Altenbach, J.M. Drossard, H. Schmickler and H. Stegelmeier, *Angew. Chem.*, 1980, **92**, 1053; *Angew. Chem., Int. Ed. Engl.*, 1980, **19**, 1016.
9. W. Lwowski and R.L. Johnson, *Tetrahedron Lett.*, 1967, 891.
10. W. Lwowski, T.J. Maricich and T.W. Mattingly, Jr, *J. Am. Chem. Soc.*, 1963, **85**, 1200.
11. S. Itoh, Y. Yokoyama, T. Okumoto, K. Saito, K. Takahashi, K. Satake, H. Takamuku, M. Kimura and S. Morosawa, *Bull. Chem. Soc. Jpn.*, 1990, **63**, 3162.
12. L.E. Chapman and R.F. Robbins, *Chem. Ind. (London)*, 1966, 1266; *Chem. Abs.*, 1966, **65**, 12170e.
13. W. Lwowski and O.S. Rao, *Tetrahedron Lett.*, 1980, 21, 727.
14. F.D. Marsh and H.E. Simmons, *J. Am. Chem. Soc.*, 1965, **87**, 3529.
15. F.D. Marsh, *US Pat.*, 3 268 512; *Chem. Abs.*, 1966, **65**, 15353q.
16. N.R. Ayyangar, M.V. Phatak and B.D. Tilak, *Indian J. Chem., Sect B*, 1978, **16B**, 547.
17. N.R. Ayyangar, M.V. Phatak, A.K. Purchit and B.D. Tilak, *Chem. Ind. (London)*, 1979, 853.
18. N.R. Ayyangar, R.J. Lahoti, B.S. Shinde and K.V. Srinivasan, *Indian J. Chem., Sect. B*, 1991, **30B**, 42.
19. N.R. Ayyangar, S.M. Kumar and K.V. Srinivasan, *Synthesis*, 1992, 499.
20. (a) N.R. Ayyangar, R.B. Bambal and A.G. Lugade, *J. Chem. Soc., Chem. Commun.*, 1981, 790; (b) N.R. Ayyangar, R.B. Bambal, D.D. Nikalje and K.V. Srinivasan, *Can. J. Chem.*, 1985, **63**, 887.
21. J.M. Photis, *J. Heterocycl. Chem.*, 1971, **8**, 167.
22. N.R. Ayyangar, R.B. Bambal and A.G. Lugade, *Heterocycles*, 1982, **18**, 77.
23. J.M. Photis, *J. Heterocycl. Chem.*, 1971, **8**, 729.
24. J.M. Photis, *J. Heterocycl. Chem.*, 1970, **7**, 1249.

25. N.R. Ayyangar, R.B. Bambal, K.V. Srinivasan, T.N.G. Row, V.G. Puranik, S.S. Tavale and P.S. Kulkarni, *Can. J. Chem.*, 1986, **64**, 1969.
26. (a) P. Casagrande, L. Pellacani and P.A. Tardella, *J. Org. Chem.*, 1978, **43**, 2725; (b) M. Barani, S. Fioravanti, M.A. Loreto, L. Pellacani and P. A. Tardella, *Tetrahedron*, 1994, **50**, 3829; (c) M. Barani, S. Fioravanti, L. Pellacani and P.A. Tardella, *Tetrahedron*, 1994, **50**, 11235.
27. R.A. Abramovich, W.D. Holcomb and S. Wake, *J. Am. Chem. Soc.*, 1981, **103**, 1525.
28. R.J. Sundberg and R.H. Smith, *Tetrahedron Lett.*, 1971, 267.
29. S. Rhovati and A. Bernou, *J. Chem. Soc., Chem. Commun.*, 1989, 730.
30. S. Tamura, H. Imaizumi, Y. Hashada and K. Matsui, *Bull. Chem. Soc. Jpn.*, 1981, **54**, 301.
31. N.R. Ayyangar, M.V. Phatak and B.D. Tilak, *J. Soc. Dyers Colour.*, 1979, **95**, 55.
32. M. Mitani, T. Tsuchida and K. Koyama, *Chem. Lett.*, 1974, 1209.
33. N. Torimoto, T. Shingaki and T. Nagai, *Bull. Chem. Soc. Jpn.*, 1978, **51**, 2983.
34. O. Meth-Cohn and S. Rhouati, *J. Chem. Soc., Chem. Commun.*, 1981, 241.
35. L.A. Paquette and D.E. Kuhla, *Tetrahedron Lett.*, 1967, 4517.
36. L.A. Paquette, D.E. Kuhla, J.H. Barrett and R.J. Haluska, *J. Org. Chem.*, 1969, **34**, 2866.
37. R.C. Bansal, A.W. McCulloch and A.G. McInnes, *Can. J. Chem.*, 1969, **47**, 2991.
38. H. Prinzbach, R. Fuchs and R. Kitzing, *Angew. Chem., Int. Ed. Engl.*, 1968, **7**, 67.
39. H. Prinzbach, H. Bingmann, H. Fritz, J. Markert, L. Knothe, W. Eberbach, J. Brokatzky-Geiger, J.C. Sekutowski and C. Krüger, *Chem. Ber.*, 1986, **119**, 616.
40. H. Prinzbach, G. Kaupp, R. Fuchs, M. Joyeux, R. Kitzing and J. Markert, *Chem. Ber.*, 1973, **106**, 3824.
41. G. Kaupp and H. Prinzbach, *Org. Photochem. Synth.*, 1976, **2**, 1.
42. H. Prinzbach, H. Babsch, H. Fritz and P. Hug, *Tetrahedron Lett.*, 1977, 1355.
43. H. Prinzbach and H. Babsch, *Heterocycles*, 1978, **11**, 113.
44. R.P. Gandhi and V.K. Chadha, *Indian J. Chem.*, 1971, **9**, 305.
45. M. Christl and H. Leininger, *Tetrahedron Lett.*, 1979, 1553.
46. M.G. Barlow, G.M. Harrison, R.N. Haszeldine, W.D. Morton, P. Shaw-Luckman and M.D. Ward, *J. Chem. Soc., Perkin Trans. 1*, 1982, 2101.
47. (a) N. Manisse and J. Chuche, *J. Am. Chem. Soc.*, 1977, **99**, 1272; (b) N. Manisse and J. Chuche, *Tetrahedron*, 1977, **33**, 2399.
48. (a) R.F. Childs and A.W. Johnson, *J. Chem. Soc. C*, 1966, 1950; (b) R.F. Childs, R. Grigg and A.W. Johnson, *Chem. Commun.*, 1966, 442.
49. G.B. Gill, D.J. Harper and A.W. Johnson, *J. Chem. Soc. C*, 1968, 1675.
50. K.H. Eckhardt, D. Hege, W. Massa, H. Perst and R. Schmidt, *Angew. Chem.*, 1981, **93**, 713; *Angew. Chem., Int. Ed. Engl.*, 1981, **20**, 699.
51. H. Perst, W. Massa, M. Lumm and G. Baum, *Angew. Chem.*, 1985, **97**, 859.
52. M. Georgarakis, H.J. Rosenkranz and H. Schmid, *Helv. Chim. Acta*, 1974, **54**, 819.
53. A. Padwa, J. Smolanoff and A. Tremper, *J. Org. Chem.*, 1976, **41**, 543.
54. S.S. Mochalov, A.N. Fedotov, A.I. Sizov and Yu.S. Shabarov, *Zh. Org. Khim.*, 1979, **15**, 1425; *Chem. Abs.*, 1979, **91**, 192930b.
55. J.W. Lown, M.H. Akhtar and W.M. Dadson, *J. Org. Chem.*, 1975, **40**, 3363.
56. M.G. Barlow, S. Culshaw, R.N. Haszeldine and W.D. Morton, *J. Chem. Soc., Perkin Trans. 1*, 1982, 2105.
57. N.R. Ayyangar, A.K. Purohit and B.D. Tilak, *Chem. Commun.*, 1981, 399.
58. L.A. Paquette, D.E. Kuhla and J.H. Barrett, *J. Org. Chem.*, 1969, **34**, 2879.
59. L.A. Paquette, J.A. Barrett and D.E. Kuhla, *J. Am. Chem. Soc.*,1969, **91**, 3616.
60. M. Mahendran and A.W. Johnson, *J. Chem. Soc. C*, 1971, 1237.

61. L.A. Paquette and D.E. Kuhla, *J. Org. Chem.*, 1969, **34**, 2885.
62. K. Saito and K. Takahashi, *Heterocycles*, 1979, **12**, 263.
63. K. Saito, H. Kojima, T. Okudaira and K. Takahashi, *Bull. Chem. Soc. Jpn.*, 1983, **56**, 175.
64. K. Saito, M. Kozaki and K. Takahashi, *Heterocycles*, 1990, **31**, 1491.
65. J.E. Baldwin and R.A. Smith, *J. Am. Chem. Soc.*, 1965, **87**, 4819.
66. R.A. Smith, J.E. Baldwin and I.C. Paul, *J. Chem. Soc. B*, 1967, 112.
67. L.A. Paquette, D.E. Kuhla, J.H. Barrett and L.M. Leichter, *J. Org. Chem.*, 1969, **34**, 2888.
68. W.S. Murphy and J.P. McCarthy, *Chem. Commun.*, 1968, 1155.
69. K. Saito, A. Yoshino, H. Watanabe and K. Takahashi, *Heterocycles*, 1992, **34**, 497.
70. J.R. Wiseman and B.P. Chong, *Tetrahedron Lett.*, 1969, 1619.
71. (a) K. Satake, R. Okuda, M. Hashimoto, Y. Fujiwara, I. Watadani, H. Okamoto, K. Kimura and S. Morosawa, *Chem. Commun.*, 1991, 1154; (b) K. Satake, M. Saitoh, M. Kimura and S. Morosawa, *Heterocycles*, 1994, **38**, 769; (c) K. Satake, R. Okuda, M. Hashimoto, Y. Fujiwara, H. Okamoto, M. Kimura and S. Morosawa, *J. Chem. Soc., Perkin Trans. 1*, 1994, 1753.
72. O. Sterner, B. Steffan and W. Steglich, *Tetrahedron*, 1987, **43**, 1075.
73. G. Schaden, *Chem. Ber.*, 1973, **106**, 1038.
74. G. Schaden, *Chem. Ber.*, 1973, **106**, 2084.
75. G. Schaden, in C.E.R. Jones and C.A. Cramers (Eds.), *Proceedings of 3rd International Symposium on Analytical Pyrolysis, 1976*, Elsevier, Amsterdam, 1976, p. 289; *Chem. Abs.*, 1977, **87**, 151885q.
76. M. Appl and R. Huisgen, *Chem. Ber.*, 1959, **92**, 2961.
77. R. Smalley and H. Suschitzky, *J. Chem. Soc., Suppl.*, 1964, 5922.
78. M. Ogata, H. Matsumoto and H. Kano, *Tetrahedron*, 1969, **25**, 5205.
79. R.J. Sundberg, M. Brenner, S.R. Suter and B.P. Das, *Tetrahedron Lett.*, 1970, 2715.
80. R.J. Sundberg, S.R. Suter and M. Brenner, *J. Am. Chem. Soc.*, 1972, **94**, 513.
81. R.J. Sundberg and R.W. Heintzelman, *J. Org. Chem.*, 1974, **39**, 2546.
82. B.A. DeGraff, D.W. Gillespie and R.J. Sundberg, *J. Am. Chem. Soc.*, 1974, **96**, 7491.
83. F.P. Tsui, Y.H. Chang, T.M. Vogel and G. Zon, *J. Org. Chem.*, 1976, **41**, 3381.
84. J.H. Boyer and C.-C. Lai, *J. Chem. Soc., Perkin Trans. 1*, 1977, 74.
85. B. Nay, E.F.V. Scriven, H. Suschitzky, D.R. Thomas and S.E. Carroll, *Tetrahedron Lett.*, 1977, 1811.
86. (a) R. Purvis, R.K. Smalley, W.A. Strachan and H. Suschitzky, *J. Chem. Soc., Perkin Trans. 1*, 1978, 191; (b) R.K. Smalley, W.A. Strachan and H. Suschitzky, *Synthesis*, 1974, 503.
87. A.K. Schrock and G.B. Schuster, *J. Am. Chem. Soc.*, 1984, **106**, 5228, and references cited therein.
88. C.J. Shields, D.R. Chrisope, G.B. Schuster, A.J. Dixon, M. Poliakoff and J.J. Turner, *J. Am. Chem. Soc.*, 1987, **109**, 4723.
89. T.-Y Liang and G.B. Schuster, *J. Am. Chem. Soc.*, 1987, **109**, 7803.
90. Y.Z. Li, J.P. Kirby, M.W. George, M. Poliakoff and G.B. Schuster, *J. Am. Chem. Soc.*, 1988, **110**, 8092.
91. (a) N. Soundararajan and M.S. Platz, *J. Org. Chem.*, 1990, **55**, 2034; (b) V. Senthil and N. Soundararajan, *Rapid Commun. Mass Spectrom.*, 1994, **8**, 755; *Chem. Abs.*, 1994, **121**, 300525.
92. U. Henriksen and O. Buchardt, *Tetrahedron Lett.*, 1990, **31**, 2443.
93. C.G. Younger and R.A. Bell, *J. Chem. Soc., Chem. Commun.*, 1992, 1359.
94. R.A. Odum and G. Wolf, *J. Chem. Soc., Chem. Commun.*, 1973, 360.

95. F. Bosold, G. Boche and W. Kleemiss, *Tetrahedron Lett.*, 1988, **29**, 1781.
96. M.A. Berwick, *J. Am. Chem. Soc.*, 1971, **93**, 5780.
97. S.E. Carroll, B. Nay, E.F.V. Scriven and H. Suschitzky, *Tetrahedron Lett.*, 1977, 943.
98. Y. Ohba, S. Kubo, M. Nakai, A. Nagai and M. Yoshimoto, *Bull. Chem. Soc. Jpn.*, 1986, **59**, 2317.
99. M. Ogata, H. Kano and H. Matsumoto, *Chem. Commun.*, 1968, 397.
100. M.F.G. Stevens, A.C. Mair and J. Reisch, *Photochem. Photobiol.*, 1971, **13**, 441; *Chem. Abs.*, 1971, **75**, 88275a.
101. M.F.G. Stevens and A.C. Mair, *J. Chem. Soc. C*, 1971, 2317.
102. A.E. Bliss, T.B. Brown, M.F.G. Stevens and C.K. Wong, *J. Pharm. Pharmacol.*, 1979, **31**, Suppl. (Br. Pharm. Conf. 1979) 66pp; *Chem. Abs.*, 1980, **92**, 140360s.
103. R.A. Mustill and A.H. Rees, *J. Org. Chem.*, 1983, **48**, 5041.
104. T.B. Brown, P.R. Lowe, C.H. Schwalbe and M.F.G. Stevens, *J. Chem. Soc., Perkin Trans. 1*, 1983, 2485.
105. Y. Ohba, S. Kubo, T. Nishiwaki and N. Aratani, *Heterocycles*, 1984, **22**, 457.
106. R. Purvis, R.K. Smalley, H. Suschitzky and M.A. Alkhader, *J. Chem. Soc., Perkin Trans. 1*, 1984, 249.
107. M. Azadiardakani, S.M. Salem, R.K. Smalley and D.I. Patel, *J. Chem. Soc., Perkin Trans. 1*, 1985, 1121.
108. C.J. Shields, D.E. Falvey, G.B. Schuster, O. Buchardt and P.E. Nielsen, *J. Org. Chem.*, 1988, **53**, 3501.
109. H. Takeuchi and K. Koyama, *J. Chem. Soc., Perkin Trans. 1*, 1982, 1269.
110. R.A. Odum and M. Brenner, *J. Am. Chem. Soc.*,1966, **88**, 2074.
111. M. Masaki, K. Fukui and J. Kita, *Bull. Chem. Soc. Jpn.*, 1977, **50**, 2013.
112. R.J. Sundberg, B.P. Das and R.H. Smith, *J. Am. Chem. Soc.*, 1969, **91**, 658.
113. J.I.G. Cadogan and M.J. Todd, *J. Chem. Soc. C*, 1969, 2808.
114. F.R. Atherton and R.W. Lambert, *J. Chem. Soc., Perkin Trans. 1*, 1973, 1079.
115. T. De Boer, J.I.G. Cadogan, H.M. McWilliam and A.G. Rowley, *J. Chem. Soc., Perkin Trans. 2*, 1975, 554.
116. J.I.G. Cadogan and R.K. Mackie, *J. Chem. Soc. C*, 1969, 2819.
117. J.I.G. Cadogan, D.J. Sears and D.M. Smith, *Chem. Commun.*, 1968, 1107.
118. J.I.G. Cadogan, D.J. Sears, D.M. Smith and M.J. Todd, *J. Chem. Soc. C*, 1969, 2813.
119. E. Vogel, R. Erb, G. Lenz and A.A. Bothner-by, *Liebigs Ann. Chem.*, 1965, **682**, 1.
120. L.A. Paquette, *J. Am. Chem. Soc.*, 1963, **85**, 4053.
121. L.A. Paquette, *J. Am. Chem. Soc.*, 1964, **86**, 4096.
122. S. Bátori, R. Gompper, J. Meier and H.-U. Wagner, *Tetrahedron*, 1988, **44**, 3309.
123. V. Nair, *J. Org. Chem.*, 1972, **37**, 802.
124. D.J. Anderson and A. Hassner, *J. Am. Chem. Soc.*, 1971, **93**, 4339.
125. D.J. Anderson and A. Hassner, *J. Org. Chem.*, 1973, **38**, 2565.
126. A. Hassner and D.J. Anderson, *J. Org. Chem.*, 1974, **39**, 3070
127. D.J. Anderson, A. Hassner and D.Y. Tang, *J. Org. Chem.*, 1974, **39**, 3076.
128. (a) M. Anderson and A.W. Johnson, *Proc. Chem. Soc.*, 1964, 263; (b) M. Anderson and A.W. Johnson, *J. Chem. Soc.*, 1965, 2411.
129. T.J. van Bergen and R.M. Kellogg, *J. Org. Chem.*, 1971, **36**, 978.
130. U. Goeckel, U. Hartmannsgruber, A. Steigel and J. Sauer, *Tetrahedron Lett.*, 1980, **21**, 595.
131. A.R. Katritzky, J.M. Aurrecoechea, K. Quian, A.E. Koziol and G.J. Palenik, *Heterocycles*, 1987, **25**, 387.
132. K. Satake, H. Saito, K. Kimura and S. Morosawa, *Chem. Commun.*, 1988, 1121.

133. W.A. Nasutavicus, S.W. Tobey and F. Johnson, *J. Org. Chem.*, 1967, **32**, 3325.
134. J.W. Streef and H.C. van der Plas, *Tetrahedron Lett.*, 1979, 2287.
135. J.W. Streef, H.C. van der Plas, A. van Veldhuisen and K. Goubits, *Recl. Trav. Chim. Pays-Bas*, 1984, **103**, 225.
136. J.W. Streef and H.C. van der Plas, *Heterocycles*, 1985, **23**, 2715.
137. N.A. Kravchenko, L.G. Menchikov and I.A. Cherkasov, *Bioorg. Khim.*, 1991, **17**, 283.
138. R.J. Sundberg amd K.B. Sloan, *J. Org. Chem.*, 1973, **38**, 2052.
139. D.W. Jones, *Chem. Commun.*, 1973, 67.
140. M. Nitta, K. Shibata and H. Miyano, *Heterocycles*, 1989, **29**, 253.
141. A. Mannschreck, G. Rissmann, F. Vögtle and D. Wild, *Chem. Ber.*, 1967, **100**, 335.
142. J.W. Streef, H.C. van der Plas, N. Nieman and C.H. Stam, *Recl. Trav. Chim. Pays-Bas*, 1985, **104**, 166.
143. R.A. Odum and B. Schmall, *Chem. Commun.*, 1969, 1299.
144. J.W. Streef, H.C. van der Plas, Y.Y. Wei, J.P. Declercq and M. Vanmeerssche, *Heterocycles*, 1987, **26**, 685.
145. M. Anderson and A.W. Johnson, *J. Chem. Soc.*, 1966, 1075.
146. P.J. Brignell, E. Bullock, U. Eisner, B. Gregory, A.W. Johnson and H. Williams, *J. Chem. Soc.*, 1963, 4819.
147. P.J. Brignell, U. Eisner and H. Williams, *J. Chem. Soc.*, 1965, 4226.
148. B. Gregory, E. Bullock and T.-S. Chen, *Can. J. Chem.*, 1979, **57**, 44.
149. W.F. Dittmar, J. Sauer and A. Streigel, *Tetrahedron Lett.*, 1969, 5171.
150. T. Sano, Y. Horiguchi and Y. Tsuda, *Heterocycles*, 1978, **9**, 731.
151. T. Sano, Y. Horiguchi, S. Kambe and Y. Tsuda, *Heterocycles*, 1981, **16**, 363.
152. T. Sano, Y. Horiguchi and Y. Tsuda, *Chem. Pharm. Bull.*, 1990, **38**, 3283.
153. L. Ghosez, A. Demoulin, M. Henriet, E. Sonveaux, M. Van Meerssche, G. Germain and J.P. Declercq, *Heterocycles*, 1977, **7**, 895.
154. W. Tueckmantel, *Liebigs Ann. Chem.*, 1944, 1165.
155. S. Murata, R. Yoshidome, Y. Satoh, N. Kato and H. Tomioka, *J. Org. Chem.*, 1995, **60**, 1428.

CHAPTER 12

Azepinones

Examples of the three types of azepinones are known but the most common by far are [2H]azepin-2-ones. All may be regarded as azatropones and have aroused interest for that reason. Whether or not tropolones and tropones should be regarded as 'aromatic' has been much discussed.[1-3] The present view is that the word 'aromatic' should be applied with some reservation.[4-6] Just as pyridines exhibit fewer of the ideal 'aromatic' properties (both physical and chemical) of benzene, so one might expect aza analogues of tropone and tropolone to behave even less like benzene in these respects. On the whole, that prediction seems to be correct (see below).

I. [2H]AZEPIN-2-ONES

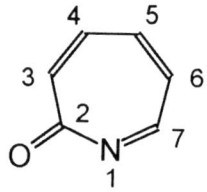

PREPARATION

This group has proved more accessible than the other two types. Treatment of the diones **1** and **2**, obtainable by Schmidt reaction on quinones (Chapter 10), with either trimethyl- or triethyloxonium fluoroborate gave the products **3** and **4** (R = CH$_3$ or C$_2$H$_5$) in fairly good yield (50–70%).[7] Azepin-2-ones have also been obtained from a range of [1*H*]azepines (**5**, R^1 to R^4 = H or CH_3) by oxygenation immediately after photolytic treatment of aryl azides in diethylamine.[8] The products (**6**, R^1 to R^4 = H or CH_3) (see Table 1) were separated by chromatography and isolated in moderate yields (20–36%).

The only α-hydroxy ketones (azatropolones) known have been made by an ingenious protocol developed by Sano and co-workers.[9] Thus, irradiation of phenylacetylene and the pyrrolinedione **7** gave the cyclobutene adduct **8**, which was converted thermally into the hydroxy ketone **9** in 60% yield. On the other hand, irradiation of **8** gave the isomeric product **10** but in only 5% yield. It was plausibly proposed that the latter result was due to a photo-induced ring-opening/reclosure of **8** (via **11** and **12**) to **13**, which would provide **10** thermally.

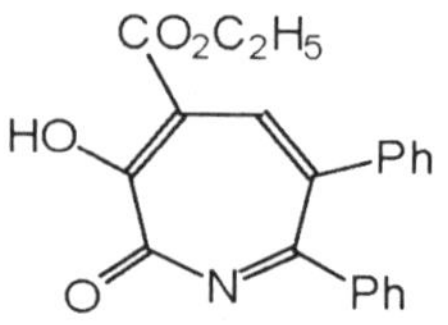

(9) 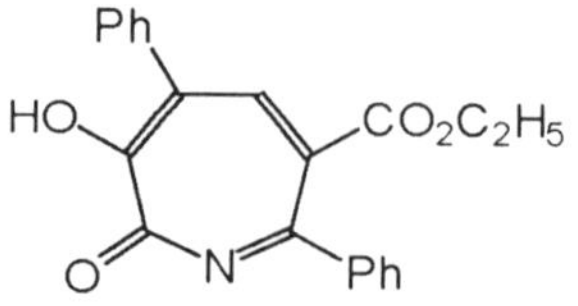 (10)

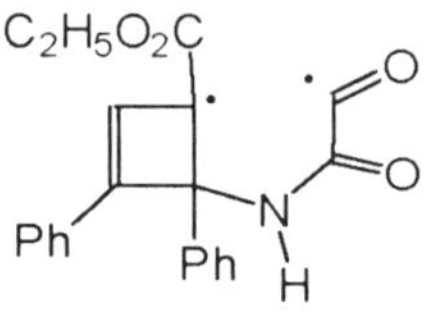

(11) 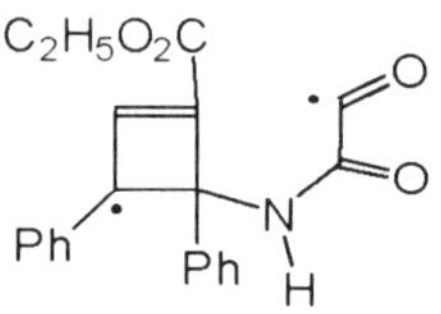 (12)

The structure of **9** was confirmed by X-ray analysis of the *O*-methyl ether of a 7-*p*-bromophenyl analogue.[10]

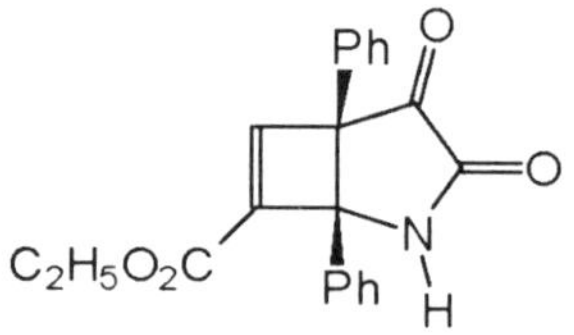

(13) 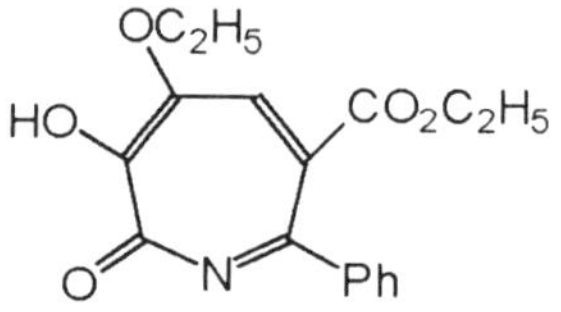 (14)

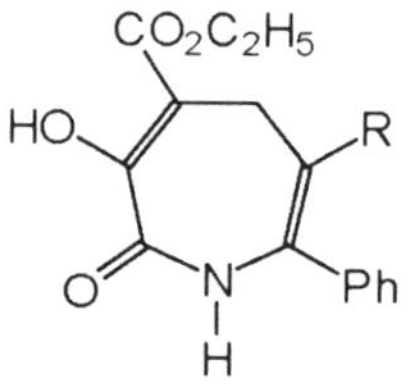

(15) 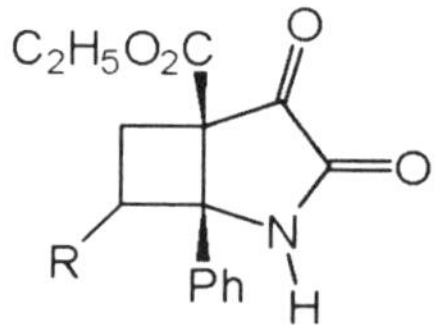 (16)

The tendency for rearrangement (e.g. to **10**) referred to above was known to intrude seriously in some cases.[11] For example, the procedure employing ethoxyacetylene yielded only 20% of the rearranged azepinone **14**.[11] Accordingly, a more robust synthesis was developed which commenced with photo-addition of the pyrrolidinedione **7** with alkenes rather than alkynes. This produced dihydroazepinones (Chapters 9 and 10)[12,13] (**15**, R = Ph, C_2H_5, OAc, OC_2H_5, SPh and H) after treatment of the cyclobutanes **16** (R = Ph, C_2H_5, OAc, OC_2H_5, SPh and H) with DBU.[11] DDQ dehydrogenation of **15** was capricious, but the *O*-methyl ethers of **15** underwent slow but moderate yielding reaction with DDQ; the products were the *O*-methyl ethers of the azepinones **17** (X = CH_3, R = Ph, C_2H_5, OAc, OC_2H_5, SPh, H). All except **17** (X = CH_3, R = H) were easily purified and appeared to be yellow gums; however, some were later found to be crystalline.[14]

(17) **(18)** **(19)** **(20)**

Less highly substituted azepinones were made by two modifications. Decarboxylation of **15** (R = Ph or OC_2H_5) using $CaCl_2$/DMSO yielded the diones **18** (R = Ph, OC_2H_5), which were dehydrogenated using DDQ, producing **19** (R = Ph, OC_2H_5).[15] Alternatively, eliminative ring expansion of **20**[17] (hydride reduction products from **16**) as O-acetates or mesylates (X = Ac or OMs) with DBU gave the dihydroazepinones **21** (R = Ph, OC_2H_5, vinyl), which DDQ

(21) **(22)**

(23) **(24)**

converted into the fully unsaturated products (**22**, R = Ph, OC_2H_5, vinyl) in variable but satisfactory yields.[16] Finally, there is a report[18] that the β-lactam **23** reacted with sodium hydride to give 40% of the azepinone **24**.

PROPERTIES

On the whole, azepin-2-ones are solids of reasonable stability; the 3-hydroxy and 3-methoxy derivatives are yellow. Fairly extensive physical data[7–17] are available and serve to illustrate that these compounds behave as cyclic ketones with multiple unsaturation; there are no striking features suggesting abnormal electron delocalisation.[7] Thus peripheral hydrogen atoms resonate between about δ 6.5 and 8.0, M − CO peaks are generally seen in the mass spectra and the only X-ray analysis reported[10] reveals that the heterocyclic ring is non-planar and bond lengths are alternate. Infrared stretching frequencies attributable to carbonyl groups are seen in the region 1660–1695 cm^{-1}, usual for $\alpha\beta$-unsaturated ketones.

Since most azepin-2-ones seen so far are fairly heavily substituted, one should not be surprised at the foregoing; a final assessment of 'aromaticity' or the lack of it awaits the discovery of simple analogues. Judging by chemical reactivity, these substances do not exhibit especially stabilising features, indeed the opposite is the case. 3-Hydroxyazepin-2-ones (e.g. **17**, X = H) are extremely reactive in protic solvents,[9,14] generating pyridine esters (**25**, R = Ph, OC$_2$H$_5$, OAc, C$_2$H$_5$, SPh, H) in methanol. The O-ethyl analogues (**17**, X = C$_2$H$_5$) also

(25)

underwent ring contraction to pyridine esters,[14] but usually this required acid or base. The mechanism has been discussed;[14] compared with tropolones which, in general, require more forcing conditions, these azepin-2-one analogues are far less stable. The 3-hydroxy compounds **17** (X = H) give yellow-green colorations with iron(III) chloride and are sufficiently acidic to methylate with diazomethane.[9,14]

It is entirely to be expected that the azepinones **3** and **4** were hydrolytically converted back to their precursors (**1** and **2**).[7] Instability in solution prevented the measurement of pK_a values.[14] One has to conclude that the introduction of a nitrogen atom to this ring system has had a considerable perturbing effect and that these 'azotropolones' seem unlikely to exhibit the interesting features associated with their carbocyclic analogues.

Azepinones

Table 1. [2H]Azepin-2-ones

Structure	M.p. (°C)	Spectroscopic evidence	Ref.
7-N(C₂H₅)₂ azepin-2-one	80–81	UV, ¹H NMR	8
3-CH₃, 7-N(C₂H₅)₂	84–85	UV, ¹H NMR	8
3-C₂H₅, 7-N(C₂H₅)₂	61–62	UV, ¹H NMR, MS, IR	8
3-iC₃H₇, 7-N(C₂H₅)₂	80–81	UV, ¹H NMR	8
3-Ph, 7-N(C₂H₅)₂	116-118	UV, ¹H NMR	8
4-CH₃, 7-N(C₂H₅)₂	77–78	UV, ¹H NMR	8
3-OH, 7-Ph	133–138	IR, UV, MS, ¹H NMR	15
3,4-(CH₃)₂, 7-N(C₂H₅)₂	Liquid	¹H NMR	8
3-OCH₃, 6-OC₂H₅, 7-Ph	130–133	IR, UV, MS, ¹H, ¹³C NMR	14

Table 1 (continued)

Structure	M.p. (°C)	Spectroscopic evidence	Ref.
HO-, OC₂H₅, O, N, Ph (azepinone)	Yellow gum	IR, UV, MS, ¹H NMR	15
CH₃, CH₃, O, N, N(C₂H₅)₂	115–116	¹H NMR	8
CO₂C₂H₅, OC₂H₅, O, N, Ph	109–110	IR, UV, MS, ¹H, ¹³C NMR	16
CO₂C₂H₅, Ph, O, N, Ph	Colourless gum	IR, UV, MS, ¹H, ¹³C NMR	16
CO₂C₂H₅, HO, O, N, Ph	Yellow gum 187–189	UV IR, MS, ¹H, ¹³C NMR	11 14
CH₃, CH₃, O, N, N(C₂H₅)₂	110–112	¹H NMR	8
CO₂C₂H₅, HO, Ph, O, N, Ph	151–154	IR, UV, MS, ¹H, ¹³C NMR	9, 14

Table 1 (continued)

Structure	M.p. (°C)	Spectroscopic evidence	Ref.
[azepinone with CH$_3$O, CO$_2$C$_2$H$_5$, Ph, Ph substituents]	131–134	IR, ^{1}H, ^{13}C NMR	9, 14
[azepinone with HO, CO$_2$C$_2$H$_5$, OC$_2$H$_5$, Ph substituents]	92–95	UV IR, MS, ^{1}H, ^{13}C NMR	11 14
[azepinone with CH$_3$O, CO$_2$C$_2$H$_5$, OC$_2$H$_5$, Ph substituents]	Yellow gum 72–73	UV IR, MS, ^{1}H, ^{13}C NMR	11 14
[azepinone with HO, CO$_2$C$_2$H$_5$, Ph, 4-BrC$_6$H$_4$ substituents]	153–154 155–156	IR, UV MS, ^{1}H, ^{13}C NMR	10 14
[azepinone with CH$_3$O, CO$_2$C$_2$H$_5$, Ph, 4-BrC$_6$H$_4$ substituents]	133–136	IR, UV, X-ray MS, ^{1}H, ^{13}C NMR	10 14
[azepinone with CH$_3$O, CO$_2$C$_2$H$_5$, OCOCH$_3$, Ph substituents]	Yellow gum	UV IR, MS, ^{1}H NMR	11 14
[azepinone with CH$_3$O, CO$_2$C$_2$H$_5$, SPh, Ph substituents]	Yellow gum	UV IR, MS, ^{1}H NMR	11 14

Table 1 (continued)

Structure	M.p. (°C)	Spectroscopic evidence	Ref.
(3-CH$_3$O, 4-CO$_2$C$_2$H$_5$, 6-C$_2$H$_5$, 7-Ph azepinone)	Yellow gum	UV IR, MS, ^{1}H NMR	11 14
(3-HO, 4-Ph, 6-CO$_2$C$_2$H$_5$, 7-Ph azepinone)	191–194	UV, IR, MS, ^{1}H NMR	9
(3-CH$_3$O, 4-Ph, 6-CO$_2$C$_2$H$_5$, 7-Ph azepinone)	102–105	IR, UV, ^{1}H NMR MS	9 14
(3-HO, 4-C$_2$H$_5$O, 6-CO$_2$C$_2$H$_5$, 7-Ph azepinone)	179–181	IR, UV, MS, ^{1}H, ^{13}C NMR	14
(3-CH$_3$O, 4-C$_2$H$_5$O, 6-CO$_2$C$_2$H$_5$, 7-Ph azepinone)	60–63	IR, UV, MS, ^{1}H NMR	14
(3-CH$_3$, 4-OCH$_3$, 6-CH$_3$, 7-CH$_3$ azepinone)	52	^{1}H NMR	7
(3-CH$_3$, 4-OC$_2$H$_5$, 6-CH$_3$, 7-CH$_3$ azepinone)	71–72	^{1}H NMR	7

Table 1 (continued)

Structure	M.p. (°C)	Spectroscopic evidence	Ref.
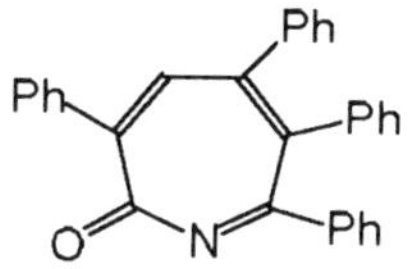 (Ph, Ph, Ph, Ph azepinone structure)	52	IR, ^{1}H NMR, MS	18
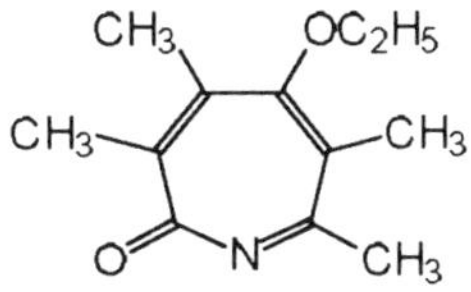 (CH_3, CH_3, OC_2H_5, CH_3, CH_3 azepinone)	68.5–70	^{1}H NMR	7
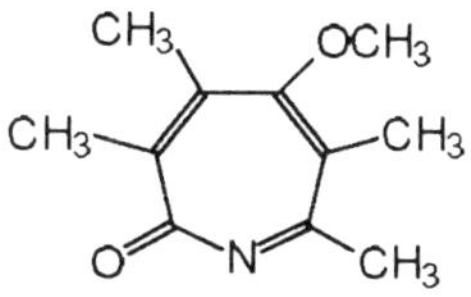 (CH_3, CH_3, OCH_3, CH_3, CH_3 azepinone)	84–85	^{1}H NMR	7

II. [3H]AZEPIN-3-ONES

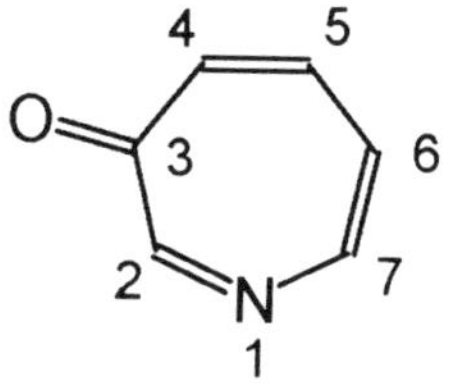

The only examples are to be found in the work of Sano and co-workers.[7,9,14] When compounds **16** (R = Ph, OC_2H_5, OAc) were converted into the lactim ethers **26** (R = Ph, OC_2H_5, OAc), the latter reacted with tin(IV) chloride giving

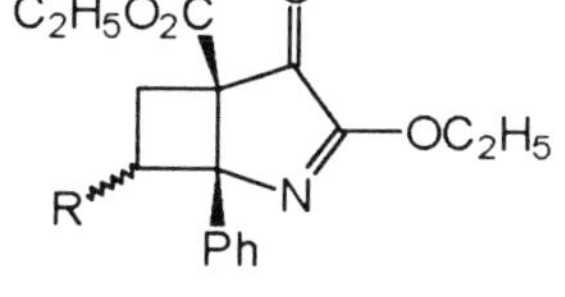

(26) 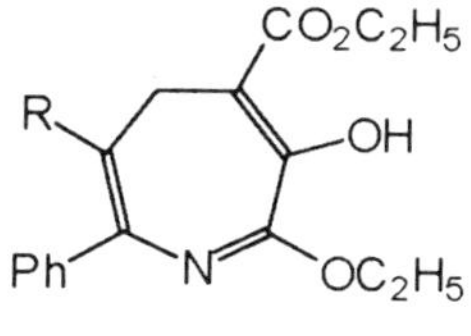 (27) 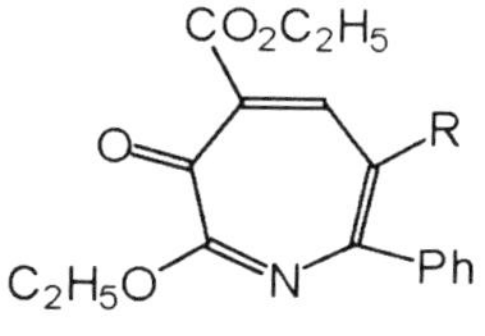 (28)

the ring-expanded products **27** (R = Ph, OC_2H_5, OAc), which DDQ dehydrogenation transformed into the [3*H*]azepin-3-ones **28** (R = Ph, OC_2H_5, OAc).[7] It was later reported[14] that **28** (R = H) could also be obtained in this way. Alternatively, diazomethane methylation of the hydroxy ketones **10**, **14** and **19** (R = OC_2H_5) gave roughly equal amounts of the azepin-3-ones **29** (R = Ph, OC_2H_5) and **30**[9,14] and the corresponding azepin-2-ones **31** (R = Ph, OC_2H_5) and **32**. Such behaviour is typical of tropolones, but **9** with diazomethane gave exclusively one product (**17**, X = CH_3, R = Ph) in 70% yield.[9] This result would appear to be attributable to hydrogen bonding involving the ester group at C-4.

(29)　　　(30)

(31)　　　(32)

The chemistry of [3*H*]azepin-3-ones has not been studied.

Table 2. [3*H*]Azepin-3-ones

Structure	M.p. (°C)	Spectroscopic evidence	Ref.
	79–81	IR, UV, MS, ^{1}H, ^{13}C NMR	14
	Yellow gum	IR, UV, ^{1}H NMR	22, 14

Table 2 (continued)

Structure	M.p. (°C)	Spectroscopic evidence	Ref.
(3-oxo-2-methoxy-4-ethoxy-6-ethoxycarbonyl-7-phenyl-azepine)	Yellow gum	IR, UV, MS, ^{1}H NMR	14
(3-oxo-2-methoxy-4-phenyl-6-ethoxycarbonyl-7-phenyl-azepine)	133–135	IR, UV, ^{1}H NMR MS	9, 14
(3-oxo-2-ethoxy-4-ethoxycarbonyl-6-phenyl-7-phenyl-azepine)	Yellow gum	IR, UV, ^{1}H NMR ^{13}C NMR, MS	22, 14
(3-oxo-2-ethoxy-4-ethoxycarbonyl-6-ethoxy-7-phenyl-azepine)	Yellow gum	IR, UV, ^{1}H NMR ^{13}C NMR, MS	22, 14
(3-oxo-2-ethoxy-4-ethoxycarbonyl-6-acetoxy-7-phenyl-azepine)	Yellow gum	IR, UV, ^{1}H NMR MS	22, 14

III. [4H]AZEPIN-4-ONES

Until recently, the only references to [4H]azepin-4-ones appeared during 1972. First, it was disclosed that if the oxygenation of [1H]azepines (**5**, $R^2 = R^3 = R^4 = H$, $R^1 = H$ or ^{i}Pr) previously mentioned (Section I) was carried out in the presence of small amounts of copper(II) ion, the products included azepin-4-ones (**33**, R = H or ^{i}Pr).[8] Yields were low.

(**33**) (**34**) (**35**)

The dione **34** reacted with triethyloxonium fluoroborate giving extremely low yields (<5%) of the azepin-4-one **35**.[7,19] One has to suspect that these compounds were reactive. The latter view is supported by very recent matrix photolysis work.[20] In concentrated nitrogen matrices at 10–20 K, irradiation ($\lambda = 280 \pm 10$ nm) of both dichloro- and dibromo-4-azidophenols (**36**, X = Cl,

(**36**) (**37**) (**38**)

Br) showed evidence of the presence of azepin-4-ones (**37**, R = Cl, Br). Thus, IR bands at 1625 cm^{-1} (C=N) and 1679, 1672, 1669 cm^{-1} (C=O) were seen to persist up to near room temperature. In less concentrated matrices, evidence ($\nu_{C=C=N}$ 1887 cm^{-1}) for the established[20] didehydroazepine intermediates **38** (X = Cl, Br), but not for ketones, was seen. Intramolecular hydrogen transfer in concentrated matrices was adduced to explain the appearance of **37**. At present this seems unlikely to provide a synthetic route to the [4H]azepin-4-ones.[21]

Table 3. [4H]Azepin-4-ones

Structure	M.p. (°C)	Spectroscopic evidence	Ref.
(2-N(C₂H₅)₂, 3-C₂H₅ azepinone)	Liquid	IR, UV, ¹H NMR	8
(2-N(C₂H₅)₂, 3-iC₃H₇ azepinone)	Liquid	¹H NMR, UV	8
(2-N(C₂H₅)₂, 3-CH₃, 7-CH₃ azepinone)	57–58 69	IR UV, MS, ¹H NMR	7 19

REFERENCES

1. (a) D.M.G. Lloyd, *Carbocyclic Non-Benzenoid Aromatic Compounds*, Elsevier, Amsterdam, 1966, p. 117; (b) G.M. Badger, *Aromatic Character and Aromaticity*, Cambridge University Press, Cambridge, 1969.
2. (a) W.N. Hubbard, C. Katz, G.B. Guthrie and G. Waddington, *J. Am. Chem. Soc.*, 1952, **74**, 4456; (b) R.B. Turner, W.R. Meador, W. Von E. Doering, L.H. Knox, J.R. Mayer and D.W. Wiley, *J. Am. Chem. Soc.*, 1957, **79**, 4127.
3. E.J. Forbes, M.J. Gregory, T.A. Hamor and D.J. Watkin, *J. Chem. Soc., Chem. Commun.*, 1966, 114.
4. M.J.S. Dewar and N. Trinajstic, *Croat. Chem. Acta*, 1970, **42**, 1; *Chem. Abs.*, 1970, **72**, 110617a.
5. D.J. Bertelli and T.G. Andrews, Jr, *J. Am. Chem. Soc.*, 1969, **91**, 5280.
6. (a) D.J. Bertelli, T.G. Andrews, Jr, and P.O. Crews, *J. Am. Chem. Soc.*, 1969, **91**, 5286; (b) D.J. Bertelli and T.G. Andrews, Jr, *Tetrahedron Lett.*, 1967, **45**, 4467.
7. E.J. Moriconi and I.A. Maniscalco, *J. Org. Chem.*, 1972, **37**, 208.
8. R.J. Sundberg, S.R. Suter and M. Brenner, *J. Am. Chem. Soc.*, 1972, **94**, 513.
9. T. Sano, Y. Horiguchi and Y. Tsuda, *Heterocycles*, 1978, **9**, 731.
10. Y. Tsuda, M. Kaneda, T. Sano, Y. Horiguchi and Y. Iitaka, *Heterocycles*, 1979, **12**, 1423.

11. T. Sano, Y. Horiguchi and Y. Tsuda, *Heterocycles*, 1979, **12**, 1427.
12. T. Sano and Y. Tsuda, *Heterocycles*, 1976, **4**, 1229.
13. T. Sano, Y. Tsuda, H. Ogura, K. Furuhata and Y. Iitaka, *Heterocycles*, 1976, **4**, 1233.
14. T. Sano, Y. Horiguchi and Y. Tsuda, *Chem. Pharm. Bull.*, 1990, **38**, 3283.
15. Y. Horiguchi, T. Sano and Y. Tsuda, *Heterocycles*, 1985, **23**, 1509.
16. T. Sano, Y. Horiguchi, K. Tanaka and Y. Tsuda, *Chem. Pharm. Bull.*, 1985, **33**, 5197.
17. T. Sano, K. Tanaka, Y. Horiguchi and Y. Tsuda, *Heterocycles*, 1985, **23**, 813.
18. K. Narasimhan, P.R. Kumar and T. Selvi, *Heterocycles*, 1984, **22**, 2751.
19. R.G. Cooke and I.M. Russell, *Aust. J. Chem.*, 1972, **25**, 2421.
20. I.R. Dunkin, A. El Ayeb and M.A. Lynch, *J. Chem. Soc., Chem. Commun.*, 1994, 1695.
21. I.R. Dunkin, personal communication.
22. T. Sano, Y. Horiguchi, S. Kambe and Y. Tsuda, *Heterocycles*, 1981, **16**, 363.

Index

acetates, 468
acetic acid, 451
acetoxy groups, 498
2-acetoxy[3H]azepine, 451
N-acetyl azepanes, anodic oxidation, 25
acetyl-substituted 1,3-dihydro[2H]azepinone, 451
acetylenes, 306
    Diels–Alder reaction, 491
    ring closure, 326
acetylenic ester, ring expansion, 413
acrylonitrile, 133, 292
acyclic azido ester, 124
acyclic haloamino acid, 11
N-acyl-[1H]azepine, 494
N-acyl iminium salts, 25
acylation, 24, 30, 289
    N-acylation, 133
    thiocaprolactam, 144
N-acyliminium ion, 26
acyloins, 219, 283
Adam's platinum oxide, 16
adipic acid derivatives, cyclisation, 275
adipimide, 135, 275
adipodialdehyde, reductive cyclisation, 10
adipoyl chloride, 276, 362
Alcaligenes faecalis, 137
alcohol, 304, 524
aliphatic amine, 524
    azepane-2,7-dione, 276
alkenamine, 20
alkenes, 141
    [2 + 2]cycloaddition, 467
2-alkenyl-N-alkyl azepane, 138
3-alkenyl-2-ethoxyazepane, 26
alkenylazepane, 23–4

2-alkenylazepane, 23, 29, 138
2-alkoxy[3H]azepine, 524, 525
N-alkoxyalkylsuccinimide, photolysis, 406
alkyl imidates, 126
2-alkyl-Δ^1-pyrrolinium salts, 290
alkyl-3,4,5,6-tetrahydro[2H]azepine, 325
2-alkyl-3,4,5,6-tetrahydro[2H]azepine, 17
alkylation, 30
    N-alkylation, 24, 140
    O-alkylation, 452
    5-phenyl[3H]azepine, 580
    thiocaprolactam, 144
2-alkylazepanes, 15
2-alkylbenzothiazole, 412
N-alkylcaprolactam, 138
    anodic oxidation, 135
2-alkylcyclohexanone, 121
3-alkylideneazepane, 211
2-alkylindazoles, irradiation, 451
N-alkyllactams, 138
alkyllithium, 323
N-alkylthioalkylsuccinimides, photolysis, 406
alkyne cyclisations, iminium ion, 211
alkynes, 141
    cycloadditions, 423
    treatment, 144
alkynyl borane, 138
alkynylalkyl-2-ethoxyazepane, 26
alkynylamine, 138
2-alkynylazepane, 26, 144
alkynylsilane, 18
allenic amine, 20
allenic structure, 18
allyl alkoxides, lithium salts, 126

607

allyl enol ethers, Claisen rearrangement, 261
N-allyl-2-propenoic acid amides, cyclisation, 361
(2Z,4E)-α-allyldiene-nitrile, 430
allylic peroxidation, 126
S-allylthioimidate, 144
N-allylthiolactam, 144
alumina, 12, 123
aluminium chloride, 25, 130
aluminium hydride, 23
amide reduction, 305
amidrazone, 527
amines, 9
  non-cyclic, 305
  primary, 10, 450
  tertiary, 305
ω-amino acid, 30
amino acid chloride, chloroalkenyl-protected, 219
α-amino acid derivative, 10
amino alcohols, 12
3-amino-azepane, 24
2-amino[3H]azepine, 529
  hydrolysis, 451
3-amino[3H]azepine, 525
2-amino-1-azetine, 478
2-amino-7-bromo-5-phenyl[3H]azepine hydrobromide, 409
3-amino-caprolactam, 117, 131
6-amino-6-desoxy-D-glucose, 12
amino diacids, 143
6-amino-6,6-dimethylhexanoic acid, 123
7-amino-3,4,5,6-tetrahydro[2H]azepine, 327
2-amino-6,7,8,9-tetrahydro[5H]pyrazino[2,3-d]azepine, 222
1-aminoacryl derivatives, 414
3-aminoalkylcaprolactam, 121
7-aminoalkylcaprolactam, 121
aminoalkynes, 211
6-aminocaproic acid, 143
aminocaprolactam, 124, 128
3-aminocaprolactam, 120, 137
  catalytic reduction, 140
  L-lysine conversion, 143
(R)-aminocaprolactam, 137
D-aminocaprolactam racemase, 137
2-aminocyclohexanone oxime, 120
6-aminohexanoic acid, 143
  cyclisation, 123
aminomethylation–desilylation, 314

ammonia, 9, 124, 137
1-anilino-amino-1,5,6,7-tetrahydro[4H]azepin-4-one, 396
annulated rings, 222
annulations, caprolactam nucleus, 140–2
anodic oxidation, 15, 25, 135
antimony pentachloride, 134
aqueous alkali, 11
arenes, 25, 130
aromatic azides, photolysis, 578
aryl azides, photolysis, 593
aryl Grignard reagents, 428
aryl isocyanates, 2,3,4,5-tetrahydro[1H]azepine, reaction, 289
2-aryl pyrroline-4,5-dione, 467
aryl sulphimide, 130
7-aryl-1,3-dihydro[2H]azepin-2-one, 450
  irradiation, 454
2-aryl-3,4,5,6-tetrahydro[2H]azepine, 17
N-arylation, 25
2-arylazepane, 12
2-arylcyclohexanone, 122
aryllithium, 16
arylmagnesium bromide, 252
7-arylmethyl-caprolactam, 122
2-arylmethylcyclohexanone, 121
2-aryloxy[3H]azepines, 524
2-arylpiperidinium methylide, 21
aza-Cope rearrangement, 451
aza-Wittig reaction, 24, 327
1-aza[3.2.1]bicyclooctanone, 21
2-azabicyclo[3.2.0]hept-6-ene, 416
2-azabicyclo[3.2.0]hepta-2,6-diene, 530
azabicyclo[3.2.0]heptane-3,4-dione, 579
2-azabicyclo[3.2.0]heptane-3,4-dione, 467
2-azabicyclo[3.2.0]heptene, 454
2-azabicyclo[3.2.1]oct-3-ene, 432
1-azabicyclo[5,4,0]undecane, 26
azabicycloheptane-3,5-diones, irradiation, 405
azadienes, chromium complex reaction, 446
azanorcaradiene structure, 450
[4H]azapine, 469
azaquadricyclanes, UV irradiation, 491
azatropolone, 468, 593
azatropone, 592
azepan-2-one, 117–237
  see also caprolactam
azepan-3-one, 211–16

Index 609

azepan-4-one, 216–37, 219, 282–3
  organometal reagents, 222
N-azepan-4-one, selenium dioxide oxidation, 222
azepanes, 9–109
  catalytic hydrogenation, 432
  2,3-dihydro[1H]azepine, 416
  2-substituted, 25
  3-substituted, 14
  N-substituted, cycloaddition reactions, 498
  N-unsubstituted, N-alkylation, 24
azepane diester, 10
azepane-2,3-dione, 251–9
azepane-2,4-dione, 260–6
azepane-2,5-dione, 266–74
azepane-2,6-dione, 274–5
azepane-2,7-dione, 275–80
azepane-3,4-dione, 281
azepane-3,5-dione, 281–2
azepane-3,6-dione, 282
azepane-4,5-dione, 282–3
azepane ester, 11
azepane halide, dehydrohalogenation, 305
azepane-3-keto ester, 211
azepane-4-keto ester, 211
azepane ring, annulation to, 26
azepanedione, 251–84
azepanyl phosphonate, 16
azepin-2-one, 602
  catalytic hydrogenation, 473
[2H]azepin-2-one, 592–601
[3H]azepin-3-one, 601–3
[4H]azepin-4-one, 604–5
azepine, 487–587
  catalytic hydrogenation, 16
  reduction, 16–17
[1H]azepine, 430, 488–521
  oxygenation, 593, 604
  N-substituted, 491
[1H]azepine-N-carboxylate, hydrolysis, 488
[2H]azepine, 521–2
[3H]azepine, 414, 523–77
[4H]azepine, 431, 469, 527, 577–87
azepinone, 592–605
azepinopyrrolopyrimidine, 253
azepinothiazolidine compounds, 12
azetidinium ion, 22
azeto[1,2-a]pyrrole, 411, 478
  hydrolysis, 410

azide
  acid treatment, 325
  Curtius rearrangement, 395, 400
  ion, 137
  ring closure, 326
azido ester, 489
  m-xylene, 490
  o-xylene, 490
  p-xylene, 490
azido ketone, photolysis of, 260
azido-sugar derivatives, 124
2-azidoethylthiocyclohexenone, 361
azidonitrile, 489
1,1-azidosulphide, 127
aziridine, 306, 324
  flash vacuum pyrolysis, 429
trans-aziridine, ring expansion, 493
aziridinium salt, 22
azirine aldehyde, 495

base cyclisation, 20
Beauveria sulfurescens, 28, 136
Beckmann fragmentation, 119
Beckmann rearrangement, 14, 15, 22, 118, 119, 120, 121, 123, 260, 360, 445
  azepane-2,5-dione, 267
  anti cyclohexen-2-one oxime, 359
  methane sulphonate, 326
  syn-oxime, 379, 442
  PPA, 260
bengamide A, 132
bengamide B, 132
bengamide E, 132
bengamides, 131
benzene, 16
benzeneseleninic anhydride, 136, 277
benzenesulphonyl azide, 489
benzofuran, 495
benzofuroazepine, 253
benzofuroazepinone, 261
benzofuroxan, UV irradiation, 485
benzonitrile, photochemical reaction, 446
benzophenone, 139
benzoquinone, Schmidt reaction, 268, 482
benzothiazinoazepine, 253
N-benzoyl, microbiological oxygenation of, 221
N-benzoyl-4-methylazepane, 221
N-benzoylation, 133
N-benzoylazepane, 28, 212
N-benzoylcaprolactam, CS_2 reaction, 140
benzyl-cobactin T, 132

5-benzyl-1-methyl-1,5,6,7-tetrahydro[4H]azepin-4-one, 396
7-benzyl-1-methyl-1,5,6,7-tetrahydro[4H]azepin-4-one, 396
1-benzyl-1,3,4,7-tetrahydro[2H]azepin-2-one, 368, 369
1-benzyl-4-vinylazetidinone, 372
benzylamine, 10
N-benzylazepan-4-one, 221
N-benzylcaprolactam, oxidation of, 136
N-benzylpiperid-4-one, diazomethane ring expansion, 217
benzyne, 25
bicyclic isoxazole, 524
[3.2.0] bicyclic ring, 496
[2.2.1]bicyclic system, 491
bis(methylthiomethyl)amine, 19
bisallylamine, 220
biscyclopropapyrrolidine, thermal rearrangements, 413
borane–dimethyl sulphoxide, 15
boric acid, 324
boron compounds, 134
bromination, 222, 485
5-bromo isomer, bromination, 222
3-bromocaprolactam, 372
 dehydrobromination, 138
7-p-bromophenyl, 594
butane-1,4-diol, 10
N-tert-butoxycarbonyl, 218
t-butyl-formamidine, 25
tert-butyl hypochlorite, 138
1-tert-butylazepine, 493
(S)-(−)-5-tert-butylcaprolactam, 125

C–C bond, 17–19, 25
C–N bond, 19–20
C-2, substitution at, 138–9
C-2 deprotonation, 25
C-3, 14, 252
 alkylation, 139
 substitution at, 139–40
Candida humicola, 137
caprolactams, 14, 117–237, 288, 289, 323, 451
 azepane-2,3-diones from, 251
 C-3 acylated, 139
 chlorination, 138
 hydrolysis, 143
 nitration, 140
 oxidation, 275, 276
 phosgene on, 322

 reduction, 13–16
 7-substituted, 121
 N-substituted, 136, 139
L-caprolactam, 137
L-caprolactam hydrolase, 137
caprolactam acetal, 23
caprolactam imine, 129–30
caprolactam thione, 130–1
caprolactin A, 132
caprolactin B, 132
1-carbethoxy[1H]azepine, 498, 499
N-carbomethoxyazepane, electrochemical oxidation, 213
carbon, 128
carbon tetrachloride, 137
carbonyl group, 315
 reduction, 222
carbonyl sulphide elimination, 22
carbonylation
 palladium-catalysed, 124, 382
 ring-expansion, 372
7-carboxyalkyl-caprolactam, 143
7-ω-carboxyalkyl-caprolactam, 121
catalysis, 23
catalytic hydrogenation, 12, 16, 414, 445, 451
 azepanes, 432
 azepin-2-one, 473
 2,3-dihydro[1H]azepine, 416
catalytic reduction, 140, 221
chalciporone, 521
Chalciporus piperatus, 521
charcoal, 17
chiral amino ester, reductive cyclisation, 124
chiral aminocaprolactam L-Boc-, 124
chiral β-keto ester, 15, 122
chiral imidazolidinone, 12
chiral lactams, LiAlH$_4$ reduction of, 14
chiral oxaziridine → caprolactam synthesis, 125
chiral oxime, 14
chloramine, 449
chlorination, 138, 453
4-chloro compounds, 476
1-chloro-1-nitrosocyclohexane, 126–7
6-chloro-1,2,3,7-tetrahydro[4H]azepin-4-one, 394
chloroalkenyl, 219
3-chlorocaprolactam, 136, 137
N-chlorocaprolactam, 138
2-chlorocyclohexanone, 121

Index

chloroiminium salts, 138
4-chloromethyl-1,4-dihydropyridine, 427
4-chloromethyldihydropyridine, 493
3-[N-(2-chloroprop-2-enyl)propanoic acid, Friedel–Crafts cyclisation, 393–4
chlorosilanes, reductive silylation, 430
chromium carbonyl carbene, 135
chromium complexes, 446
cinnamalaniline, 291
cinnamyl bromide, 307
circinatin, 132
Claisen rearrangement, 126
  allyl enol ether, 261
  vinyl-substituted oxazolidine, 368
Clarke–Eschweiler methylation, 18
Clemmensen reduction, 213
Cobactin T, 132
cobactins, 132
cobalt salts, 135
cobalt-catalysed carbonylation/ring-expansion, 372
cobalt-catalysed oxidation, 135
Cope rearrangement
  cyclopropyl imidate, 368
  cis-2,2-dimethyl-3-isobutenylcyclopentyl isocyanate, 373
  cis-divinylaziridine, 429
  cis-iminovinylcyclopropane, 423
copper chelate, 211
copper(II), 133
CS_2 reaction, 140
Cu^{2+} complexes, 24
CuCrO, 10
Curtius rearrangement
  azides, 395, 400
  cis-2-vinylcyclopropyl carbonylazide, 451
1-cyano-[1H]azepines, 489
cyano ester, 268
4-cyano-4-phenyl-1-methylazepane, 13
2-cyanocyclohexanone, 121
cyanodihydroazepine, 427
cyanogen bromide, 28
cyclic β-keto ester, 217
cyclic vinylaziridine, ring expansion, 414
cyclisation
  adipic acid derivatives, 275
  N-allyl-2-propenoic acid amide, 361
  dinitrile, 327
  nitrile ester, 408
  tertiary amine, 305

cyclising reactions, 123–5
cyclo-adducts, 499
cycloaddition, 252, 476, 527, 579
  alkenes, 467
  alkynes, 423
  N-azepines, 498
  cyclopentadiene, 485
3-cycloalkylcaprolactam, 121
7-cycloalkylcaprolactam, 121
cyclobutene adduct, 593
cyclohexa-2,4-diene-1-ylidenesulphonamide, 495
1,4-cyclohexadiene, 491
cyclohexane, 125
cyclohexane-1,3-diones, ring expansion, 260
cyclohexanone, ring expansion, 127
cyclohexanone oxime, mesyl derivative, 22
anti-cyclohexen-2-one oxime, Beckmann rearrangement, 359
cyclohexenone, Schmidt rearrangement, 380
cyclohexenone oxime, O-methyl ether, 23
1-cyclohexyl-amino-1,5,6,7-tetrahydro-[4H]azepin-4-one, 396
cyclolysine, 131
cyclopentadiene, cycloaddition reaction, 485
cyclopropene, 527, 579
cyclopropyl imidate, Cope rearrangement, 368

de-ethoxycarbonylation, 403
decarboxylation, boric acid, 324
dehydration, 304
  alcohol, 2,3,6,7-tetrahydro[1H]azepine, 315
dehydrobromination, 138, 485
dehydrogenation, dichlorodicyanobenzoquinone, 594, 602
dehydrohalogenation, 305
deoxygenation, nitrobenzene, 525
deprotonation, 139, 452
desulphurisation, 121
di-tert-butyl benzene, 489
di-tert-butyl[1H]azepine, 527
dialdehyde, reductive amination of, 220
2,2-dialkylazepane, 15
2,6-dialkylphenols, sodium salts, 449
1,6-diaminohexane, 10

1,8-diazabicyclo[5.4.0]undec-7-ene, 467
diazo ester ring expansion, 217
α-diazo ketones, 213
diazoacetate ring expansion, 211
diazoacetic ester ring expansion, 218
diazoalkene, 141
diazomethane methylation, 468, 602
diazomethane ring expansion, 217
diazothiolactam, 144
dibromo-4-azidophenol, 604
4,6-dibromo compound, 476
trans-3,4-dibromo-2,3,4,7-tetrahydro-[1H]azepine-2,7-dione, 408
1,5-dibromopentane, 11
dichloro-4-azidophenols, 604
4,6-dichloro compounds, 476
6,6-dichloro-N-tosylazepan-4-one, 394
dichlorodicyanobenzoquinone, 468, 595
dehydrogenation, 594, 602
3,3-dichlorolactam, 136
dicycanocyclobutene, 429
didehydroazepine, 604
Dieckmann cyclisation, 26, 140–1, 211, 216, 217
cyano ester, 268
Diels–Alder reaction, 416, 452
acetylenes, 491
dienamides, 369
dienamine derivatives, flash vacuum pyrolysis, 474
cis,cis dienes, nitrogen nucleophile, 426
diethylaluminium chloride, 23
diethylaluminium iodide, 23
2-diethylamino-5-phenyl[3H]azepine, 529
diethylaminoazepine, 451
1,6-difunctional hexane, 9
α,ω-dihalo-monocarboxylic acids, 10
1,6-dihalohexanes, 10
1,2-dihydro[3H]azepin-3-one, 474–8
1,3-dihydro[2H]azepin-2-one, 415, 449–67
photochemistry, 453
1,5-dihydro[2H]azepin-2-one, 467–72
1,7-dihydro[2H]azepin-2-one, 472–3
2,3-dihydro[4H]azepin-4-one, 478–9
3,4-dihydro[2H]azepin-2-one, 473
3,6-dihydro[2H]azepin-2-one, 474
2,3-dihydro[1H]azepine, 412–22, 451–2
catalytic hydrogenation, 416
2,3-dihydro[1H]azepine-2,3-dione, 482
2,5-dihydro[1H]azepine, 423–5
2,5-dihydro[1H]azepine-2,5-dione, 268, 482–4

2,7-dihydro[1H]azepine, 426
2,7-dihydro[1H]azepine-2,7-dione, 485–6
3,4-dihydro[2H]azepine, 442–4, 531
3,6-dihydro[2H]azepine, 444
photochemical irradiation, 446
4,5-dihydro[1H]azepine, 427–41, 578
4,5-dihydro[3H]azepine, 445–6
5,6-dihydro[2H]azepine, 444–5
trans-3,4-dihydro-3,4-dibromo-1-methylazepine-2,7-dione, 485
3,6-dihydro-3,3,6,6-tetramethyl-[2H]azepin-2-one, 373
dihydroazatropolone, 403
dihydroazepine, 17
dihydroazepinedione, 481–6
dihydroazepines, 412–46
reduction, 16–17
2,3-dihydroazepine, 414
dihydroazepinone, 128, 449–79, 594, 595
4,5-dihydroisoxazole-5-spirocyclobutane, flash vacuum thermolysis, 396
dihydropyridine, 428, 577
1,2-dihydropyridine, 527
1,4-dihydropyridine, ring expansion, 527
1,4-dihydropyridine derivatives, ring contraction, 432
dihydropyridine ethyl ester, 428
dihydropyridine methyl ester, 428
diisobutylaluminium hydride, 16, 23
2,5-diisopropyl[3H]azepine, 528
3,6-diisopropyl[3H]azepine, 528
diketene silyl acetal, 19
2,3-diketoazepane, 14
dimethoxyamine, 127
dimethoxybenzazepinone, ozonolysis, 126
dimethyl acetylenecarboxylate, 368
dimethyl acetylenedicarboxylate, 290, 395–6, 412, 426, 452, 472, 476, 579
dimethyl sulphate, 138
salt formation, 139
dimethyl sulphoxide, 291
sodium cyanide, 428
cis-2,2-dimethyl-3-isobutenylcyclopentyl isocyanate, Cope rearrangement, 373
2-dimethylamino[3H]azepine, 445
2,3-dimethylbut-2-ene, 252
7,7-dimethylcaprolactam, 119, 123
2,2-dimethylcyclohexanone oxime, 119
dimethylformamide, 291
dinitrile, 327, 528
dioxopyrrolidine, 252
N,α-diphenyl nitrone, 499

1,4-diphenyl-3,3-dichloro-1,3,4,7-
 tetrahydro[2H]azepin-2-one, 368
3,3-diphenyl-2,3,4,7-
 tetrahydro[1H]azepine-2,7-dione,
 408
3,4-diphenylcyclopentadienone, 526
4,5-*trans*-diphenyltetrahydroazepine, 291
1,3-dipolar cycloadditions, 429
1,2-disubstituted azepane, 25
2,7-disubstituted azepane, 25
4,4-disubstituted azepane, 13
2,5-disubstituted azepine, 489
3,6-disubstituted [1H]azepine, 490
2,2-disubstituted N-hydroxyazepane, 17
cis-divinylaziridine, Cope rearrangement,
 429
2,3-double bond, 498
4,5-double bond, 498, 499
double bond substitution, position 6, 291

electrochemical oxidation, 213
electrophiles, 25
electrophilic cyclisation, Friedel–Crafts
 type, 219
eliminative ring expansion, 468
enamine lactam, ring expansion, 368
enaminopyrrolidone, 472
endocyclic imine, 126
enol ether, 323
esters, ring closure, 326
ethanol, 15
ethanolysis, N-methyl-*cis*,*cis*-muconamate,
 485
ethers, saturated, 290
ethoxyazepanes, 25, 136
7-ethoxy[4H]azepines, 579
2-ethoxy-3,5,7-trimethyl[3H]azepine, 526
2-ethoxycarbonylcyclohexanone, 121
ethyl acrylate, 431
ethyl chloroformate, 289
ethyl propiolate, 476
1-ethyl-1,3,4,5-tetrahydro[2H]azepin-2-
 one, 361
2-ethylamino[3H]azepines, 525
3-ethylcaprolactam, 121
4-ethylcaprolactam, 121
7-ethylcaprolactam, 121
N-ethylcaprolactam, 125
ethylene ketal, 221

Favorskii rearrangement, 21, 143
fission, 27–8

flash vacuum pyrolysis
 aziridine, 429
 dienamine derivatives, 474
flash vacuum thermolysis, 220
 4,5-dihydroisoxazole-5-
 spirocyclobutanes, 396
formaldehyde, 18
formic acid, 18
N-formylazepanes, anodic oxidation of,
 25
fragmentation, 21
free-radical ring expansion, 218
Friedel–Crafts cyclisation, 130, 219
 3-[N-(2-chloroprop-2-enyl)propanoic
 acid, 393–4
 N-tosylchloroalkenol, 212
fulvene, 496
furan amine, 12
furo[2,3-b]pyridine, 432
furoazepine, 253
Fusarium, 212

Grignard addition, 14
Grignard alkylation, 2,3,6,7-
 tetrahydro[1H]azepine, 315
Grignard reagent, 16, 17, 138, 288
 aryl, 428
 caprolactam, 323

halides, elimination of, 138
α-halo esters, radical-like cyclisations, 18
α-halo ketones, radical-like cyclisations,
 18
ω-halo-tertiary amines, 13
1,6-haloamines, cyclisation of, 11–13
halogenation, 136–8
α-halolactams, Favorskii rearrangement
 of, 21
heptynylhydroxylamines, 327
hetero-Cope rearrangement, 424
hexadecylaminocaproic acid, 143
N-hexadecylcaprolactam, 143
hexafluoro[1H]azepine, 493
hexahydroazepines, 9–109
hexane, 9
hexane-1,2,6-triol, 10
hexane-1,6-diol, 10
HIV-1 protease, inhibitors, 374
Hofmann degradation, 21
Hofmann elimination reaction, 223
homobenzomorphan, 14

2,2-homodisubstituted azepanes, 16
7-hydrazino-3,4,5,6-tetrahydro-
 [2*H*]azepine, 328, 330
hydroboration/carbon monoxide
 insertion, 220
hydrogen, 10, 124
hydrogen bromide, 528
hydrogen catalyst, 268
hydrogen cyanide, 252
hydrogen peroxide oxidation, 144
hydrogenated carbon, 128
hydrolysis, 143, 451
 2-amino[3*H*]azepine, 451
 [1*H*]azepine-*N*-carboxylate, 488
 azeto[1,2*a*]pyrrole, 410
hydroxy esters, 16
hydroxy ketone, 593, 602
3-hydroxy-*N*-arylazepane, 11
6-hydroxy-1,7-dihydro[2*H*]azepinone, 368
3-hydroxy-*N*-methyl-azepane, 252
3-hydroxyazepane, 213
3-hydroxyazepin-2-one, 596
N-hydroxyazepine, 373
5-hydroxycaprolactam, 268
N-hydroxyimide, 2-nitrocyclohexanone
 conversion, 277
hydroxylamine, ring expansion, 325
hydroxylamine–alkyne cyclisations, 17
hydroxylamine carbonates, 22
7-hydroxylamino-3,4,5,6-
 tetrahydro[2*H*]azepine, 331
(−)-2-hydroxypinan-3-one, 11

imidate, 126
imidazoazepine, 261
imide, 485
imidic esters, 467
imidoyl chloride, 138
imine, 128
 spiro cyclisation, 142
iminium ion, 17, 18, 141
 cyclisation, 12, 125, 211
iminium salt, 23, 144
imino thioether, 127
iminophosphorane, 276, 362, 413
cis-iminovinylcyclopropanes,
 Cope rearrangement, 423
immonium salt, 316
indoloazepine, 253
2-indolyl-2-cyclohexanone, 123
infrared stretching frequencies, 261
inhibitors, HIV-1 protease, 374

intramolecular cyclisation, 268
intramolecular hydroamination, 19
iodine azide, 430
iodine isocycanate, 491
iodoalkenylamine, 382
 palladium catalysed carbonylation of,
 124
3-iodocaprolactam, 137
irradiation, 416
 2-alkylindazoles, 451
 7-aryl-1,3-dihydro[2*H*]azepin-2-one, 454
 azabicycloheptane-3,5-dione, 405
 4-phenyl-5-nitrocyclohexene, 373
 7-isobutyl-1,3,4,7-tetrahydro[2*H*]azepin-
 2-one, 369
isocyano-esters, 11
isomeric product, irradiation of, 593
isophorone, 442
 anti-oxime, 359, 445
N-isopropylazepane-2,7-dione, 277
N-isopropylcaprolactam, 276
4,4′-isopropylidene, 121
5,5′-isopropylidenebiscaprolactam, 121
isothiazolo[2,3-*a*]azepine, 30
isothiazolo[5,4-*d*]azepine, 222
isoxazole, 30, 396
isoxazolin-5-one, 317
isoxazoloazepine, 254
isoxazolo[3,4-*d*]azepine derivative, 222

keto azide, 327
β-keto ester, 16, 23, 25, 217
 chiral substituted, 122
keto hydrazones, reductive cyclisation, 12
ketoazepanes, reduction, 13–16
4-ketoazepane, 16
6-ketohexanoic esters, 124
3-ketomolinate, 212
4-ketomolinate, 221
ketones, ring closure, 326

lactam, 526
 reduction, 13
β-lactam, 595
lactam acetal, 292
lactim ether, 138, 601
Lawesson's reagent, 131
lead tetra acetate, 283
(*S*)-leucine, 369
LiAlH$_4$ reduction, 13, 14, 22, 136, 222, 252
 1,3-dihydro[2*H*]azepin-2-one, 415
 2,3-dihydro[1*H*]azepine, 451–2

LiNEt$_2$/HMPT, 139
lithium diisopropylamine, 139
lithium salts, 126
lithium triphenylsilylacetylide, 24, 142, 396
lutidine, 372
L-lysine, 117, 137
L-(S)-lysine, 137

maleic anhydride, 476
malonates, 23
malononitrile, 23
manganese(II), 133
manganese(III) tetraphenylporphyrin, 125
manganese-catalysed oxidation, 135
manganic acetylacetonate, 276
Mannich reaction, 314
mass spectra, 596
matrix photolysis, 604
Meerwein's reagent, 469
Meldrum's acid, 23, 138, 474, 475
Meldrum's acid derivative, 29, 30
(−)-menthone oxime, 120
mercury(II) acetate, 128
mesoionic oxazolone, 429
metal complexes, 133–5
methane sulphonate
  Beckmann rearrangement, 326
  ring expansion, 331
methanesulphonic acid, 122
methanol, 128
N-methoxy-carbonyl-3-piperidone, 211
6-methoxy-1,2,3,7-tetrahydro[4H]azepin-4-one, 394
methoxyazepane, 25
2-methoxy[3H]azepine, 526
N-methoxycaprolactam, 127
1-methoxycyclopentene, 446
N-methyl-caprolactam, 129
O-methyl ether, 23
  X-ray analysis, 594
6-methyl-5-heptenenitrile, 119
methyl L-lysinate, 14
methyl lysinate hydrochloride, 124
N-methyl-L-lysine, 143
2-methyl-N-methylazepane, 213
N-methyl-cis,cis-muconamate, ethanolysis, 485
1-methyl-1,3,4,5-tetrahydro[2H]azepin-2-one, 360
1-methyl-1,5,6,7-tetrahydro[4H]azepin-4-one, 396

7-methyl-3,4,5,6-tetrahydro[2H]azepine, 327
3-methyl-3-vinyl-2-dimethylaminoazirine, ring expansion, 579
methylation, diazomethane, 468
5-methylazepan-4-one, 221
N-methylazepane, 24
N-methylazepane-4-carboxylic acid, 28
1-methylazepine-2,7-dione, 408
(S)-2-methylazepine-2-carboxylic acid, 12
3-methylcaprolactam, 119
7-methylcaprolactam, 119, 121
N-methylcaprolactam, 138, 289
  with dimethyl sulphate, 139
  with lithium triphenylsilylacetylide, 142
(S)-(−)-5-methylcaprolactam, 120
2-methylcyclohexanone, 121
(+)-4-methylcyclohexanone oxime, 120
methylene azepanes, 21
methylene-2,2-bicyclohexanone, 121
methylene caprolactam, 124
7,7'-methylenebiscaprolactam, 121
syn/anti-N-methylnitrones, 360
N-methylpyrrolidone, 396
6-methyltetrahydropyridazin-3-one, ring-expansion/rearrangement, 395
methylthio-2,3,4,5-tetrahydro[1H]azepine, 289
5-methylthioazabicycloheptanes, 381
N-methylthiocaprolactam, 144
Michael reaction, 24, 133, 140
microbial enzymes, 143
microbiological oxygenation, 221
microorganisms, 281
molinate, 212, 213, 221
molybdenum, 369
molybdenum catalyst, 305
ω-monoalkylamino-allyltrimethylsilanes, 314
muscaflavin, 415
mushroom, 521
mycobactin, 131, 132

NaBH$_4$ reduction, 221, 261
nickel, 12
nickel salts, 23
nickel-aluminium, 124
nitration, 140
nitrene, 523
  insertion, 16, 488

nitrile ester
 cyclisation, 408
 reduction, 124
nitrile ylid, 429
nitrobenzene, deoxygenation, 525
3-nitrocaprolactam, 140
2-nitrocyclohexanone, N-hydroxyimide
 conversion, 277
nitroethane, 125
nitrogen nucleophiles, reaction with
 cis,*cis* dienes, 426
nitrones, 17, 327
N-nitroso ketone, 218
N-nitrosoazepan-3-one, 212
N-nitrosoazepan-4-one, 221
N-nitrosoazepane, 25, 221
nitrosobenzene, 525
N-nitrosobenzene, 499
N-nitrosocaprolactam, 133
NMR spectra, 476
norchalciporyl propionate, 522
nylon-6, 117

Ordram, 213, 281
organoaluminium reagents, 22
organolanthanide cyclisation, 20
organometal reagents, 222
oxaziridines, 125
oxazolidine, vinyl-substituted, 368
oxazoline derivatives, 494
oxazolo-azepines, 222
oxidation, 135–6
 caprolactams, 275, 276
 electrochemical, 213
 selenium dioxide, 222
oxidative coupling, phenols, 450
oxime, 379
 irradiation of, 125
 photolysing, 118
anti-oxime
 isophorone, 359, 445
 1,3,4,5-tetrahydro[2H]azepin-2-one,
  379
syn-oxime
 Beckmann rearrangement, 379, 442
 1,5,6,7-tetrahydro[2H]azepin-2-one, 359
oxime sulphonates, 22
ω-oximino esters, 124
oximinocaprolactam, 129
oxygenation, 135
 [1H]azepines, 593, 604
ozonolysis, 126

P_2S_5, N-methylcaprolactam, 289
palladised carbon, 128
palladium, 16, 17, 140
 chemistry, 18
palladium catalysed carbonylation, 124,
 382
palladium(II) catalysis, 144
Pd(OAc)$_2$, 498
Pd^{2+} complexes, 24
PdCl$_2$(PhCN)$_2$/CO/MeOH/Et$_3$N, 20
peptidomimetic azepine, 374
peracetic acid, 276
peracid-catalysed oxidation, 135
Periconia circinata, 132
N-phenacylcaprolactam,
 1,7-photocyclisation, 140
phenols, oxidative coupling, 450
phenyl azide, 524
 photolysis, 451
 thermolysis, 451
phenyl hydrazone, 252
phenyl selenide, ring expansion, 212
4-phenyl-4-isopropyl-[4H]azepine, 580
4-phenyl-5-nitrocyclohexene, irradiated,
 373
6-phenyl-2,3,4,7-tetrahydro[1H]azepine,
 306
4-phenyl-2,3,6,7-tetrahydro[1H]azepine-
 2,7-dione, 409
2-phenyl-2-vinylaziridine, ring expansion,
 306
phenylacetylene, 133, 593
2-phenylamino[3H]azepine, 523
2-phenylazepan-4-one, flash vacuum
 thermolysis, 220
2-phenylazepane, 23
N-phenylazepane, 28
5-phenyl[3H]azepine, alkylation, 580
phenylmagnesium bromide, 23
N-phenylmaleimide, spiro cyclisation,
 142
phenylnitrene, 525
phosgene, 322
photochemical [2 + 2] cycloaddition, 252
photochemical inducement, 20
photochemical irradiation, 3,6-dihydro-
 [2H]azepine, 446
photochemical reaction, benzonitrile, 446
photochemically induced rearrangement,
 496
photochemically induced ring expansion,
 277

Index   617

photochemistry, 1,3-dihydro[2*H*]azepin-2-one, 453
1,7-photocyclisation, 140
photolysis, 16, 118, 476
  N-alkoxyalkylsuccinimide, 406
  N-alkylthioalkylsuccinimide, 406
  aromatic azides, 578
  aryl azides, 593
  azido ketone, 260
  matrix, 604
  phenyl azide, 451
  N-substituted succinimide, 266
(*S*)-3-*N*-phthalimidoadipimide, 136
(*S*)-3-*N*-phthaloylcaprolactam, 136
pipecolic acids, 143
piperid-4-one, 217
piperidine, 293
piperidone, 219
pKα values, 24
platinum, 16
platinum(II), 134
polyhydroxycaprolactam, 124
polyphosphate ester, 118
polyphosphoric acid, 118, 121
  Beckmann rearrangements, 260
position 6, double bond substitution, 291
potassium cyanide, 291, 427, 495
potassium hydroxide, 493
potassium iodide, 11
primary amines, 450
2-propylazepane, 12, 16
N-propylazepane, 24
3-propylcaprolactam, 121
7-propylcaprolactam, 121
protic solvents, 596
O-protonation, 476
Pt^{2+} complexes, 24
2-pyranone compounds, 414
pyrazoles, 30
pyrazolo[3,4-*c*]azepine, 473
pyrazoloazepine, 254
  structure, 222
pyridine ester, 596
pyridoazepines, 253
pyridocoumarin, 253
pyridopyrroloazepine, 253
pyrimido[1,6-*a*]azepine, 29
pyrimido[3,4-*b*]azepine, 29
pyrimido[4,5-*b*]azepine, 222, 413
  derivatives, 142
  ring system, 142

N-pyrrole, 491
pyrrole acrylate, 431
pyrrole derivatives, ring expansion, 413
pyrrolidine ring expansion, 21
pyrrolinedione, 593
pyrrolo[2,3-*b*]pyridine, 432
pyrroloazepine, 29

quinolo[2,3-*b*]azepine system, 142
quinones, Schmidt reaction, 592

racemic aminocaprolactam, 124
Raney nickel, 10, 16, 118, 144, 289
reduction
  caprolactams, 136
  2,3,6,7-tetrahydro[1*H*]azepine, 315
reductive amination, 220
reductive coupling, 314
reductive silylation, chlorosilanes, 430
regiospecific results, 119
rhodium(III), 134
rhodium acetate, 141, 144
ring closure
  acetylenes, 326
  [3*H*]azepine, 528
  azides, 326
  esters, 326
  ketones, 326
ring contraction, 21, 23, 142–3, 293, 530
  azepan-4-one, 223
  1,4-dihydropyridine derivatives, 432
  2,3,4,5-tetrahydro[1*H*]azepine, 292
ring expansion, 16, 21–3, 24, 142–3, 290, 324, 530, 602
  acetylenic ester, 413
  cyclic vinylaziridine, 414
  cyclohexane-1,3-dione, 260
  diazo ester, 217
  diazoacetate, 212
  diazoacetic ester, 218
  diazomethane, 217
  1,5-dihydro[2*H*]azepin-2-one, 468
  1,4-dihydropyridine, 527
  eliminative, 595
  enamine lactam, 368
  free-radical, 218
  hydroxylamines, 325
  methane sulphonate, 331
  method, 306
  3-methyl-3-vinyl-2-dimethylaminoazirine, 579
  6-methyltetrahydropyridazin-3-one, 395

phenyl selenide, 212
photochemically induced, 277
protocols, 126
pyrrole derivatives, 413
2,3,4,7-tetrahydro[1H]azepine, 305
thermally induced, 493
thioacetals, 322–3
trans-aziridine, 493
ring formation, non-cyclic amine, 305
ring fragmentation, 21
ruthenium, 140
ruthenium catalyst, 370
ruthenium halide, 10
ruthenium tetraoxide
  oxidations, 135
  with sodium periodate, 276

salt formation, 139
salts, 416
samarium iodide, 25
Schmidt reaction, 15, 121–3, 223, 442
  benzoquinone, 268
  1,4-benzoquinone, 482
  cyclohexenone, 380
  quinone, 592
  1,3,4,5-tetrahydro[2H]azepin-2-one, 360
1,2,3-selenadiazole rings, 222
selenium dioxide oxidation, 222
[2,3]sigmatropic shift, 1-benzyl-4-vinylazetidinone, 372
silylamine, 18
silylation, 430
six-membered ring, 488
sodium, 314
sodium azide, 121, 122, 323
sodium borohydride, 15, 128, 136, 252
sodium cyanide, dimethyl sulphoxide, 428
sodium hydride, 11
sodium methoxide, 138
sodium periodate, with ruthenium tetraoxide, 276
sodium salts, 2,6-dialkylphenols, 449
spiro cyclisation, 142
spirooxaziridine, 125
  ring opening, 125
spirooxazolidine, flash vacuum thermolysis, 220
Sporotrichum sulfurescens, 212, 221
Stevens rearrangement, 21
styrene, 133

styrylamine, 20
N-styrylcaprolactam, 133
succinimides, 266
sulphonation, 24
1-sulphonyl-2-cyano[1H]azepine, 495
sulphuryl chloride, 137

tertiary amine, 305
tetrabutylammonium perrhenate(VII), 118
tetracarbonyl hydridoferrate, 10
tetracyanoethylene, 452
1,3,6,7-tetrahydro-3,3,6,6-tetramethyl-[2H]azepin-2-one, 374
tetrahydroazepin-2-one, 128
1,2,3,7-tetrahydro[4H]azepin-4-one, 393–4
1,2,6,7-tetrahydro[3H]azepin-3-one, 392–3
1,3,4,5-tetrahydro[2H]azepin-2-one, 359–67
  anti-oxime, 379
  Schmidt rearrangement, 360
1,3,4,7-tetrahydro[2H]azepin-2-one, 367–71
1,3,6,7-tetrahydro[2H]azepin-2-one, 372–9
1,5,6,7-tetrahydro[2H]azepin-2-one, 360, 379–92
  syn-oxime, 359
1,5,6,7-tetrahydro[4H]azepin-4-one, 395–9
2,3,5,6-tetrahydro[4H]azepin-4-one, 399–400
3,5,6,7-tetrahydro[4H]azepin-4-one, 400
tetrahydroazepines, 288–353
  reduction, 16–17
2,3,4,5-tetrahydro[1H]azepine, 288–303
2,3,4,5-tetrahydro[1H]azepine-2,3-dione, 403–4, 405
2,3,4,5-tetrahydro[1H]azepine-2,5-dione, 406–7
2,3,4,7-tetrahydroazepine, 17
2,3,4,7-tetrahydro[1H]azepine, 304–13
2,3,4,7-tetrahydro[1H]azepine-2,7-dione, 408–9
2,3,6,7-tetrahydro[1H]azepine, 314–21
2,3,6,7-tetrahydro[1H]azepine-2,7-dione, 409–10
2,3,6,7-tetrahydro[2H]azepine, 17
2,3,4,5-tetrahydro[1H]azepine-4,5-dione, 410

2,5,6,7-tetrahydro[1H]azepine-2,5-dione, 407–8
3,4,5,6-tetrahydro[2H]azepine, 17, 289, 321–53
3,4,5,6-tetrahydro[2H]azepine-4,5-dione, 411
tetrahydroazepinedione, 403–11
tetrahydroazepinone, 359–400, 430
tetrahydropyridinium salt, expansion of, 22
6,7,8,9-tetrahydro[5H]pyrido[2,3-d]-azepin-2-one, 317
tetrahydroxyazepane, 12
tetramethyl[1H]azepine-3,4,5,6-tetracarboxylate, 496
3,3,7,7-tetramethylcaprolactam, 127
2,4,6,7-tetraphenyl[3H]azepine, 527
tetrazole corazole, 330
tetrazolodiazocine, 223
thermal means, 16
thermal rearrangement, 491, 530
  biscyclopropapyrrolidine, 413
thermolysis, phenyl azide, 451
thiazolidines, fused, 361
thiazoloazepine, 222, 253
thioacetals, 322–3
thioacetamide, 222
thiocaprolactams, 130–1, 144
thiolactone, 381
thiophenocaprolactam, desulphurisation, 121
thiophenocyclohexanone, 121
Thorpe–Ziegler cyclisation, 217
TiCl$_3$, 17
tin(IV) chloride, 127, 467
tosyl analogue, 129
tosyl isocyanate, 129
N-tosylazepane, 11
  microbiological oxygenation, 221
N-tosylchloroalkenols, Friedel–Crafts cyclisation, 211
tosyloxy quinuclidene, 21
trialkylaluminium reagents, 16
trialkylaluminium/diisobutylaluminium hydride, 22
1,2,4-triazines, 527, 579
1,2,4-triazoles, fused, 330
triazoloazepinone, 254
tributylphosphine, 10
tributyltin chloride, 25
triethyloxonium fluoroborate, 592
triethyloxonium tetrafluoroborate, 452

trifluoroacetic acid, 127
3,5,7-trimethyl-1,3-dihydro[2H]azepin-2-one, 449, 526
4,6,7-trimethyl-2,3,4,5-tetrahydro[1H]azepine-2,5-dione, 406
trimethyloxonium fluoroborate, 592
2,4,6-trimethylphenol, 449
trimethylsilyl chloride, 314
  acyloin reaction, 219
trimethylsilyl iodide, 381
trimethylsilyl polyphosphate, 118
trimethylsilyl triflate, 127
trimethylsilylation, 133
N-trimethylsilylcaprolactam, 139
2,4,6-triphenyl-N-benzylpyridinium tetrafluoroborate, 527
triphenylaluminium, 127
triphenylphosphine, 10, 24, 127, 327
Triton B, 133
tropolones, aromatic, 592
tropones, aromatic, 592

UV irradiation, 135, 495, 524, 527, 530
  azaquadricyclane, 491
  benzofuroxan, 485

Vilsmeier reaction, 140
  N-methylcaprolactam, 139–40
Vilsmeier reagent, 430
vinyl-magnesium bromide, 17
vinyl-substituted oxazolidine, Claisen rearrangement, 368
4-vinylazepane, 21
vinylazetidines, 372
vinylaziridines, 423
cis-2-vinylcyclopropyl carbonylazide, Curtius rearrangement, 451
2-vinylpyrrolidines, 306, 307
VOCl$_2$ complexes, 134

Wagner–Meerwein rearrangement, 218
Wolff–Kishner reduction, 21, 30, 213, 223

X-ray analysis, 474, 476, 596
  O-methyl ether, 594
xylene
  azido ester, 490
  refluxing, 430

zinc(II), 133
zinc chloride, 135